Board Review Series

GROSS ANATOMY
3rd edition

Look for all of the titles in this series:

Board Review Series

GROSS ANATOMY
3rd edition

Kyung Won Chung, Ph.D.

David Ross Boyd Professor and Vice Chairman
Department of Anatomical Sciences
College of Medicine
University of Oklahoma Health Sciences Center
Oklahoma City, Oklahoma

Williams & Wilkins

BALTIMORE • PHILADELPHIA • HONG KONG
LONDON • MUNICH • SYDNEY • TOKYO

A WAVERLY COMPANY

Williams & Wilkins

Copyright © 1995
Williams & Wilkins
Rose Tree Corporate Center
Building II, Suite 5025
1400 North Providence Road
Media, Pa 19063-2043 USA

Library of Congress Cataloging-in-Publication Data

Chung, Kyung Won.
  Gross anatomy/Kyung Won Chung.—3rd ed.
    p.   cm.—(Board review series)
  Includes bibliographical references and index.
  ISBN 0-683-01563-X
  1. Human anatomy—Outlines, syllabi, etc.   2. Human anatomy—
Examinations, questions, etc.   I. Title.   II. Series.
  [DNLM: 1. Anatomy—examination questions.   2. Anatomy—programmed
instruction.   QS 18 C559g 1995]
QM31.C54   1995
611—dc20
DNLM/DLC
for Library of Congress   94-23875
                                                                          CIP

Printed in the United States of America

96 97 98
2 3 4 5 6 7 8 9 10

Dedication

To My Wife, Young Hee,

and

My Sons, Harold and John

Contents

Preface to the Third Edition

This concise review of human anatomy is designed for medical and dental students, and it is intended primarily to help these students prepare for the United States Medical Licensing Examination Step 1 as well as other examinations. It presents the essentials of human anatomy in the form of condensed descriptions and simple illustrations. The text is tightly outlined with related Board-type questions following each section. I have attempted to include all Board-relevant information without introducing a vast amount of material or entangling students in a web of details. However, because this book is in summary form, students are encouraged to consult a standard textbook for a comprehensive study of difficult concepts and fundamentals.

Organization

As with the previous editions, the third edition begins with a brief introduction to the skeletal, muscular, nervous, and circulatory systems. Following are chapters on regional anatomy, which include the upper limb, lower limb, thorax, abdomen, perineum and pelvis, back, and head and neck.

I believe that anatomy is a visual science of the configuration of the body, and thus the success of learning and understanding it depends largely on the quality of dissection and the illustrations of the human structure. Many of the illustrations are simple schematic drawings used to enhance the student's understanding of the descriptive text. A few of the illustrations are more complex, attempting to exhibit important anatomical relationships. The considerable number of tables of muscles will prove particularly useful as a summary and review. In addition, several summary charts for muscle innervation and action, cranial nerves, autonomic ganglia, and foramina of the skull are included in order to highlight and summarize pertinent aspects of the system.

Test questions at the end of each chapter are designed to emphasize important information and hence lead to a better understanding of the material. These questions also serve as a self-evaluation to help the student uncover areas of weakness. Answers and explanations are provided after the questions.

Features of the new edition

• The upper and lower extremities have been organized systemically in order to present the continuity of the musculoskeletal, nervous, and vascular systems.

• Questions reflect the guidelines set forth beginning in 1991 by the National Board of Medical Examiners. K-type questions have been eliminated despite their educational value. Test questions measure anatomical knowledge as well as the student's ability to solve clinically oriented problems.

• A Comprehensive Examination serves as a practice exam and self-assessment tool to help the student diagnose weaknesses prior to beginning a review of the subject. It also serves as a self-examination upon completion of the review book prior to the Board examination.

• Some illustrations have been rearranged and new ones have been added.

• Roentgenograms and computer tomograms are used in the test questions to aid in the study of anatomical structures and their relationships.

• The practical application of anatomical knowledge is included throughout the book as clinical considerations. More clinical considerations have been added and expanded for this edition.

Kyung Won Chung

Acknowledgments

I wish to express my sincere thanks to the many medical students, colleagues, and friends who have made valuable suggestions regarding the preparation of this new edition. My appreciation is extended to Ms. Diane Abeloff, Medical Illustrator, for her book, *Medical Art, Graphics for Use,* which was used for illustrations with a little modification in some cases, and to Shawn C. Schlinke, M.D., and Ms. Laura Barton for their excellent illustrations and full cooperation. I am deeply indebted to Ms. Susan Kelly for her critical advice and constructive suggestions for improvement of the text and to Dr. Ae Son Om for her copyediting during the preparation of this manuscript. Finally, I greatly appreciate and enjoy the privilege of working with the Williams & Wilkins staff—Ms. Elizabeth Nieginski, Acquisitions Editor; Ms. Jane Velker, Senior Managing Editor; Ms. Joan Scullin, Editorial Assistant; Ms. Beth Goldner, Editorial Coordinator; and Ms. Mary Durkin, Manuscript Editor—for their constant guidance, enthusiasm, and unfailing support throughout the production of this book.

1

Introduction

Skeleton and Joints

I. Bones

– are calcified connective tissue consisting of cells (**osteocytes**) in a matrix of ground substance and collagen fibers.
– act as levers on which muscles act to produce the movements permitted by joints.
– serve as a **reservoir** for **calcium** and **phosphorus.**
– contain internal soft tissue, the **marrow,** where blood cells are formed.
– are classified, according to shape, into long, short, flat, irregular, and sesamoid bones; they also are classified according to their developmental history (i.e., endo-chondral and membranous bones).

A. Long bones

– are longer than they are wide; they include the clavicle, humerus, radius, ulna, femur, tibia, fibula, metacarpals, and phalanges.
– develop by **replacement of hyaline cartilage plate.**
– have a shaft (**diaphysis**) and two ends (**epiphyses**).

1. Diaphysis

– is the central region and is composed of a thick collar of compact bone surrounded by the periosteum.
– contains the **marrow cavity** and the **metaphysis,** which is the more recently developed part of the bone adjacent to the epiphyseal disk.

2. Epiphyses

– are composed of a **trabecular bony meshwork** surrounded by a thin layer of compact bone.
– have articular surfaces that are covered by hyaline cartilage.

B. Short bones

– are found only in the **wrist** and **ankle** and are approximately cuboid shaped.
– are composed of **spongy bone and marrow** surrounded by a thin outer layer of dense compact bone.

C. Flat bones

– include the ribs, sternum, scapulae, and bones in the vault of the skull.
– consist of **two layers of compact bone separated by spongy bone** and marrow space (**diploë**).

1

– have articular surfaces that are covered with fibrocartilage.
– grow by **replacement of connective tissue.**

D. Irregular bones

– include bones of mixed shapes, such as bones of the **skull, vertebrae,** and **coxa.**
– contain mostly **spongy bone** enveloped by a **thin outer layer of dense compact bone.**

E. Sesamoid bones

– develop in certain **tendons** and serve to reduce friction on the tendon, thus protecting it from excessive wear.
– are commonly found where tendons cross the ends of long bones in the limbs, as in the knee and the wrist.

II. Joints

– are places of union between two or more bones.
– are innervated as follows: The nerve supplying a joint also supplies the muscles that move the joint and the skin covering the insertion of such muscles (**Hilton's law**).
– are classified on the basis of their structural features into fibrous, cartilaginous, and synovial types.

A. Fibrous joints (synarthroses)

– are barely movable or nonmovable, and are found in these forms:

1. Sutures

– are connected by fibrous connective tissue.
– are found between the flat bones of the skull.

2. Syndesmoses

– are connected by fibrous connective tissue.
– occur as the inferior tibiofibular and tympanostapedial syndesmoses.

B. Cartilaginous joints

1. Synchondroses (primary cartilaginous joints)

– are united by hyaline cartilage.
– **permit no movement** but growth in the length of the bone.
– include epiphyseal cartilage plates (the union between the epiphysis and the diaphysis of a growing bone) and spheno-occipital and manubriosternal synchondroses.

2. Symphyses (secondary cartilaginous joints)

– are joined by a plate of fibrocartilage and are slightly movable joints.
– include the **pubic symphysis** and the **intervertebral disks.**

C. Synovial joints (diarthrodial joints)

– permit a great degree of free movement.
– are characterized by four features: **joint cavity, articular cartilage, synovial membrane** (which produces synovial fluid), and **articular capsule.**
– are classified according to axes of movement into plane, hinge, pivot, ellipsoidal, saddle, and ball-and-socket joints.

1. **Plane joints**
   - have flat articular surfaces and are limited in movement by the articular capsule.
   - allow simple **gliding or sliding movement** between two flat surfaces.
   - occur in the proximal tibiofibular, intercarpal, intermetacarpal, carpometacarpal, sternoclavicular, and acromioclavicular joints.

2. **Hinge (ginglymus) joints**
   - resemble door hinges and allow movement around one axis (**uniaxial**) at right angles to the bones.
   - allow movements of **flexion** and **extension** only.
   - occur in the elbow, knee, ankle, and interphalangeal joints.

3. **Pivot (trochoid) joints**
   - are formed by a central bony pivot turning within a bony ring.
   - allow movement around one longitudinal axis (**uniaxial**).
   - allow **rotation** only.
   - occur in the superior and inferior radioulnar joints and in the atlantoaxial joint.

4. **Ellipsoidal (condyloid) joints**
   - have reciprocal elliptical convex and concave articular surfaces.
   - allow movement in two directions (**biaxial**) at right angles to each other.
   - allow flexion and extension movements and abduction and adduction movements but *no axial rotation*.
   - occur in the **wrist** (radiocarpal), **atlanto-occipital**, and **metacarpophalangeal joints**.

5. **Saddle (sellar) joints**
   - resemble a saddle on a horse's back and allow movement in several directions.
   - allow flexion and extension, abduction and adduction, and circumduction, but *no axial rotation*.
   - occur in the **carpometacarpal joint of the thumb**.

6. **Ball-and-socket (spheroidal) joints**
   - are formed by the reception of a globular head into a cup-shaped cavity and allow movement in many directions (**multiaxial**).
   - allow flexion and extension, abduction and adduction, medial and lateral rotations, and circumduction.
   - occur in the **shoulder** and **hip joints**.

Muscular System

I. Muscle

- consists predominantly of **contractile cells**.
- produces the movements of various parts of the body by contraction.
- appears in three types: skeletal, cardiac, and smooth muscles.

A. Skeletal muscle

- is under voluntary control and makes up about 40% of the total body mass.
- has two attachments, an **origin** and an **insertion**. The origin is usually the more fixed and proximal attachment; the insertion is the more movable and distal attachment.

– is enclosed by **epimysium,** a thin layer of connective tissue. Smaller bundles of muscle fibers are surrounded by **perimysium.** Each muscle fiber is enclosed by **endomysium.**

B. Cardiac muscle

– is known as **myocardium** and forms the middle layer of the heart.
– is innervated by the autonomic nervous system but contracts spontaneously without any nerve supply.
– responds to increased demands by increasing the size of its fiber; this is known as **compensatory hypertrophy.**

C. Smooth muscle

– is generally arranged in two layers, **circular** and **longitudinal,** in the walls of many visceral organs.
– is innervated by the autonomic nervous system, regulating the size of the lumen of a tubular structure.
– undergoes rhythmic contractions called **peristaltic waves** in the walls of the gastrointestinal tract, uterine tubes, ureters, and other organs.

II. Structures Associated with Muscles

A. Tendons

– are **fibrous bands of dense connective tissue** that always have one end attached to muscle and the other end blending with the fibrous connective tissue of the structure to which they attach (usually bone).
– are supplied by sensory fibers extending from muscle nerves.

B. Bursae

– are **flattened sacs of synovial membrane** that contain a viscid fluid that moistens the bursa wall to facilitate movement by minimizing friction.
– are found where a tendon rubs against a bone, ligament, or other tendon.
– are prone to fill with fluid when infected or injured.

C. Synovial tendon sheaths

– are **tubular sacs** wrapped **around the tendons;** they are similar to bursae in their fundamental structure and are filled with synovial fluid.
– occur where tendons pass under ligaments or retinacula and through osseofibrous tunnels, thus facilitating movement by reducing friction.
– have linings, like synovial membrane, that respond to infection by forming more fluid and by proliferating more cells, causing adhesions and thus restriction of movement of the tendon.

D. Aponeuroses

– are **flat fibrous sheets** or **expanded broad tendons** that attach to muscles and serve as the means of origin or insertion of a flat muscle.

E. Ligaments

– are **fibrous bands or sheets** connecting bones or cartilage or are folds of peritoneum serving to support and strengthen joints, muscles, and visceral structures.

F. Fascia

– is a **fibrous sheet that envelops the body** under the skin and invests the muscles.
– may **limit the spread of pus** and extravasated fluids such as urine and blood.

1. **Superficial fascia**
   - is a **loose connective tissue** between the dermis and the deep (investing) fascia.
   - contains fat, cutaneous vessels, and nerves.
   - has a fatty superficial layer and a membranous deep layer.

2. **Deep fascia**
   - is a **sheet of fibrous tissue** that invests the muscles and helps to support them by serving as an elastic sheath or stocking, providing origins and insertions for muscles and forming retinacula and fibrous sheaths for tendons.
   - forms potential **pathways for infection** or extravasation of fluids.
   - has no sharp distinction from epimysium.

Nervous System

I. Divisions of the Nervous System

- The nervous system is divided anatomically into the **central nervous system** (CNS), consisting of the brain and spinal cord, and the **peripheral nervous system** (PNS), consisting of 12 pairs of cranial nerves and 31 pairs of spinal nerves.
- The nervous system is divided functionally into the **somatic nervous system,** which controls primarily voluntary activities, and the **visceral (autonomic) nervous system,** which controls primarily involuntary activities.
- It is composed of **neurons** and **neuroglia,** which are non-neuronal cells such as astrocytes, oligodendrocytes, and microglia.
- It controls and integrates the activity of various parts of the body.

II. Neurons

- are the structural and functional units of the nervous system (neuron doctrine).
- are specialized for the reception, integration, transformation, and transmission of information.

A. Components of neurons

- Neurons consist of **cell bodies** (perikaryon or soma) and their processes, **dendrites** and **axons.**

1. **Dendrites** (dendron means "tree") are usually short and highly branched and **carry impulses toward the cell body.**

2. **Axons** are usually single and long, have fewer branches (collaterals), and **carry impulses away from the cell body.**

B. Classification of neurons

1. **Unipolar neurons**
   - have only one process, which divides into a central branch that functions as an axon, and a peripheral branch that serves as a dendrite.
   - are sensory neurons of the PNS (e.g., cerebrospinal ganglion cells).

2. **Bipolar neurons**
   - have two processes, one dendrite and one axon, and are found in the **olfactory epithelium,** in the **retina,** and in the **inner ear.**

3. **Multipolar neurons**
   - have several dendrites and one axon and are most common in the CNS (e.g., motor cells in anterior and lateral horns of the spinal cord, autonomic ganglion cells).

C. **Other components of the nervous system**

1. **Cells that support neurons**
   - include **Schwann cells** and **satellite cells** in the PNS.
   - are called **neuroglia** in the CNS and are composed mainly of three types: **oligodendrocytes, astrocytes,** and **microglia.**

2. **Myelin**
   - is the fat-like substance forming a sheath around certain nerve fibers.
   - is formed by **Schwann cells** in the PNS and **oligodendrocytes** in the CNS.

3. **Synapses**
   - are the **sites of functional contact** of a neuron with another neuron, an effector (muscle, gland) cell, or a sensory receptor cell.
   - are classified by the site of contact as **axodendritic, axoaxonic,** or **axosomatic** (between axon and cell body).
   - subserve the transmission of nerve impulses, commonly from the axon terminals (presynaptic elements) to the receiving cell's plasma membranes (postsynaptic elements). In most cases, the impulse is transmitted by means of neurotransmitters released into a **synaptic cleft** that separates the presynaptic from the postsynaptic membrane.

III. Central Nervous System

A. **Brain**
   - is enclosed within the cranium, or brain case.
   - has a **cortex,** which is the outer part of the cerebral hemispheres and is composed of **gray matter,** which consists largely of the nerve cell bodies.
   - has an interior part composed of **white matter,** which consists largely of axons forming tracts or pathways, and ventricles, which are filled with cerebrospinal fluid (CSF).

B. **Spinal cord**
   - is **cylindrical,** occupies approximately the upper two-thirds of the vertebral canal, and is enveloped by the meninges.
   - has cervical and lumbar enlargements for the nerve supply of upper and lower limbs, respectively.
   - has **centrally located gray matter,** in contrast to the cerebral hemispheres, and peripherally located white matter.
   - has a conical end known as the **conus medullaris.**
   - grows more slowly than the vertebral column during fetal development, and hence its terminal end gradually shifts to a higher level.
   - ends at the level of L2 (or between L1 and L2) in the adult and at the level of L3 in the newborn.

C. **Meninges**
   - consist of three layers of connective tissue membranes (**pia, arachnoid,** and **dura mater**) that surround and protect the brain and spinal cord.

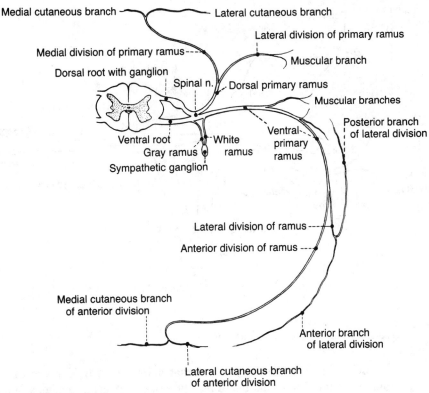

Figure 1-1. Typical spinal nerve.

   – contain the **subarachnoid space,** which is the interval between the arachnoid and pia mater, filled with CSF.

IV. Peripheral Nervous System

A. Cranial nerves

   – consist of **12 pairs.**
   – are connected to the brain rather than to the spinal cord.
   – have **motor fibers** with cell bodies located within the CNS and **sensory fibers** with cell bodies that form sensory ganglia located outside the CNS.
   – emerge from the ventral aspect of the brain (except for the trochlear nerve [cranial nerve IV]).
   – contain all four functional components of the spinal nerves (GSA, GSE, GVA, GVE) and an additional three components (SSA, SVA, SVE) [see Nervous System IV C; Chapter 8].

B. Spinal nerves (Figure 1-1)

   – consist of **31 pairs:** 8 cervical, 12 thoracic, 5 lumbar, 5 sacral, and 1 coccygeal.
   – are formed from dorsal and ventral roots; each dorsal root has a ganglion that is within the intervertebral foramen.
   – are connected with the sympathetic chain ganglia by **rami communicantes.**
   – contain sensory fibers with cell bodies in the dorsal root ganglion (GSA and GVA); motor fibers with cell bodies in the anterior horn of the spinal cord (GSE); and motor fibers with cell bodies in the lateral horn of the spinal cord (only segments between T1 and L2) [GVE].

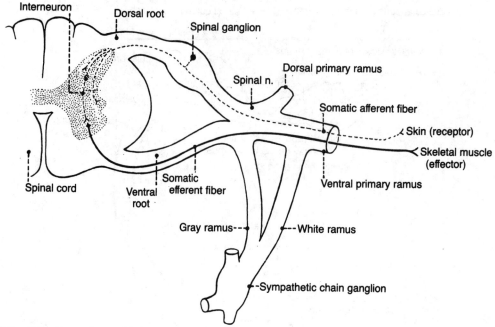

Figure 1-2. General somatic afferent and efferent nerves.

   – are divided into the **ventral and dorsal primary rami.** The ventral primary
      rami enter into the formation of plexuses (i.e., cervical, brachial, and lumbosa-
      cral); the dorsal primary rami innervate the skin and deep muscles of the back.

C. **Functional components in peripheral nerves** (Figures 1-2 and 1-3)

   1. **General somatic afferent (GSA) fibers**

      – transmit pain, temperature, touch, and proprioception from the body to the
         CNS.

   2. **General somatic efferent (GSE) fibers**

      – carry motor impulses to skeletal muscles of the body.

   3. **General visceral afferent (GVA) fibers**

      – convey sensory impulses from visceral organs to the CNS.

   4. **General visceral efferent (GVE) fibers (autonomic nerves)**

      – transmit motor impulses to smooth muscle, cardiac muscle, and glandular
         tissues.

   5. **Special somatic afferent (SSA) fibers**

      – convey special sensory impulses of vision, hearing, and equilibration to the
         CNS.

   6. **Special visceral afferent (SVA) fibers**

      – transmit smell and taste sensations to the CNS.

   7. **Special visceral efferent (SVE) fibers**

      – conduct motor impulses to the muscles of the head and neck.
      – arise from branchiomeric structures, such as muscles for mastication, mus-
         cles for facial expression, and muscles for elevation of the pharynx and
         movement of the larynx.

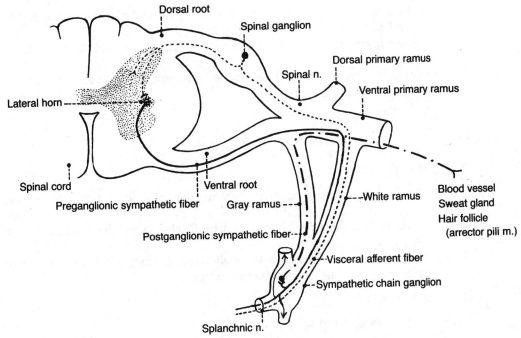

Figure 1-3. General visceral efferent (autonomic) and afferent nerves.

V. Autonomic Nervous System

– is divided into the **sympathetic** (thoracolumbar outflow) and **parasympathetic** (craniosacral outflow) systems and is composed of two neurons, preganglionic and postganglionic, which are GVE neurons.

A. Sympathetic nerve fibers (see Figure 1-3)

– have preganglionic nerve cell bodies that are located in the lateral horn of the thoracic and upper lumbar levels (L2 or L1–L3) of the spinal cord.

– have **preganglionic fibers** that pass through ventral roots, spinal nerves, and white rami communicantes. These fibers enter adjacent sympathetic chain ganglia, where they synapse or travel up or down the chain to synapse in remote ganglia or run further through the splanchnic nerves to synapse in collateral ganglia, located along the major abdominal blood vessels.

– have **postganglionic fibers** from the chain ganglia that return to spinal nerves by way of gray rami communicantes and supply the skin with secretory fibers to sweat glands, motor fibers to smooth muscles of the hair follicles (arrectores pilorum), and vasomotor fibers to the blood vessels.

– function primarily in **emergencies,** preparing individuals for fight or flight and thus increase heart rate, inhibit gastrointestinal motility and secretion, and dilate pupils and bronchial lumen.

B. Parasympathetic nerve fibers

– comprise the preganglionic fibers that arise from the brainstem (cranial nerves III, VII, IX, and X) and sacral part of the spinal cord (second, third, and fourth sacral segments).

– are, with few exceptions, characterized by **long preganglionic fibers** and **short postganglionic fibers.**

– are distributed to the walls of the visceral organs and glands of the digestive system but not to the skin or to the periphery.

– decrease heart rate, increase gastrointestinal peristalsis, and stimulate secretory activity.

– function primarily in **homeostasis,** tending to promote quiet and orderly processes of the body.

Circulatory System

I. Vascular System

– functions to transport vital materials between the external environment and the internal fluid environment of the body.

– consists of the **heart** and **vessels** (arteries, capillaries, veins) that transport blood through all parts of the body.

– includes the **lymphatic vessels,** a set of channels that begin in the tissue spaces and return excess tissue fluid to the bloodstream.

A. Circulatory loops

1. Pulmonary circulation

– pumps blood from the right ventricle to the lungs through the pulmonary arteries and returns it to the left atrium of the heart through the pulmonary veins.

2. Systemic circulation

– pumps blood from the left ventricle through the aorta to all parts of the body and returns it to the right atrium through the superior and inferior vena cavae and the cardiac veins.

B. Fetal circulation (Figure 1-4)

1. The fetus

– blood supply is oxygenated in the placenta rather than in the lungs.

– has three shunts that partially bypass the lungs and liver:

a. Foramen ovale

– is an opening in the **septum secundum.**

– usually closes functionally at birth, but anatomic closure occurs later.

– shunts blood from the right atrium to the left atrium, partially bypassing the lungs (pulmonary circulation).

b. Ductus arteriosus

– is derived from the sixth aortic arch and connects the bifurcation of the pulmonary trunk.

– closes functionally soon after birth, but anatomic closure requires several weeks.

– becomes the **ligamentum arteriosum,** which connects the left pulmonary artery (at its origin from the pulmonary trunk) to the concavity of the arch of the aorta.

– shunts blood from the pulmonary trunk to the aorta, partially bypassing the lungs (pulmonary circulation).

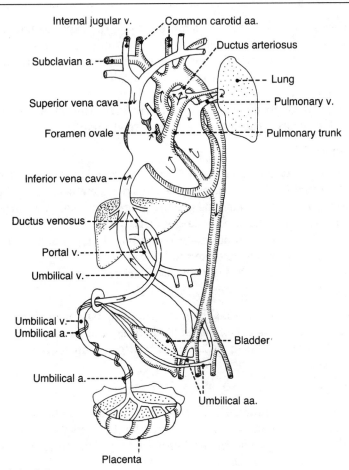

Figure 1-4. Fetal circulation.

c. Ductus venosus

– shunts oxygenated blood from the umbilical vein (returning from the placenta) to the inferior vena cava, without bypassing the liver (portal circulation).

– joins the left branch of the portal vein to the inferior vena cava and is obliterated to become the **ligamentum venosum** after birth.

2. Umbilical arteries

– carry blood to the placenta for reoxygenation before birth.

– become **medial umbilical ligaments** after birth, after their distal parts have atrophied.

3. Umbilical veins

– carry highly oxygenated blood from the placenta to the fetus.

– The right vein is obliterated during the embryonic period; the left is obliterated to form the **ligamentum teres hepatis** after birth.

C. Heart

– is a hollow, muscular, **four-chambered organ** that pumps blood through the pulmonary and systemic circulations.

– receives the venous blood from the body in the right atrium and then passes it into the right ventricle, which pumps it to the lungs for oxygenation.

– receives the oxygenated blood from the lungs in the left atrium and passes it to the left ventricle, which pumps it through the arteries to supply the tissues of the body.

– is regulated in its pumping rate and strength by the autonomic nervous system, which controls a pacemaker (the **sinoatrial node**).

D. Blood vessels

– carry blood to the lungs, where carbon dioxide is exchanged for oxygen.

– carry blood to the intestines, where nutritive materials in fluid form are absorbed, and to the endocrine glands, where hormones pass through the vessel walls and are distributed to target cells.

– transport the waste products of tissue fluid to the kidneys, intestines, lungs, and skin, where they are excreted.

– are of four types: arteries, veins, capillaries, and sinusoids.

1. Arteries

– carry blood away from the heart and distribute it to all parts of the body.

– have thicker and stronger walls than those of veins.

– consist of three main types: **elastic arteries, muscular arteries,** and **arterioles.**

2. Veins

– carry blood toward the heart from all parts of the body.

– consist of the **pulmonary veins,** which return oxygenated blood to the heart from the lungs, and the **systemic veins,** which return deoxygenated blood to the heart from the rest of the body.

– contain valves that prevent the reflux of blood. Valves are numerous in the veins of the limbs, but are absent in the small veins and in the venae cavae as well as in the hepatic, renal, uterine, and ovarian veins.

– **Venae comitantes** are veins or pairs of veins that closely accompany an artery in such a manner that the pulsations of the artery aid venous return; they are often found in the medium-sized vessels of the arm, forearm, and leg.

3. Capillaries

– are composed of endothelium and its basement membrane and connect the arterioles to the venules.

– are the **exchange sites** where oxygen and nutritive materials from oxygenated blood diffuse across the endothelial wall of the arteriolar end of the capillary into the tissue spaces, whereas metabolic waste products and carbon dioxide diffuse from the tissue spaces into the blood through the wall of the venous end.

– are absent in the cornea, epidermis, and hyaline cartilage.

– may not be present in some areas where the arterioles and venules have direct connections. These **arteriovenous anastomoses** (AV shunts) bypass the capillaries and are especially numerous in the skin of the nose, lips, fingers, and ears, where they conserve body heat.

4. Sinusoids
- are wider and more irregular than capillaries.
- substitute for capillaries in the liver, spleen, red bone marrow, carotid body, adenohypophysis, suprarenal cortex, and parathyroid glands.
- have walls that consist largely of phagocytic cells.
- form a part of the **reticuloendothelial system,** which is concerned chiefly with phagocytosis and antibody formation.

II. Lymphatic System
- provides an important **immune mechanism** for the body.
- is involved in the **metastasis** of cancer cells.

A. Lymphatic vessels
- serve as one-way drainage toward the heart and return lymph to the bloodstream through the **thoracic duct,** the largest lymphatic vessel.
- are not generally visible in dissections but are the major route by which carcinoma metastasizes.
- function to **absorb large protein molecules** and transport them to the bloodstream because the molecules cannot pass through the walls of the blood capillaries back into the blood.
- carry lymphocytes from lymphatic tissues to the bloodstream.
- have valves to ensure the flow of lymph away from the tissues and toward the venous system.
- are constricted at the sites of valves, showing a knotted or beaded appearance.
- are absent in the brain, spinal cord, eyeballs, bone marrow, splenic pulp, hyaline cartilage, nails, and hair.

B. Lymphatic capillaries
- begin blindly in most tissues, collect tissue fluid, and join to form large collecting vessels that pass to regional lymph nodes.
- are wider than blood capillaries.
- **absorb lymph** from tissue spaces and transport it back to the venous system.
- are called **lacteals** in the villi of the small intestine, where they absorb emulsified fat.

C. Lymph nodes
- are organized collections of lymphatic tissue permeated by lymphatic channels.
- have fibroelastic capsules from which connective tissue trabeculae extend into the substance of the organ.
- have a **cortex** (outer part), which contains collections of lymphatic cells called **germinal centers,** and a **medulla** (inner part), which contains cords of lymphatic cells.
- produce **lymphocytes** and **plasma cells.**
- **trap bacteria** drained from an infected area and contain reticuloendothelial cells and phagocytic cells (**macrophages**) that ingest these bacteria.
- serve as **filters.** (Thus, the cancer cells in lymph vessels migrate or metastasize to lymph nodes and tend to remain within them, proliferating and gradually destroying them.)
- are hard and often palpable when there is a metastasis and are enlarged and tender during infection.

D. Lymph

- is a clear, watery fluid that is collected from the intercellular spaces. It contains no cells until lymphocytes are added in its passage through the lymph nodes.
- contains constituents similar to those of blood plasma (e.g., proteins, fats, lymphocytes).
- often contains fat droplets (called **chyle**) when it comes from intestinal organs.
- is filtered by passing through several lymph nodes before entering the venous system.

Review Test

Directions: Each of the numbered items or incomplete statements in this section is followed by answers or by completions of the statement. Select the ONE lettered answer or completion that is BEST in each case.

1. Which of the following structures is a fibrous sheet or band that covers the body under the skin and invests the muscles?

(A) Tendon
(B) Fascia
(C) Synovial tendon sheath
(D) Aponeurosis
(E) Ligament

2. Which of the following statements concerning cranial nerves is true?

(A) They consist of 31 pairs
(B) They are connected to the brain
(C) All but the trochlear nerve emerge from the dorsal aspect of the brain
(D) They have motor fibers with cell bodies located in the nerve outside the CNS
(E) They are connected to the sympathetic chain ganglia by gray rami communicantes

3. Which blood vessel or group of vessels below carries richly oxygenated blood to the heart?

(A) Superior vena cava
(B) Pulmonary arteries
(C) Pulmonary veins
(D) Ascending aorta
(E) Cardiac veins

Directions: Each of the numbered items or incomplete statements in this section is negatively phrased, as indicated by a capitalized word such as NOT, LEAST, or EXCEPT. Select the ONE lettered answer or completion that is BEST in each case.

4. All of the following statements concerning neurons are correct EXCEPT

(A) their axons carry impulses toward the cell bodies
(B) they consist of cell bodies, axons, and dendrites
(C) they are unipolar, bipolar, or multipolar in shape
(D) they are the basic structural and functional units of the nervous system
(E) they are specialized for the reception, integration, and transmission of information

5. All of the following statements concerning fetal circulation are true EXCEPT

(A) the right atrium receives oxygenated blood from the placenta
(B) right atrial pressure is higher than left atrial pressure
(C) blood flows from the pulmonary artery to the aorta through the ductus arteriosus
(D) the foramen ovale shunts blood from the left atrium to the right atrium, partially bypassing the lungs
(E) the ductus venosus shunts oxygenated blood from the umbilical vein to the inferior vena cava, bypassing the liver

Directions: The set of matching questions in this section consists of a list of four to twenty-six lettered options (some of which may be in figures) followed by several numbered items. For each numbered item, select the ONE lettered option that is most closely associated with it. To avoid spending too much time on matching sets with large numbers of options, it is generally advisable to begin each set by reading the list of options. Then, for each item in the set, try to generate the correct answer and locate it in the option list, rather than evaluating each option individually. Each lettered option may be selected once, more than once, or not at all.

Questions 6–10

Match each example or description below with the appropriate type of joint.

(A) Syndesmoses
(B) Synchondroses
(C) Hinge joints
(D) Ellipsoidal joints
(E) Ball-and-socket joints

6. Radiocarpal and metacarpophalangeal joints

7. Joints that allow flexion and extension only

8. Fibrous joints

9. Joints that allow movement in many directions

10. Primary cartilaginous joints; contain epiphyseal cartilage plates

Answers and Explanations

1–B. The fascia is a fibrous sheet or band that covers the body under the skin and invests the muscles.

2–B. The cranial nerves consist of 12 pairs; some have sensory fibers with cell bodies located in the nerve outside the CNS. All cranial nerves except the trochlear nerve emerge from the ventral aspect of the brain. The trochlear nerve emerges from the posterior aspect of the midbrain.

3–C. Pulmonary veins return oxygenated blood to the heart from the lungs. Pulmonary arteries carry deoxygenated blood from the heart to the lungs for oxygen renewal. The aorta carries oxygenated blood from the left ventricle to all parts of the body.

4–A. The dendrites of neurons carry impulses to the cell bodies; axons carry impulses away from the cell bodies.

5–D. The foramen ovale shunts blood from the right atrium to the left atrium, partially bypassing the lungs.

6–D. The wrist (radiocarpal) and metacarpophalangeal joints are ellipsoidal (condyloid) joints, which allow flexion and extension, and abduction and adduction.

7–C. Hinge (ginglymus) joints allow flexion and extension only. The elbow and interphalangeal joints are hinge joints.

8–A. Syndesmoses are fibrous joints. The tympanostapedial and inferior tibiofibular joints are syndesmoses.

9–E. Ball-and-socket (spheroidal) joints allow movement in many directions (multiaxial). The shoulder and hip joints are ball-and-socket joints.

10–B. Synchondroses are united by hyaline cartilage and contain epiphyseal cartilage plates.

2
Upper Limb

Bones and Joints

I. Bones of the Upper Limb (Figure 2-1)

A. Clavicle (collarbone)
- forms the **girdle of the upper limb** with the scapula.
- is the first bone to begin ossification during fetal development, but it is the last one to complete ossification, at about age 21 years.
- is the only long bone to be ossified intermembranously.
- Medial two-thirds are convex forward; lateral one-third is flattened with a marked concavity.
- articulates with the sternum at the **sternoclavicular joint** and with the scapula at the **acromioclavicular joint.**
- is a **commonly fractured bone.**

B. Scapula (shoulder blade)
- is a **triangular** flat bone.

1. Spine of the scapula
- is a triangular-shaped process and continues laterally as the **acromion.**
- divides the dorsal surface of the scapula into the **upper supraspinous and lower infraspinous fossae.**
- provides an origin for the deltoid muscle and an insertion for the trapezius muscle.

2. Acromion
- is the lateral end of the spine and articulates with the clavicle.
- provides an origin for the deltoid muscle and an insertion for the trapezius muscle.

3. Coracoid process
- provides the origin of the coracobrachialis and biceps brachii muscles and the insertion of the pectoralis minor muscle.
- gives attachment to the coracoclavicular, coracohumeral, and coracoacromial ligaments and the costocoracoid membrane.

4. Scapular notch
- is bridged by the superior transverse scapular ligament and is converted into a foramen, which permits passage of the **suprascapular nerve.**

5. **Glenoid cavity**

– articulates with the head of the **humerus.**

– is deepened by a fibrocartilaginous lip **(glenoid labrum).**

6. **Supraglenoid and infraglenoid tubercles**

– provide origins for the tendons of the long heads of the biceps brachii and triceps brachii muscles, respectively.

C. **Humerus** (see Figure 2-1)

1. **Head of the humerus**

– has a smooth, rounded, articular surface and articulates with the scapula at the **glenohumeral joint.**

2. **Anatomical neck of the humerus**

– is an indentation distal to the head of the humerus and provides for the attachment of the articular capsule.

3. **Greater tubercle**

– lies on the lateral side of the humerus, just lateral to the anatomical neck.

– provides attachments for the supraspinatus, infraspinatus, and teres minor muscles.

4. **Lesser tubercle**

– lies on the anterior medial side of the humerus, just distal to the anatomical neck.

– provides an insertion for the subscapularis muscle.

5. **Intertubercular (bicipital) groove**

– lies between the greater and lesser tubercles.

– lodges the tendon of the long head of the biceps brachii muscle, which forces the humeral head medially into the glenohumeral joint.

– is spanned by the **transverse humeral ligament,** which restrains the tendon of the long head of the biceps brachii.

– provides insertions for the pectoralis major on its lateral lip, the teres major on its medial lip, and the latissimus dorsi on its floor.

6. **Surgical neck of the humerus**

– is a narrow area distal to the tubercles and a **common site of fracture.**

– is in contact with the axillary nerve and the posterior humeral circumflex artery.

7. **Deltoid tuberosity**

– is a V-shaped roughened area on the lateral aspect of the midshaft and marks the insertion of the deltoid muscle.

8. **Spiral groove**

– is a groove for the **radial nerve,** separating the origin of the lateral head of the triceps above and the origin of the medial head below.

9. **Trochlea**

– is shaped like a spool or pulley and has a deep depression between two margins; this depression articulates with the **trochlear notch of the ulna.**

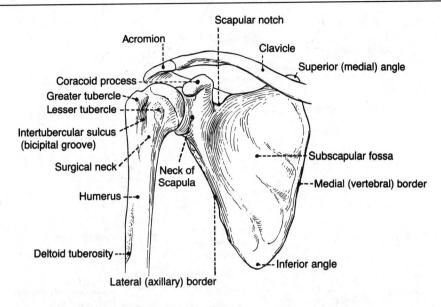

Anterior view

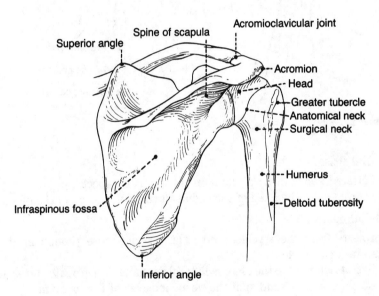

Posterior view

Figure 2-1. Pectoral girdle and humerus.

10. Capitulum
– is globular in shape and articulates with the **head of the radius.**

11. Olecranon fossa
– is located above the trochlea on the posterior aspect of the humerus.
– houses the olecranon of the ulna on full extension of the forearm.

12. Coronoid fossa
– is located above the trochlea on the anterior aspect of the humerus.
– receives the coronoid process of the ulna on flexion of the elbow.

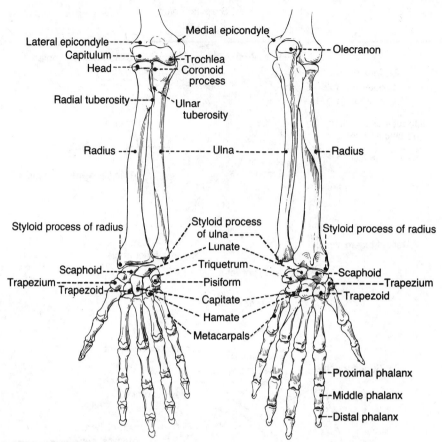

Figure 2-2. Bones of the forearm and hand.

13. Radial fossa

- is located above the capitulum on the anterior aspect.
- is occupied by the head of the radius during full flexion of the elbow joint.

14. Medial epicondyle

- projects from the trochlea and is larger and more prominent than the lateral epicondyle.
- gives attachment to the ulnar collateral ligament, the pronator teres muscle, and the common tendon of the flexor muscles of the forearm.

15. Lateral epicondyle

- projects from the capitulum and provides the origin of the supinator and extensor muscles of the forearm.

D. Radius (see Figure 2-2)

- is shorter than the ulna and is situated lateral to the ulna.
- is characterized by displacement of the hand dorsally and radially when fractured at its distal end (**Colles' fracture**).

1. Head (proximal end)

- articulates with the **capitulum of the humerus** and the **radial notch of the ulna**.

2. Distal end

– articulates with the **proximal row of carpal bones,** including the scaphoid, lunate, and triquetrous bones but excluding the pisiform bone.

3. Tuberosity

– is an oval prominence just distal to the neck and gives attachment to the biceps brachii tendon.

4. Styloid process

– is located on the distal end of the radius and is about 1 cm distal to that of the ulna.
– is palpable in the **anatomical snuff-box** (which is located between the extensor pollicis longus and brevis tendons) and provides insertion of the brachioradialis muscle.

E. Ulna (see Figure 2-2)

1. Olecranon

– is the curved projection on the back of the elbow and gives attachment to the triceps tendon.

2. Coronoid process

– is located below the trochlear notch and gives attachment to the brachialis.

3. Trochlear notch

– receives the trochlea of the humerus.

4. Radial notch

– accommodates the head of the radius.

5. Head (distal end)

– articulates with the articular disk of the **distal radioulnar joint** and contains a styloid process.

F. Carpal bones (see Figure 2-2)   *S L T P*

– are arranged in two rows of four:

1. Proximal row (lateral to medial): scaphoid, lunate, triquetral, and pi-siform

– except for the pisiform, articulate with the radius and the articular disk (the ulna has no contact with the carpal bones).   *T T C H*

2. Distal row (lateral to medial): trapezium, trapezoid, capitate, and hamate

G. Metacarpals

– are miniature long bones consisting of **bases** (proximal ends), **shafts** (bodies), and **heads** (distal ends).
– Heads form the knuckles of the fist.

H. Phalanges

– are miniature long bones consisting of **bases, shafts,** and **heads.**
– There are three in each finger and two in the thumb.
– The heads of the proximal and middle phalanges form the knuckles.

II. Joints and Ligaments of the Upper Limb (see Figures 2-1 and 2-2)

A. Acromioclavicular joint

- is a plane joint between the acromion and the lateral border of the clavicle.
- Articular surfaces are covered by fibrous cartilages.
- Stability is provided by the **coracoclavicular ligament,** which consists of the conoid and trapezoid ligaments.

B. Sternoclavicular joint

- is a double plane (gliding) joint, united by the fibrous capsule, which is reinforced by the anterior and posterior sternoclavicular, interclavicular, and costoclavicular ligaments.

C. Shoulder (glenohumeral) joint

- is a **multiaxial ball-and-socket (spheroidal) joint** between the glenoid cavity of the scapula and the head of the humerus and allows abduction and adduction, flexion and extension, and circumduction and rotation.
- Both articular surfaces are covered with hyaline cartilage.
- capsule lies deep to the tendon of the **rotator (musculotendinous) cuff** and is attached to the margin of the glenoid cavity and to the anatomical neck of the humerus.
- cavity is deepened by the fibrocartilaginous **glenoid labrum** and communicates with the subscapular bursa.
- is innervated by the axillary, suprascapular, and lateral pectoral nerves.
- receives blood from branches of the suprascapular, anterior and posterior humeral circumflex, and scapular circumflex arteries.
- Inferior (or anterior) dislocation stretches the fibrous capsule, avulses the glenoid labrum, and may injure the axillary nerve.

1. Rotator (musculotendinous) cuff

- contributes to the stability of the shoulder joint by keeping the head of the humerus pressed into the glenoid fossa.
- is formed by the tendons of the **subscapularis, supraspinatus, infraspinatus, and teres minor** muscles.

2. Bursae around the shoulder

- form a **lubricating mechanism** between the rotator cuff and the coracoacromial arch during movement of the shoulder joint.

a. Subacromial bursa

- lies between the acromion and the coracoacromial ligament superiorly and the supraspinatus muscle inferiorly.
- usually communicates with the subdeltoid bursa but normally does not communicate with the synovial cavity of the glenohumeral joint.
- facilitates the movement of the deltoid muscle over the joint capsule and the supraspinatus tendon.

b. Subdeltoid bursa

- lies between the deltoid muscle and coracoacromial arch superiorly and the tendon of the supraspinatus muscle inferiorly.
- usually communicates with the subacromial bursa.

c. Subscapular bursa

– lies between the tendon of the subscapularis and the neck of the scapula.
– communicates with the synovial cavity of the shoulder joint.

3. Ligaments of the shoulder

a. Glenohumeral ligament

– extends from the supraglenoid tubercle to the upper part of the lesser tubercle of the humerus (**superior glenohumeral ligament**), to the lower anatomical neck of the humerus (**middle glenohumeral ligament**), and to the lower part of the lesser tubercle of the humerus (**inferior glenohumeral ligament**).

b. Transverse humeral ligament

– extends between the greater and lesser tubercles and holds the tendon of the long head of the biceps in the intertubercular groove.

c. Coracohumeral ligament

– extends from the coracoid process to the greater tubercle of the humerus.

d. Coracoacromial ligament

– extends from the coracoid process to the acromion.

D. Elbow joint

– forms a **hinge (ginglymus) joint** between the capitulum of the humerus and the head of the radius (**humeroradial joint**) and between the trochlea of the humerus and the trochlear notch of the ulna (**humeroulnar joint**), and allows flexion and extension.
– also includes the **proximal radioulnar (pivot) joint** within a common articular capsule, which allows supination and pronation.
– is innervated by the musculocutaneous, median, radial, and ulnar nerves.
– Its blood supply is derived from the anastomosis formed by branches of the brachial artery and recurrent branches of the radial and ulnar arteries.
– is reinforced by the following ligaments:

1. Annular ligament

– is a fibrous band that forms nearly four-fifths of a circle around the head of the radius; the remainder is formed by the **radial notch.**
– attaches to the anterior and posterior lips of the radial notch of the ulna, and to the origin of the supinator muscle.
– forms a collar around the head of the radius and thus serves as a restraining ligament, preventing withdrawal of the head of the radius from its socket.
– fuses with the radial collateral ligament and blends with the articular capsule of the elbow joint.

2. Radial collateral ligament

– is a fibrous thickening that extends from the lateral epicondyle to the anterior and posterior margins of the radial notch of the ulna and the annular ligament of the radius.

3. Ulnar collateral ligament

– is **triangular** and is composed of anterior, posterior, and oblique bands.

– extends from the medial epicondyle to the coronoid process and the olecranon of the ulna.

E. Proximal radioulnar joint

– forms a **pivot (trochoid) joint** in which the head of the radius articulates with the radial notch of the ulna, and allows pronation and supination.

F. Distal radioulnar joint

– forms a **pivot joint** between the head of the ulna and the ulnar notch of the radius, and allows pronation and supination.

G. Wrist (radiocarpal) joint

– is an **ellipsoidal (condyloid) joint** formed superiorly by the radius and the articular disk and inferiorly by the proximal row of carpal bones (scaphoid, lunate, and triquetrous), exclusive of the pisiform. (The ellipsoidal surface of the carpal bones fits into the concave surface of the radius and articular disk.)
– allows flexion and extension, abduction and adduction, and circumduction.
– Articular capsule is strengthened by radial and ulnar collateral ligaments and dorsal and palmar radiocarpal ligaments.

H. Midcarpal joint

– is an articulation between the proximal and distal rows of carpal bones.
– forms an **ellipsoidal joint** by fitting the hamate bone and the head of the capitate bone into the concavity of the scaphoid, lunate, and triquetral bones.
– also forms a plane joint by joining the scaphoid with the trapezium and trapezoid bones.

I. Carpometacarpal joints

– form **saddle (sellar) joints** between the trapezium and the base of the first metacarpal bone.
– also form **plane joints** between the carpal bones and the medial four metacarpal bones.

J. Metacarpophalangeal joints

– are **ellipsoidal joints** and are supported by a palmar ligament and two collateral ligaments.

K. Interphalangeal joints

– are **hinge joints** and are supported by a palmar ligament and two collateral ligaments.

Cutaneous Nerves, Superficial Veins, and Lymphatics

I. Cutaneous Nerves of the Upper Limb (Figure 2-3)

A. Supraclavicular nerve

– arises from the cervical plexus (C3, C4) and innervates the skin over the upper pectoral, deltoid, and outer trapezius areas.

B. Medial brachial cutaneous nerve

– arises from the medial cord of the brachial plexus and innervates the medial side of the arm.

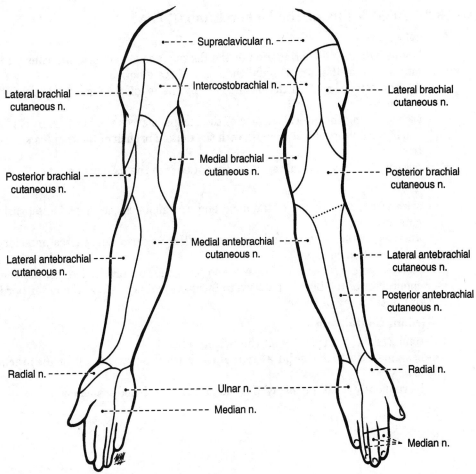

Figure 2-3. Cutaneous nerves of the upper limb.

C. Medial antebrachial cutaneous nerve

– arises from the medial cord of the brachial plexus and innervates the medial side of the forearm.

D. Lateral brachial cutaneous nerve

– arises from the axillary nerve and innervates the lateral side of the arm.

E. Lateral antebrachial cutaneous nerve

– arises from the musculocutaneous nerve and innervates the lateral side of the forearm.

F. Posterior brachial and antebrachial cutaneous nerves

– arise from the radial nerve and innervate the posterior sides of the arm and forearm, respectively.

G. Intercostobrachial nerve

– is the lateral cutaneous branch of the second intercostal nerve and emerges from the second intercostal space by piercing the intercostal and serratus anterior muscles.

– may communicate with the medial brachial cutaneous nerve.

II. Superficial Veins of the Upper Limb (Figure 2-4)

A. Cephalic vein

- begins as a radial continuation of the dorsal venous network and runs on the lateral side of the forearm and the front of the elbow.
- is often connected with the basilic vein by the median cubital vein in front of the elbow.
- ascends along the lateral surface of the biceps, pierces the brachial fascia, and lies in the deltopectoral triangle with the deltoid branch of the thoracoacromial trunk.
- pierces the clavipectoral fascia and empties into the axillary vein.

B. Basilic vein

- arises from the dorsum of the hand and accompanies the medial antebrachial cutaneous nerve.
- ascends on the posteromedial surface of the forearm and passes anterior to the medial epicondyle.
- pierces the deep fascia of the arm and joins the two brachial veins, the venae comitantes of the brachial artery, to form the axillary vein at the lower border of the teres major muscle.

C. Median cubital vein

- connects the cephalic vein to the basilic vein.
- lies superficial to the **bicipital aponeurosis,** which separates it from the brachial artery.
- is frequently used for intravenous injections and blood transfusions.

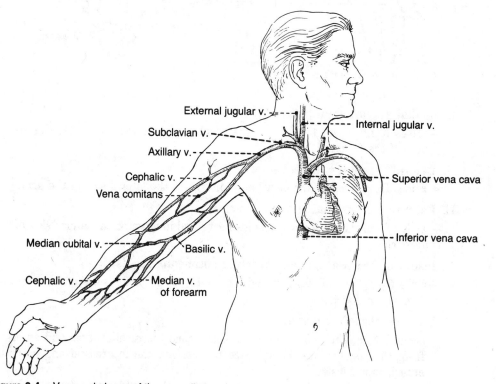

Figure 2-4. Venous drainage of the upper limb.

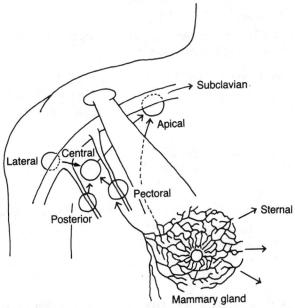

Figure 2-5. Lymphatic drainage of the breast and axillary lymph nodes.

D. Median antebrachial vein

– arises in the palmar venous network, ascends on the front of the forearm, and terminates in the median cubital or the basilic vein.

E. Dorsal venous network

– receives dorsal digital veins by means of dorsal metacarpal veins.
– also receives palmar digital veins by means of intercapitular and palmar metacarpal veins.
– Its radial part is continued proximally as the **cephalic vein,** and its ulnar part is continued proximally as the **basilic vein.**

III. Superficial Lymphatics and Axillary Lymph Nodes

A. Lymphatics of the finger

– drain into the plexus on the dorsum and palm of the hand.

B. Medial group of lymphatic vessels

– accompanies the basilic vein, passes through the cubital or supratrochlear nodes, and ascends to enter the lateral axillary nodes.

C. Lateral group of lymphatic vessels

– accompanies the cephalic vein and drains into the lateral axillary nodes and also into the **deltopectoral (infraclavicular) nodes.** The deltopectoral nodes drain into the apical nodes.

D. Axillary lymph nodes (Figure 2-5)

1. Central nodes

– lie near the base of the axilla between the lateral thoracic and subscapular veins; receive lymph from the lateral, pectoral, and posterior groups of nodes; and drain into the apical nodes.

2. **Lateral nodes**
   - lie posteromedial to the axillary veins, receive lymph from the upper limb, and drain into the central nodes.

3. **Subscapular (posterior) nodes**
   - lie along the subscapular vein and drain lymph from the posterior thoracic wall and the posterior aspect of the shoulder to the central nodes.

4. **Pectoral nodes**
   - lie along the inferolateral border of the pectoralis minor muscle and drain lymph from the anterior and lateral thoracic walls, including the breast.

5. **Apical nodes**
   - lie at the apex of the axilla medial to the axillary vein and above the upper border of the pectoralis minor muscle, receive lymph from all of the other axillary nodes (and occasionally from the breast), and drain into the subclavian trunks.

Muscles and Fasciae of the Upper Limb

I. Fasciae of Pectoral and Axillary Regions

A. Clavipectoral fascia
- extends between the coracoid process, the clavicle, and the thoracic wall.
- envelops the subclavius and pectoralis minor muscles.

B. Costocoracoid membrane
- is a part of the clavipectoral fascia between the first rib and the coracoid process and covers the deltopectoral triangle.
- is pierced by the **cephalic vein,** the **thoracoacromial artery,** and the **lateral pectoral nerve.**

C. Pectoral fascia
- covers the pectoralis major muscle, is attached to the sternum and clavicle, and is continuous with the axillary fascia.

D. Axillary fascia
- is continuous anteriorly with the pectoral and clavipectoral fasciae, laterally with the brachial fascia, and posteromedially with the fascia of the latissimus dorsi and serratus anterior muscles.

E. Axillary sheath
- is a fascial prolongation of the prevertebral layer of the cervical fascia into the axilla, enclosing the axillary vessels and the brachial plexus.

II. Breast and Mammary Gland (Figure 2-6)

A. Breast
- consists of mammary gland tissue, fibrous and fatty tissue, blood and lymph vessels, and nerves.
- extends from the second to sixth ribs and from the sternum to the midaxillary line.

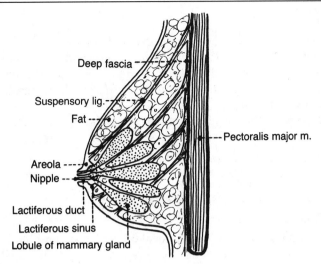

Figure 2-6. Breast.

– has glandular tissue, which lies in the superficial fascia.
– is supplied by the anterior perforating branches of the internal thoracic artery and the lateral mammary branches of the lateral thoracic artery.
– is innervated by the anterior and lateral cutaneous branches of the second to sixth intercostal nerves.
– has **suspensory ligaments (Cooper's ligaments),** which are strong fibrous processes that support the breast and run from the dermis of the skin to the deep layer of the superficial fascia through the breast.
– The **nipple** usually lies at the level of the fourth intercostal space.
– The **areola** is a ring of pigmented skin around the nipple.
– has an **axillary tail,** which is a normal extension of the mammary gland into the axilla.
– More than one pair of breasts (**polymastia**) and more than one pair of nipples (**polythelia**) may be present.

B. Mammary gland

– is a **modified sweat gland** located in the superficial fascia.
– has 15 to 20 lobes of glandular tissue, each of which opens by a **lactiferous duct** onto the tip of the nipple. Each duct enlarges to form a **lactiferous sinus,** which serves as a reservoir for milk during lactation.

C. Lymphatic drainage (see Figure 2-5)

– drains mainly (75%) to the **axillary nodes,** more specifically to the **pectoral nodes** (including drainage of the nipple).
– follows the perforating vessels through the pectoralis major muscle and the thoracic wall to enter the **parasternal (internal thoracic) nodes,** which lie along the internal thoracic artery.
– also drains to the apical nodes and may connect to lymphatics draining the opposite breast and to lymphatics draining the anterior abdominal wall.
– is of great importance in view of the frequent development of cancer and subsequent dissemination of cancer cells through the lymphatic stream.

1. **Breast cancer**
   - forms a palpable mass in the advanced stage.
   - enlarges, attaches to Cooper's ligaments, and produces shortening of the ligaments, causing depression or **dimpling** of the overlying skin.
   - may also attach to and shorten the lactiferous ducts, resulting in a **retracted or inverted nipple**.
   - may invade the deep fascia of the pectoralis major muscle, so that contraction of this muscle produces a **sudden upward movement** of the whole breast.

2. **Detection and treatment of breast cancer**

   a. **Mammography**
      - is a roentgenographic examination of the breast.

   b. **Radical mastectomy**
      - is **extensive surgical removal of the breast and its related structures,** including the pectoralis major and minor muscles, axillary lymph nodes and fascia, and part of the thoracic wall.
      - may injure the long thoracic and thoracodorsal nerves.
      - may cause postoperative swelling (edema) of the upper limb as a result of lymphatic obstruction caused by removal of most of the lymphatic channels that drain the arm, or by venous obstruction caused by thrombosis of the axillary vein.

   c. **Modified radical mastectomy**
      - involves **excision of the entire breast and axillary lymph nodes,** with preservation of the pectoralis major and minor muscles. (The pectoralis minor muscle is usually retracted or severed near its insertion into the coracoid process.)

III. Axilla

A. Boundaries of the axilla

1. **Medial wall:** upper ribs and their intercostal muscles and serratus anterior muscle

2. **Lateral wall:** humerus

3. **Posterior wall:** subscapularis, teres major, and latissimus dorsi muscles

4. **Anterior wall:** pectoralis major and pectoralis minor muscles

5. **Base:** axillary fascia

6. **Apex:** interval between the clavicle, scapula, and first rib

B. Contents of the axilla

- include the axillary vasculature, branches of the brachial plexus, the long and short heads of the biceps brachii, and the coracobrachialis.

IV. Muscles of the Pectoral Region (Figure 2-7)

A. Pectoralis major

- arises from the medial half of the clavicle, the manubrium and body of the sternum, and the upper six costal cartilages.
- inserts on the lateral lip of the intertubercular groove (the crest of the greater tubercle) of the humerus.

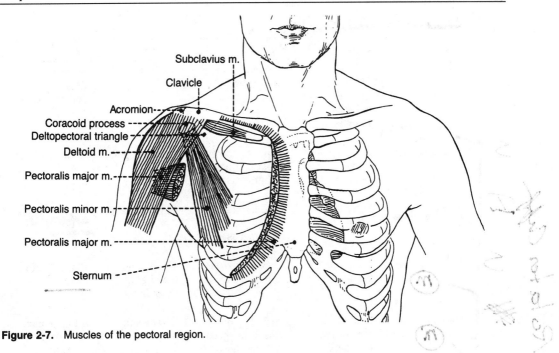

Figure 2-7. Muscles of the pectoral region.

- – is innervated by the medial and lateral pectoral nerves.
- – adducts and medially rotates the arm. The clavicular part can rotate the arm medially and flex it; the sternocostal part depresses the arm and shoulder. Its lower fibers can help extend the arm when it is flexed.
- – forms the anterior wall of the axilla, and its lateral border forms the anterior axillary fold.

B. Pectoralis minor

- – arises from the external surfaces of the **second to fifth ribs.**
- – inserts into the coracoid process and depresses the shoulder.
- – is **innervated** chiefly by the **medial pectoral nerve;** also innervated by the lateral pectoral nerve as a result of their communication.
- – is invested by the clavipectoral fascia, divides the axillary artery into three parts, forms part of the anterior wall of the axilla, and crosses the cords of the brachial plexus.

C. Subclavius

- – originates from the **junction of the first rib and its cartilage.**
- – inserts on the lower surface of the clavicle.
- – is **innervated** by the **nerve to the subclavius.**
- – assists in depressing the lateral portion of the clavicle.

D. Serratus anterior

- – arises from the external surfaces of the **upper eight ribs.**
- – is inserted on the medial border of the scapula.
- – is **innervated** by the **long thoracic nerve.**
- – **rotates the scapula upward** so that the inferior angle swings laterally and abducts the arm and elevates it above a horizontal position.

V. Muscles of the Shoulder

Muscle	Origin	Insertion	Nerve	Action
Deltoid	Lateral third of clavicle, acromion, and spine of scapula	Deltoid tuberosity of humerus	Axillary	Abducts, adducts, flexes, extends, and rotates arm medially
Supraspinatus	Supraspinous fossa of scapula	Superior facet of greater tubercle of humerus	Suprascapular	Abducts arm
Infraspinatus	Infraspinous fossa	Middle facet of greater tubercle of humerus	Suprascapular	Rotates arm laterally
Subscapularis	Subscapular fossa	Lesser tubercle of humerus	Upper and lower subscapular	Rotates arm medially
Teres major	Dorsal surface of inferior angle of scapula	Medial lip of intertubercular groove of humerus	Lower subscapular	Adducts and rotates arm medially
Teres minor	Upper portion of lateral border of scapula	Lower facet of greater tubercle of humerus	Axillary	Rotates arm laterally
Latissimus dorsi	Spines of T7–T12 thoracolumbar fascia, iliac crest, ribs 9–12	Floor of bicipital groove of humerus	Thoracodorsal	Adducts, extends, and rotates arm medially

VI. Structures of the Dorsal Scapular Region (Figure 2-8)

A. Quadrangular space

- is bounded superiorly by the teres minor and subscapularis muscles, inferiorly by the teres major muscle, medially by the long head of the triceps, and laterally by the surgical neck of the humerus.
- transmits the **axillary nerve** and the **posterior humeral circumflex vessels.**

B. Triangular space

- is bounded superiorly by the teres minor muscle, inferiorly by the teres major muscle, and laterally by the long head of the triceps.
- contains the **circumflex scapular vessels.**

C. Triangular interval

- is formed superiorly by the teres major muscle, medially by the long head of the triceps, and laterally by the medial head of the triceps.
- contains the **radial nerve** and the **profunda brachii (deep brachial) artery.**

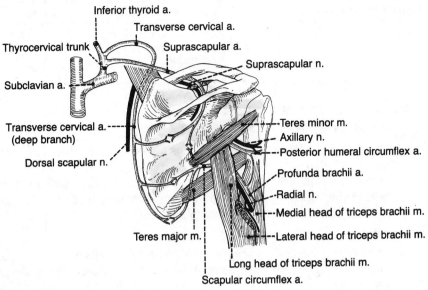

Figure 2-8. Structures of the dorsal scapular region.

D. Triangle of auscultation

– is bounded by the upper border of the latissimus dorsi muscle, the lateral border of the trapezius muscle, and the medial border of the scapula; its floor is formed by the rhomboid major muscle.
– is most prominent when the shoulders are drawn forward.
– is the site at which breathing sounds are heard most clearly.

VII. Muscles of the Arm

Muscle	Origin	Insertion	Nerve	Action
Coracobrachialis	Coracoid process	Middle third of medial surface of humerus	Musculocutaneous	Flexes and adducts arm
Biceps brachii	Long head, supraglenoid tubercle; short head, coracoid process	Radial tuberosity of radius	Musculocutaneous	Flexes arm and forearm, supinates forearm
Brachialis	Lower anterior surface of humerus	Coronoid process of ulna and ulnar tuberosity	Musculocutaneous	Flexes forearm
Triceps	Long head, infraglenoid tubercle; lateral head, superior to radial groove of humerus; medial head, inferior to radial groove	Posterior surface of olecranon process of ulna	Radial	Extends forearm

Muscle	Origin	Insertion	Nerve	Action
Anconeus	Lateral epicondyle of humerus	Olecranon and upper posterior surface of ulna	Radial	Extends forearm

VIII. Structures of the Arm and Forearm

A. Brachial intermuscular septa

– extend from the brachial fascia, which is a portion of the deep fascia enclosing the arm.

– consist of medial and lateral intermuscular septa, which divide the arm into the anterior fascial compartment (**flexor compartment**) and the posterior fascial compartment (**extensor compartment**).

B. Cubital fossa

– is a V-shaped interval on the anterior aspect of the elbow.

– is bounded laterally by the brachioradialis muscle and medially by the pronator teres muscle.

– Its upper limit is an imaginary horizontal line connecting the epicondyles of the humerus.

– Its floor is formed by the brachialis and supinator muscles.

– Its lower end is where the brachial artery divides into the radial and ulnar arteries, and its fascial roof is strengthened by the bicipital aponeurosis.

– contains (from lateral to medial) the **radial nerve, biceps tendon, brachial artery,** and **median nerve.**

C. Bicipital aponeurosis

– originates from the medial border of the biceps tendon.

– lies on the brachial artery and the median nerve and passes downward and medially to blend with the deep fascia of the forearm.

D. Interosseous membrane of the forearm

– is a broad sheet of **dense connective tissue** extending between the radius and the ulna.

– Its proximal border and the oblique cord form a gap through which the posterior interosseous vessels pass. (The oblique cord extends from the ulnar tuberosity to the radius, somewhat distal to its tuberosity.)

– Its distal part is pierced by the anterior interosseous vessels.

– provides extra surface area for attachment of the deep extrinsic flexor, extensor, and abductor muscles of the hand.

E. Characteristics of the arm and forearm

1. Carrying angle

– is the angle formed by the axis of the arm and forearm when the forearm is extended, because the medial edge of the trochlea projects more inferiorly than its lateral edge; thus, the long axis of the humerus lies at an angle of about 170° to the long axis of the ulna.

– carries the hand away from the side of the body in extension and pronation. (It is wider in women than in men.)

– disappears when the forearm is flexed or pronated, because the trochlea runs in a spiral direction from anterior to posterior aspects.

2. Pronation and supination

- occur at the proximal and distal radioulnar joints.
- In supination, the palm faces forward (lateral rotation); in pronation, the radius rotates over the ulna, and thus the palm faces backward (medial rotation about a longitudinal axis, in which case the shafts of the radius and ulna cross each other).
- are movements in which the upper end of the radius nearly rotates within the annular ligament.
- have unequal strengths, with supination being the stronger.

IX. Muscles of the Anterior Forearm

Muscle	Origin	Insertion	Nerve	Action
Pronator teres	Medial epicondyle and coronoid process of ulna	Middle of lateral side of radius	Median	Pronates forearm
Flexor carpi radialis	Medial epicondyle of humerus	Bases of second and third metacarpals	Median	Flexes forearm, flexes and abducts hand
Palmaris longus	Medial epicondyle of humerus	Flexor retinaculum, palmar aponeurosis	Median	Flexes hand and forearm
Flexor carpi ulnaris	Medial epicondyle, medial olecranon, and posterior border of ulna	Pisiform, hook of hamate, and base of fifth metacarpal	Ulnar	Flexes and adducts hand, flexes forearm
Flexor digitorum superficialis	Medial epicondyle, coronoid process, oblique line of radius	Middle phalanges of finger	Median	Flexes proximal interphalangeal joints, flexes hand and forearm
Flexor digitorum profundus	Anteromedial surface of ulna, interosseous membrane	Bases of distal phalanges of fingers	Ulnar and median	Flexes distal interphalangeal joints and hand
Flexor pollicis longus	Anterior surface of radius, interosseous membrane, and coronoid process	Base of distal phalanx of thumb	Median	Flexes thumb
Pronator quadratus	Anterior surface of distal ulna	Anterior surface of distal radius	Median	Pronates forearm

X. Muscles of the Posterior Forearm

Muscle	Origin	Insertion	Nerve	Action
Brachioradialis	Lateral supracondylar ridge of humerus	Base of radial styloid process	Radial	Flexes forearm

Muscle	Origin	Insertion	Nerve	Action
Extensor carpi radialis longus	Lateral supracondylar ridge of humerus	Dorsum of base of second metacarpal	Radial	Extends and abducts hand
Extensor carpi radialis brevis	Lateral epicondyle of humerus	Posterior base of third metacarpal	Radial	Extends fingers and abducts hands
Extensor digitorum	Lateral epicondyle of humerus	Extensor expansion, base of middle and digital phalanges	Radial	Extends fingers and hand
Extensor digiti minimi	Common extensor tendon and interosseous membrane	Extensor expansion, base of middle and distal phalanges	Radial	Extends little finger
Extensor carpi ulnaris	Lateral epicondyle and posterior surface of ulna	Base of fifth metacarpal	Radial	Extends and adducts hand
Supinator	Lateral epicondyle, radial collateral and annular ligaments	Lateral side of upper part of radius	Radial	Supinates forearm
Abductor pollicis longus	Interosseous membrane, middle third of posterior surfaces of radius and ulna	Lateral surface of base of first metacarpal	Radial	Abducts thumb and hand
Extensor pollicis longus	Interosseous membrane and middle third of posterior surface of ulna	Base of distal phalanx of thumb	Radial	Extends distal phalanx of thumb and abducts hand
Extensor pollicis brevis	Interosseous membrane and posterior surface of middle third of radius	Base of proximal phalanx of thumb	Radial	Extends proximal phalanx of thumb and abducts hand
Extensor indicis	Posterior surface of ulna and interosseous membrane	Extensor expansion of index finger	Radial	Extends index finger

XI. Fibrous Structures of the Hand (Figures 2-9 and 2-10)

A. Extensor retinaculum

 – is a **thickening of the antebrachial fascia** on the back of the wrist.
 – extends from the lateral margin of the radius to the styloid process of the ulna, the pisiform, and the triquetrum.
 – is crossed by the superficial branch of the radial nerve.

B. Palmar aponeurosis

 – is a triangular fibrous layer overlying the tendons in the palm.
 – is continuous with the palmaris longus tendon, the thenar and hypothenar fasciae, the flexor retinaculum, and the palmar carpal ligament.

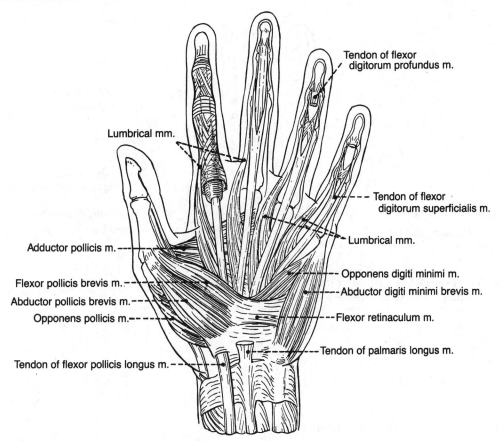

Figure 2-9. Superficial muscles of the hand.

– Its thickening, shortening, and fibrosis produce **Dupuytren's contracture,** a flexion deformity in which the fingers are pulled toward the palm.

C. **Flexor retinaculum** (see Figure 2-9)

– serves as an origin for muscles of the thenar eminence.
– forms a **carpal (osteofacial) tunnel** on the anterior aspect of the wrist.
– is attached medially to the triquetrum and pisiform bones and the hook of the hamate and laterally to the tubercles of the scaphoid and trapezium bones.
– Structures entering the palm superficial to the flexor retinaculum include the **ulnar nerve, ulnar artery, palmaris longus tendon,** and **palmar cutaneous branch of the median nerve.**

D. **Carpal tunnel**

– is formed anteriorly by the flexor retinaculum and posteriorly by the carpal bones.
– transmits the flexor tendons and the median nerve, which may be compressed.
– Structures inside the carpal tunnel include the **median nerve** and the tendons of **flexor pollicis longus, flexor digitorum profundus,** and **flexor digitorum superficialis muscles.**

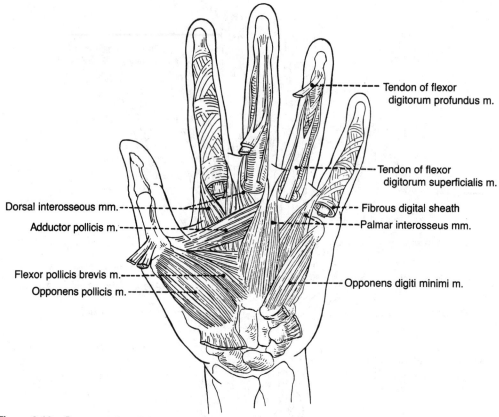

Tendon of flexor digitorum profundus m.

Tendon of flexor digitorum superficialis m.

Dorsal interosseous mm.

Adductor pollicis m.

Fibrous digital sheath

Palmar interosseus mm.

Flexor pollicis brevis m.

Opponens pollicis m.

Opponens digiti minimi m.

Figure 2-10. Deep muscles of the hand.

E. Fascial spaces of the palm

– are large fascial spaces in the hand divided by a midpalmar (oblique) septum into the **thenar space** and the **midpalmar space.**

1. Thenar space

– lies between the middle metacarpal bone and the tendon of the flexor pollicis longus muscle.

2. Midpalmar space

– lies between the middle metacarpal bone and the radial side of the hypothenar eminence.

F. Synovial flexor sheaths

1. Common synovial flexor sheath (ulnar bursa)

– envelops or contains the tendons of both the flexor digitorum superficialis and profundus muscles.

2. Synovial sheath for flexor pollicis longus (radial bursa)

– envelops the tendon of the flexor pollicis longus muscle.

G. Extensor expansion (Figure 2-11)

– is the expansion of the extensor tendon over the metacarpophalangeal joint and is referred to by clinicians as the **extensor hood.**

– provides the insertion of the lumbrical and interosseous muscles.

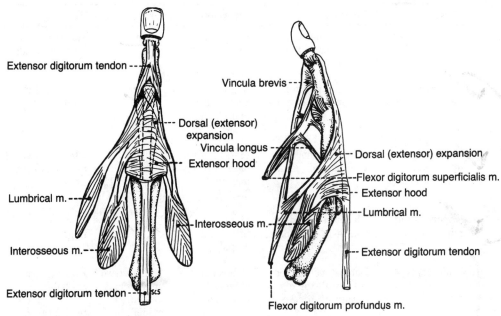

Figure 2-11.   Dorsal (extensor) expansion of the middle finger.

H. Anatomical snuff-box

– is a **triangular interval** bounded medially by the tendon of the extensor pollicis longus muscle and laterally by the tendons of the extensor pollicis brevis and abductor pollicis longus muscles.
– is limited proximally by the styloid process of the radius.
– Its floor is formed by the scaphoid and trapezium bones and is crossed by the radial artery.

XII. Muscles of the Hand

Muscle	Origin	Insertion	Nerve	Action
Abductor pollicis brevis	Flexor retinaculum, scaphoid, and trapezium	Lateral side of base of proximal phalanx of thumb	Median	Abducts thumb
Flexor pollicis brevis	Flexor retinaculum and trapezium	Base of proximal phalanx of thumb	Median	Flexes thumb
Opponens pollicis	Flexor retinaculum and trapezium	Lateral side of first metacarpal	Median	Opposes thumb to other digits
Adductor pollicis	Capitate and bases of second and third metacarpals (oblique head); palmar surface of third metacarpal (transverse head)	Medial side of base of proximal phalanx of the thumb	Ulnar	Adducts thumb

Muscle	Origin	Insertion	Nerve	Action
Palmaris brevis	Medial side of flexor retinaculum, palmar aponeurosis	Skin of medial side of palm	Ulnar	Wrinkles skin on medial side of palm
Abductor digiti minimi	Pisiform and tendon of flexor carpi ulnaris	Medial side of base of proximal phalanx of little finger	Ulnar	Abducts little finger
Flexor digiti minimi brevis	Flexor retinaculum and hook of hamate	Medial side of base of proximal phalanx of little finger	Ulnar	Flexes proximal phalanx of little finger
Opponens digiti minimi	Flexor retinaculum and hook of hamate	Medial side of fifth metacarpal	Ulnar	Opposes little finger
Lumbricals (4)	Lateral side of tendons of flexor digitorum profundus	Lateral side of extensor expansion	Median (two lateral) and ulnar (two medial)	Flex metacarpophalangeal joints and extend interphalangeal joints
Dorsal interossei (4)	Adjacent sides of metacarpal bones	Lateral sides of bases of proximal phalanges; extensor expansion	Ulnar	Abduct fingers; flex metacarpophalangeal joints; extend interphalangeal joints
Palmar interossei (3)	Medial side of second metacarpal; lateral sides of fourth and fifth metacarpals	Bases of proximal phalanges in same sides as their origins; extensor expansion	Ulnar	Adduct fingers; flex metacarpophalangeal joints; extend interphalangeal joints

Nerves of the Upper Limb

I. Brachial Plexus (Figure 2-12)

- is formed by the ventral primary rami of the lower four cervical nerves and the first thoracic nerve (C5–T1).
- has roots that pass between the scalenus anterior and medius muscles.
- is enclosed with the axillary artery and vein in the **axillary sheath,** which is formed by a prolongation of the prevertebral fascia.
- has the following subdivisions:

A. Branches from the roots

1. Dorsal scapular nerve (C5)

- pierces the scalenus medius muscle to reach the posterior cervical triangle and descends deep to the levator scapulae and the rhomboid minor and major muscles.
- innervates the rhomboids and frequently the levator scapulae muscles.

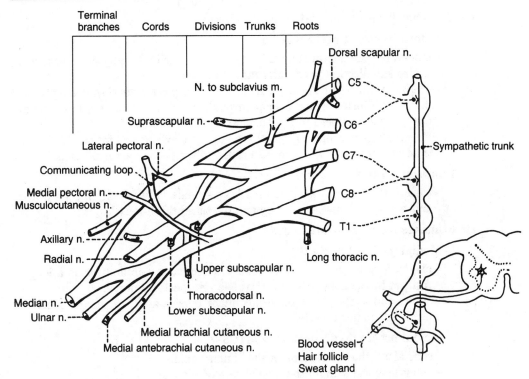

Figure 2-12. Brachial plexus.

2. Long thoracic nerve (C5–C7)

- descends behind the brachial plexus and runs on the external surface of the serratus anterior muscle, which it supplies.
- when damaged, causes **winging of the scapula** and makes elevating the arm above a horizontal position impossible.

B. Branches from the upper trunk

1. Suprascapular nerve (C5–C6)

- runs laterally across the posterior cervical triangle.
- passes through the scapular notch under the superior transverse scapular ligament, whereas the suprascapular artery passes over the ligament. (Thus, it can be said that the army [**artery**] runs over the bridge [**ligament**], and the navy [**nerve**] runs under the bridge.)
- supplies the supraspinatus muscle and the shoulder joint and then descends through the notch of the scapular neck to innervate the infraspinatus muscle.

2. Nerve to subclavius (C5)

- descends in front of the brachial plexus and the subclavian artery and behind the clavicle to reach the subclavius muscle.
- also innervates the sternoclavicular joint.
- usually branches to the **accessory phrenic nerve (C5),** which enters the thorax to join the phrenic nerve.

C. Branches from the lateral cord

1. Lateral pectoral nerve (C5–C7)

– innervates the **pectoralis major** muscle primarily; by way of a nerve loop, it also supplies the **pectoralis minor** muscle.
– sends a branch over the first part of the axillary artery to the medial pectoral nerve and forms a nerve loop through which the lateral pectoral nerve conveys motor fibers to the pectoralis minor muscle.
– pierces the costocoracoid membrane of the clavipectoral fascia.
– is accompanied by the pectoral branch of the thoracoacromial artery.

2. Musculocutaneous nerve (C5–C7)

– pierces the coracobrachialis muscle, descends between the biceps brachii and brachialis muscles, and innervates these three muscles.

D. Branches from the medial cord

1. Medial pectoral nerve (C8–T1)

– passes forward between the axillary artery and vein and forms a loop in front of the axillary artery with the lateral pectoral nerve.
– enters and supplies the pectoralis minor muscle and reaches the overlying pectoralis major muscle.

2. Medial brachial cutaneous nerve (C8–T1)

– runs along the medial side of the axillary vein.
– innervates the skin on the medial side of the arm.
– may communicate with the intercostobrachial cutaneous nerve.

3. Medial antebrachial cutaneous nerve (C8–T1)

– runs between the axillary artery and vein and then runs medial to the brachial artery.
– innervates the skin on the medial side of the forearm.

4. Ulnar nerve (C7–T1)

– runs down the medial aspect of the arm but does not branch in the brachium.

E. Branches from the medial and lateral cords: median nerve (C5–T1)

– is formed by heads from both the medial and lateral cords.
– runs down the anteromedial aspect of the arm but does not branch in the brachium.

F. Branches from the posterior cord

1. Upper subscapular nerve (C5–C6)

– innervates the upper portion of the subscapularis muscle.

2. Thoracodorsal nerve (C7–C8)

– runs behind the axillary artery and accompanies the thoracodorsal artery to enter the latissimus dorsi muscle.

3. Lower subscapular nerve (C5–C6)

– innervates the lower part of the subscapularis and teres major muscles.
– runs downward behind the subscapular vessels to the teres major muscle.

4. Axillary nerve (C5–C6)

– innervates the deltoid muscle (by its anterior and posterior branches) and the teres minor muscle (by its posterior branch).

   – gives rise to the **lateral brachial cutaneous nerve.**
   – passes posteriorly through the quadrangular space accompanied by the posterior circumflex humeral artery.
   – winds around the surgical neck of the humerus (may be injured when this part of the bone is fractured).

5. Radial nerve (C5–T1)

   – is the largest branch of the brachial plexus and occupies the musculospiral groove on the back of the humerus with the profunda brachii artery.

II. Nerves of the Arm, Forearm, and Hand (Figures 2-13 and 2-14)

A. Musculocutaneous nerve (C5–C7)

   – pierces the coracobrachialis muscle and descends between the biceps and brachialis muscles.
   – innervates all of the flexor muscles in the anterior compartment of the arm, such as the coracobrachialis, biceps, and brachialis muscles.
   – continues into the forearm as the **lateral antebrachial cutaneous nerve.**

B. Median nerve (C5–T1)

   – runs down the anteromedial aspect of the arm (has no muscular branches in the arm).
   – passes through the cubital fossa, deep to the bicipital aponeurosis and medial to the brachial artery.
   – enters the forearm between the humeral and ulnar heads of the pronator teres muscle and then passes between the flexor digitorum superficialis and the flexor digitorum profundus muscles.
   – in the cubital fossa, gives rise to the **anterior interosseous nerve,** which descends on the interosseous membrane between the flexor digitorum profundus and the flexor pollicis longus, and then passes behind the pronator quadratus, supplying these three muscles.
   – innervates all of the anterior muscles of the forearm except the flexor carpi ulnaris and the ulnar half of the flexor digitorum profundus.
   – enters the palm of the hand through the carpal tunnel deep to the flexor retinaculum, gives off a muscular branch (the **recurrent branch**) to the thenar muscles, and terminates by dividing into three **common palmar digital nerves,** which then divide into the palmar digital branches.
   – innervates the lateral two lumbricals, the skin of the lateral side of the palm, and the palmar side of the lateral three and one-half fingers, as well as the dorsal side of the index finger, middle finger, and one-half of the ring finger.

C. Radial nerve (C5–T1)

   – is the **largest branch of the brachial plexus**.
   – runs down the posterior aspect of the arm and lies in the radial groove on the back of the humerus with the profunda brachii artery.
   – gives rise to the **posterior brachial** and **antebrachial cutaneous nerves.**
   – may be damaged by a midhumeral fracture, causing paralysis of the extensor muscles of the hand.
   – passes anterior to the lateral epicondyle, between the brachialis and brachioradialis muscles, where it divides into superficial and deep branches.

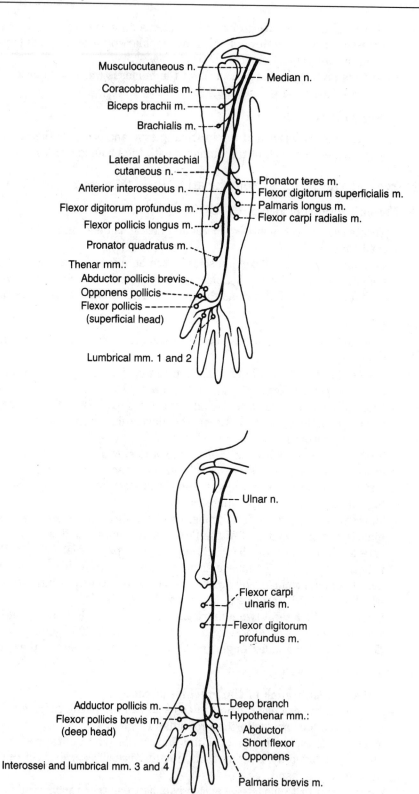

Figure 2-13. Distribution of the musculocutaneous, median, and ulnar nerves.

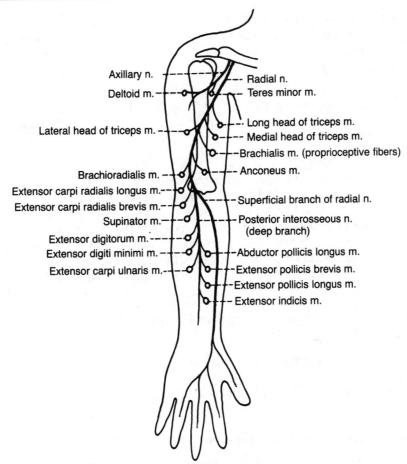

Figure 2-14. Distribution of the axillary and radial nerves.

1. Deep branch
- enters the supinator muscle, winds laterally around the radius in the substance of the muscle, and continues as the **posterior interosseous nerve** with the posterior interosseous artery.
- innervates the muscles of the back of the forearm.

2. Superficial branch
- descends in the forearm under cover of the brachioradialis muscle and then passes dorsally around the radius under the tendon of the brachioradialis.
- runs distally to the dorsum of the hand to innervate the radial side of the hand and the radial two and one-half digits over the proximal phalanx. This nerve does not supply the skin of the distal phalanges.

D. Ulnar nerve (C7–T1)
- runs down the medial aspect of the arm and behind the medial epicondyle in a groove.
- may be damaged by a fracture of the medial epicondyle and produce **funny-bone** symptoms.
- descends between and innervates the flexor carpi ulnaris and flexor digitorum profundus muscles.

– enters the hand superficial to the flexor retinaculum and lateral to the pisiform bone, and terminates by dividing into superficial and deep branches at the root of the hypothenar eminence.

1. Superficial branch

– innervates the palmaris brevis and the skin of the hypothenar eminence and terminates in the palm by dividing into **three palmar digital branches,** which supply the skin of the little finger and the medial side of the ring finger.

2. Deep branch

– arises at about the level of the pisiform bone and passes between the pisiform and the hook of the hamate, between the origins of the abductor and flexor digiti minimi brevis muscles, and then deep to the opponens digiti minimi.
– curves around the hook of the hamate, and then turns laterally to follow the course of the deep palmar arterial arch across the interossei.
– innervates the hypothenar muscles, the medial two lumbricals, all the interossei, the adductor pollicis, and usually the deep head of the flexor pollicis brevis.

III. Functional Components of the Peripheral Nerves

A. Somatic motor nerves

– include, for example, radial, axillary, median, musculocutaneous, and ulnar nerves.
– contain nerve fibers with cell bodies that are located in the following areas:

1. Dorsal root ganglia (for GSA and GVA fibers)

2. Anterior horn of the spinal cord (for GSE fibers)

3. Sympathetic chain ganglia (for sympathetic postganglionic GVE fibers)

B. Cutaneous nerves

– include, for example, medial brachial and medial antebrachial cutaneous nerves.
– contain nerve fibers with cell bodies that are located in the following areas:

1. Dorsal root ganglia (for GSA and GVA fibers)

2. Sympathetic chain ganglia (for sympathetic postganglionic GVE fibers)

Arteries of the Upper Limb

I. Branches of the Subclavian Artery (Figure 2-15)

A. Suprascapular artery

– is a branch of the **thyrocervical trunk.**
– passes over the superior transverse scapular ligament (whereas the suprascapular nerve passes under the ligament).
– anastomoses with the deep branch of the transverse cervical artery (**dorsal scapular artery**) and the circumflex scapular artery around the scapula, providing a collateral circulation.
– supplies the supraspinatus and infraspinatus muscles and the shoulder and acromioclavicular joints.

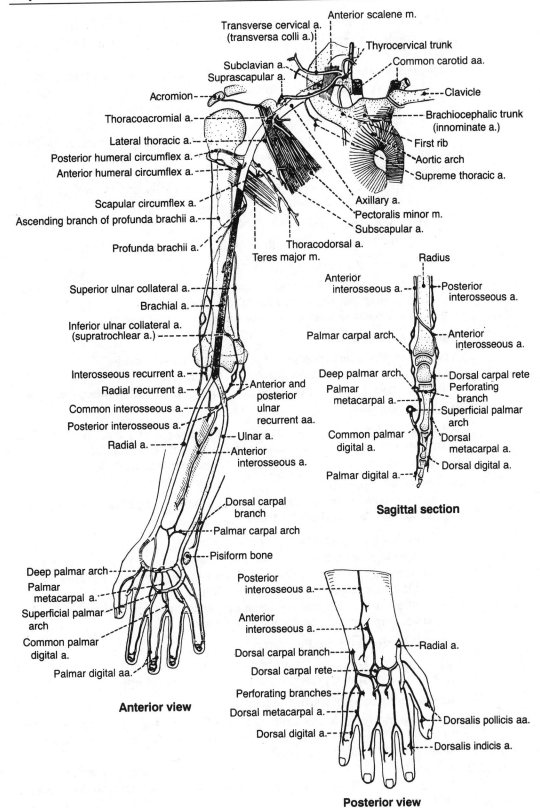

Figure 2-15. Blood supply to the upper limb.

B. Dorsal scapular or descending scapular artery

- arises from the subclavian artery but may be a deep branch of the transverse cervical artery.
- accompanies the dorsal scapular nerve and supplies the levator scapulae, rhomboids, and serratus anterior muscles.

II. Axillary Artery (see Figure 2-15)

- extends from the outer border of the first rib to the inferior border of the teres major muscle, where it becomes the **brachial artery.**
- is divided into three parts by the pectoralis minor muscle.
- is considered to be the **central structure of the axilla** and is bordered on its medial side by the axillary vein.

A. Supreme thoracic artery

- supplies the first and second intercostal spaces and adjacent muscles.

B. Thoracoacromial artery

- is a short trunk from the first or second part of the axillary artery and has pectoral, clavicular, acromial, and deltoid branches.
- pierces the costocoracoid membrane (or clavipectoral fascia).

C. Lateral thoracic artery

- runs along the lateral border of the pectoralis minor muscle.
- supplies the pectoralis major, pectoralis minor, and serratus anterior muscles and the axillary lymph nodes, and gives rise to **lateral mammary branches.**

D. Subscapular artery

- is the largest branch of the axillary artery, arises at the lower border of the subscapularis muscle, and descends along the axillary border of the scapula.
- divides into the thoracodorsal and circumflex scapular arteries.

1. Thoracodorsal artery

- accompanies the thoracodorsal nerve and supplies the latissimus dorsi muscle and the lateral thoracic wall.

2. Circumflex scapular artery

- passes posteriorly into the triangular space bounded by the subscapularis muscle and the teres minor muscle above, the teres major muscle below, and the long head of the triceps brachii laterally.
- ramifies in the infraspinous fossa and anastomoses with branches of the dorsal scapular and suprascapular arteries.

E. Anterior humeral circumflex artery

- passes anteriorly around the surgical neck of the humerus.
- anastomoses with the posterior humeral circumflex artery.

F. Posterior humeral circumflex artery

- runs posteriorly with the axillary nerve through the quadrangular space bounded by the teres minor and teres major muscles, the long head of the triceps brachii, and the humerus.
- anastomoses with the anterior humeral circumflex artery and an ascending branch of the profunda brachii artery and also sends a branch to the acromial rete.

III. Brachial Artery (see Figure 2-15)

- extends from the inferior border of the teres major muscle to its bifurcation in the cubital fossa.
- lies on the triceps brachii and then on the brachialis muscles medial to the coraco-brachialis and biceps brachii, and is accompanied by the basilic vein in the middle of the arm.
- lies in the center of the cubital fossa, medial to the biceps tendon, lateral to the median nerve, and deep to the bicipital aponeurosis.
- provides muscular branches and terminates by dividing into the radial and ulnar arteries at the level of the radial neck, about 1 cm below the bend of the elbow, in the cubital fossa.

A. Profunda brachii (deep brachial) artery

- descends posteriorly with the radial nerve and gives off an **ascending branch,** which anastomoses with the descending branch of the posterior humeral circum-flex artery.
- divides into the **middle collateral artery,** which anastomoses with the interosse-ous recurrent artery, and the **radial collateral artery,** which follows the radial nerve through the lateral intermuscular septum and ends in front of the lateral epicondyle by anastomosing with the radial recurrent artery of the radial artery.

B. Superior ulnar collateral artery

- pierces the medial intermuscular septum, accompanies the ulnar nerve behind the septum and medial epicondyle, and anastomoses with the posterior ulnar recurrent branch of the ulnar artery.

C. Inferior ulnar collateral artery

- arises just above the elbow, descends in front of the medial epicondyle, and anastomoses with the anterior ulnar recurrent branch of the ulnar artery.

IV. Radial Artery (see Figure 2-15)

- arises as the smaller lateral branch of the brachial artery in the cubital fossa and descends laterally under cover of the brachioradialis muscle, with the superficial radial nerve on its lateral side, on the supinator and flexor pollicis longus muscles.
- curves over the radial side of the carpal bones beneath the tendons of the abductor pollicis longus muscle, the extensor pollicis longus and brevis muscles, and over the surface of the scaphoid and trapezium bones.
- runs through the anatomical snuff-box, enters the palm by passing between the two heads of the first dorsal interosseous muscle and then between the heads of the adductor pollicis muscle, and divides into the **princeps pollicis artery** and the **deep palmar arch.**
- The **radial pulse** can be felt proximal to the wrist between the tendons of the brachioradialis and flexor carpi radialis muscles; it may also be palpated in the anatomical snuff-box between the tendons of the extensor pollicis longus and brevis muscles.
- gives rise to the following branches:

A. Radial recurrent artery

- arises from the radial artery just below its origin and ascends on the supinator and then between the brachioradialis and brachialis muscles.
- anastomoses with the radial collateral branch of the profunda brachii artery.

B. Palmar carpal branch

– joins the palmar carpal branch of the ulnar artery and forms the palmar carpal arch.

C. Superficial palmar branch

– passes through the thenar muscles and anastomoses with the superficial branch of the ulnar artery to complete the superficial palmar arterial arch.

D. Dorsal carpal branch

– joins the dorsal carpal branch of the ulnar artery and the dorsal terminal branch of the anterior interosseous artery to form the **dorsal carpal rete.**

E. Princeps pollicis artery

– descends along the ulnar border of the first metacarpal bone under the flexor pollicis longus tendon.
– divides into **two proper digital arteries** for each side of the thumb.

F. Radialis indicis artery

– also may arise from the deep palmar arch or the princeps pollicis artery.

G. Deep palmar arch (Figure 2-16)

– is formed by the main termination of the radial artery and usually is completed by the deep palmar branch of the ulnar artery.
– passes between the transverse and oblique heads of the adductor pollicis muscle.
– gives rise to three **palmar metacarpal arteries,** which descend on the interossei and join the common palmar digital arteries from the superficial palmar arch.

V. Ulnar Artery (see Figure 2-15)

– is the **larger medial branch of the brachial artery** in the cubital fossa.
– descends behind the ulnar head of the pronator teres muscle and lies between the flexor digitorum superficialis and profundus muscles.
– enters the hand anterior to the flexor retinaculum, lateral to the pisiform bone, and medial to the hook of the hamate bone.
– divides into the superficial palmar arch and the deep palmar branch, which passes between the abductor and flexor digiti minimi brevis muscles and runs medially to join the radial artery to complete the deep palmar arch.
– The **ulnar pulse** is palpable just to the radial side of the insertion of the flexor carpi ulnaris into the pisiform bone.
– gives rise to the following branches:

A. Anterior ulnar recurrent artery

– anastomoses with the inferior ulnar collateral artery.

B. Posterior ulnar recurrent artery

– anastomoses with the superior ulnar collateral artery.

C. Common interosseous artery

– arises from the lateral side of the ulnar artery and divides into the anterior and posterior interosseous arteries.

1. Anterior interosseous artery

– descends with the anterior interosseous nerve in front of the interosseous membrane, located between the flexor digitorum profundus and the flexor pollicis longus muscles.

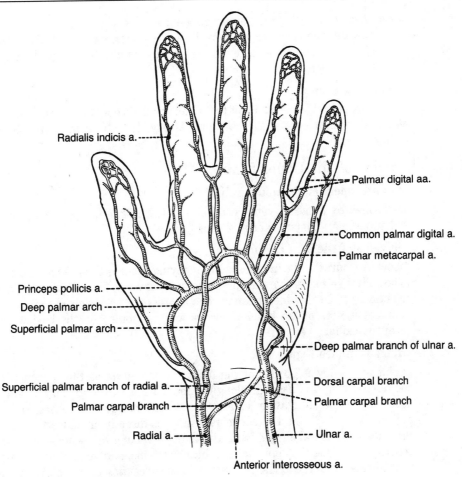

Radialis indicis a.

Palmar digital aa.

Common palmar digital a.

Palmar metacarpal a.

Princeps pollicis a.

Deep palmar arch

Superficial palmar arch

Deep palmar branch of ulnar a.

Superficial palmar branch of radial a.

Dorsal carpal branch

Palmar carpal branch

Palmar carpal branch

Radial a.

Ulnar a.

Anterior interosseous a.

Figure 2-16. Blood supply to the hand.

– perforates the interosseous membrane to anastomose with the posterior interosseous artery and join the dorsal carpal network.

2. **Posterior interosseous artery**
   – gives rise to the interosseous recurrent artery, which anastomoses with a middle collateral branch of the profunda brachii artery.
   – descends behind the interosseous membrane in company with the posterior interosseous nerve.
   – anastomoses with the dorsal carpal branch of the anterior interosseous artery.

D. **Palmar carpal branch**
   – joins the palmar carpal branch of the radial artery to form the palmar carpal arch.

E. **Dorsal carpal branch**
   – passes around the ulnar side of the wrist and joins the dorsal carpal rete.

F. **Superficial palmar arch** (see Figure 2-16)
   – is formed by the main termination of the ulnar artery and usually is completed by the superficial palmar branch of the radial artery.

– lies immediately under the palmar aponeurosis.

– gives rise to three **common palmar digital arteries,** each of which bifurcates into proper palmar digital arteries, which run distally to supply the adjacent sides of the fingers.

G. Deep palmar branch

– accompanies the deep branch of the ulnar nerve through the hypothenar muscles and anastomoses with the radial artery, thereby completing the **deep palmar arch.**

VI. Deep Veins of the Upper Limb

A. Deep and superficial venous arches

– are formed by a pair of venae comitantes, which accompany each of the deep and superficial palmar arterial arches.

B. Deep veins of the arm and forearm

– follow the course of the arteries, accompanying them as their venae comitantes. (The **radial veins** receive the dorsal metacarpal veins. The **ulnar veins** receive tributaries from the deep palmar venous arches. The **brachial veins** are the vena comitantes of the brachial artery and are joined by the basilic vein to form the axillary vein.)

C. Axillary vein (see Figure 2-4)

– begins at the lower border of the teres major muscle as the continuation of the **basilic vein** and ascends along the medial side of the axillary artery.

– continues as the **subclavian vein** at the inferior margin of the first rib.

– commonly receives the thoracoepigastric veins directly or indirectly and thus provides a collateral circulation if the inferior vena cava becomes obstructed.

– Its tributaries are the cephalic vein, brachial veins, and veins that correspond to the branches of the axillary artery, with the exception of the thoracoacromial vein.

Clinical Considerations

I. Fractures and Syndromes

A. Fracture of the clavicle

– results in upward displacement of the proximal fragment, owing to the pull of the sternocleidomastoid muscle, and downward displacement of the distal fragment, owing to the pull of the deltoid muscle and gravity.

– may be caused by the obstetrician in breech (buttocks) presentation, or it may occur when the infant presses against the maternal pubic symphysis during its passage through the birth canal.

– may cause injury to the brachial plexus (lower trunk) and a fatal hemorrhage from the subclavian vein.

– is also responsible for thrombosis of the subclavian vein, leading to pulmonary embolism.

B. Colles' fracture

– is a fracture of the lower end of the radius in which the distal fragment is displaced posteriorly.

– is called a reverse Colles' fracture (**Smith's fracture**) if the distal fragment is displaced anteriorly.

C. Inferior dislocation of the humerus

– is not uncommon because the inferior aspect of the shoulder joint is not supported by muscles.
– may damage the axillary nerve and the posterior humeral circumflex vessels.

D. Referred pain to the shoulder

– most probably indicates involvement of the phrenic nerve (or diaphragm).
– The supraclavicular nerve (C3–C4), which supplies sensory fibers over the shoulder, has the same origin as the phrenic nerve (C3–C5), which supplies the diaphragm.

E. Carpal tunnel syndrome

– is caused by **compression of the median nerve** caused by the reduced size of the osseofibrous carpal tunnel, resulting from inflammation of the flexor retinaculum, anterior dislocations of the lunate bone, arthritic changes, or inflammation of the tendon and its sheath by fibers of the flexor retinaculum.
– leads to **pain** and **paresthesia** (tingling, burning, and numbness) in the hand in the area of the median nerve.
– may also cause **atrophy of the thenar muscles** in cases of severe compression.

F. Dupuytren's contracture

– is a **disease of the palmar fascia** resulting in thickening and contracture of fibrous bands on the palmar surface of the hand and fingers.

II. Lesions of Peripheral Nerves

A. Upper trunk injury (Erb-Duchenne paralysis)

– is caused by a violent displacement of the head from the shoulder, such as results from a fall from a motorcycle or horse.
– results in a **waiter's tip hand,** in which the arm tends to lie in medial rotation due to paralysis of lateral rotator muscles.

B. Lower trunk injury (Klumpke's paralysis)

– may be caused during a difficult breech delivery (**birth palsy or obstetric paralysis**); may also be caused by a cervical rib (**cervical rib syndrome**) or by abnormal insertion or spasm of the anterior and middle scalene muscles (**scalene syndrome**).
– results in a **claw hand.**

C. Injury to the posterior cord

– is caused by the pressure of the crosspiece of a crutch, resulting in paralysis of the arm called **crutch palsy.**
– results in loss of the extensors of the arm, forearm, and hand.
– produces a **wrist drop.**

D. Injury to the long thoracic nerve

– is caused by a stab wound or during thoracic surgery.
– results in **winging of the scapula** from paralysis of the serratus anterior muscle, and inability to elevate the arm above the horizontal.

E. **Injury to the musculocutaneous nerve**
   – results in weakness of supination (biceps) and forearm flexion (brachialis and biceps).

F. **Injury to the axillary nerve**
   – is caused by a fracture of the surgical neck of the humerus or inferior dislocation of the humerus.
   – results in weakness of lateral rotation and abduction of the arm (the supraspinatus can abduct the arm but not to a horizontal level).

G. **Injury to the radial nerve**
   – is caused by a **fracture of the midshaft of the humerus.**
   – results in loss of the extensors of the forearm, hand, metacarpals, and phalanges.
   – results in loss of wrist extension, leading to **wrist drop.**
   – produces a weakness of abduction and adduction of the hand.

H. **Injury to the ulnar nerve**
   – is caused by a fracture of the medial epicondyle.
   – results in a **claw hand,** in which the ring and little fingers are hyperextended at the metacarpophalangeal joints and flexed at the interphalangeal joints.
   – results in loss of abduction and adduction of the fingers and flexion of the metacarpophalangeal joints, owing to paralysis of the palmar and dorsal interossei muscles and the medial two lumbricals.
   – results in loss of adduction of the thumb, owing to paralysis of the adductor pollicis muscle.

I. **Injury to the median nerve**
   – may be caused by a supracondylar fracture of the humerus.
   – results in loss of pronation, opposition of the thumb, flexion of the lateral two interphalangeal joints, and impairment of the medial two interphalangeal joints.
   – produces a characteristic flattening of the thenar eminence, often referred to as **ape hand.**

Summary of Muscle Actions of the Upper Limb

Movement of the Scapula
   Elevation—trapezius (upper part), levator scapulae
   Depression—trapezius (lower part), serratus anterior, pectoralis minor
   Protrusion (forward or lateral movement; abduction)—serratus anterior
   Retraction (backward or medial movement; adduction)—trapezius, rhomboids
   Anterior or inferior rotation of the glenoid fossa—rhomboid major
   Posterior or superior rotation of the glenoid fossa—serratus anterior, trapezius

Movement at the Shoulder Joint (Ball-and-Socket Joint)
   Adduction—pectoralis major, latissimus dorsi, deltoid (posterior part)
   Abduction—deltoid, supraspinatus
   Flexion—pectoralis major (clavicular part), deltoid (anterior part), coracobrachialis, biceps
   Extension—latissimus dorsi, deltoid (posterior part)
   Medial rotation—subscapularis, pectoralis major, deltoid (anterior part), latissimus dorsi, teres major
   Lateral rotation—infraspinatus, teres minor, deltoid (posterior part)

Movement at the Elbow Joint (Hinge Joint)
  Flexion—brachialis, biceps, brachioradialis, pronator teres
  Extension—triceps (medial head), anconeus

Movement at the Radioulnar Joints (Pivot Joints)
  Pronation—pronator quadratus, pronator teres
  Supination—supinator, biceps brachii

Movement at the Radiocarpal and Midcarpal Joints (Ellipsoidal Joints)
  Adduction—flexor carpi ulnaris, extensor carpi ulnaris
  Abduction—flexor carpi radialis, extensor carpi radialis longus and brevis
  Flexion—flexor carpi radialis, flexor carpi ulnaris, palmaris longus, abductor pollicis
  longus
  Extension—extensor carpi radialis longus and brevis, extensor carpi ulnaris

Movement at the Metacarpophalangeal Joint (Ellipsoidal Joint)
  Adduction—palmar interossei
  Abduction—dorsal interossei
  Flexion—lumbricals and interossei
  Extension—extensor digitorum

Movement at the Interphalangeal Joint (Hinge Joint)
  Flexion—flexor digitorum superficialis (proximal interphalangeal joint), flexor digitorum
  profundus (distal interphalangeal joint)
  Extension—lumbricals and interossei (when metacarpophalangeal joint is extended by
  extensor digitorum)
  Extension—extensor digitorum (when metacarpophalangeal joint is flexed by lumbricals
  and interossei)

Summary of Muscle Innervations of the Upper Limb

Muscles of the Anterior Compartment of the Arm: Musculocutaneous Nerve
  Biceps brachii
  Coracobrachialis
  Brachialis

Muscles of the Posterior Compartment of the Arm: Radial Nerve
  Triceps
  Anconeus

Muscles of the Posterior Compartment of the Forearm: Radial Nerve
  Superficial layer—brachioradialis; extensor carpi radialis longus; extensor carpi radialis
  brevis; extensor carpi ulnaris; extensor digitorum communis; extensor digiti minimi
  Deep layer—supinator; abductor pollicis longus; extensor pollicis longus; extensor pollicis
  brevis; extensor indicis

Muscles of the Anterior Compartment of the Forearm: Median Nerve
  Superficial layer—pronator teres; flexor carpi radialis; palmaris longus; flexor carpi ul-
  naris (ulnar nerve)*
  Middle layer—flexor digitorum superficialis
  Deep layer—flexor digitorum profundus (median nerve and ulnar nerve)*; flexor pollicis
  longus; pronator quadratus

Thenar Muscles: Median Nerve
   Abductor pollicis brevis
   Opponens pollicis
   Flexor pollicis brevis (median and ulnar nerves)*

Adductor Pollicis Muscle: Ulnar Nerve

Hypothenar Muscles: Ulnar Nerve
   Abductor digiti minimi
   Opponens digiti minimi
   Flexor digiti minimi

Lumbrical (Medial Two) and Interossei Muscles: Ulnar Nerve

Lumbrical Muscles (Lateral Two): Median Nerve

---

* Indicates exception or dual innervation

Review Test

Directions: Each of the numbered items or incomplete statements in this section is followed by answers or by completions of the statement. Select the ONE lettered answer or completion that is BEST in each case.

1. Which of the following statements concerning the brachial plexus is true?

(A) It contains nerve fibers originating only from ventral roots of spinal nerves
(B) It contains nerve fibers originating only from ventral primary rami of spinal nerves
(C) Each root is formed by fibers originating from more than one ventral primary ramus of a spinal nerve
(D) Each trunk is formed by fibers originating from more than one ventral ramus
(E) All trunks directly give rise to muscular branches

2. If the thoracoacromial trunk were ligated, which of the following branches would remain unobstructed?

(A) Acromial
(B) Pectoral
(C) Clavicular
(D) Deltoid
(E) Superior thoracic

3. The defect known as winged scapula is caused by damage to

(A) a nerve arising from the medial cord of the brachial plexus
(B) thoracodorsal nerve
(C) dorsal scapular nerve
(D) a nerve arising from the roots of the brachial plexus
(E) a nerve arising from the upper trunk of the brachial plexus

4. Which of the following statements concerning the muscles of the hand is true?

(A) The adductor pollicis muscle is innervated by the median nerve
(B) The thenar muscles are innervated by a nerve derived from the posterior cords of the brachial plexus
(C) The lumbrical muscles arise from tendons of the flexor digitorum superficialis
(D) The dorsal interosseous muscles assist flexion of the metacarpophalangeal joints and extension of the interphalangeal joints
(E) The palmar interosseous muscles abduct the fingers

5. Which of the following statements concerning the pectoralis minor muscle is true?

(A) It divides the axillary nerve into three parts
(B) It is innervated by the medial pectoral nerve and acts to elevate the shoulder
(C) It originates from the coracoid process
(D) It is invested by the clavipectoral fascia
(E) It forms the posterior wall of the axilla

6. Which of the following statements concerning the intercostobrachial nerve is true?

(A) It is a lateral cutaneous branch of the second intercostal nerve
(B) It typically communicates with the medial antebrachial cutaneous nerve
(C) It pierces the coracobrachialis muscle
(D) It is a branch of the medial cord of the brachial plexus
(E) It contains sympathetic preganglionic fibers

7. A patient presents with a severely damaged radial nerve resulting from fracture of the lower third of the humerus. The patient will experience

(A) a loss of wrist extension, leading to wrist drop
(B) a weakness in pronating the forearm
(C) a sensory loss over the ventral aspect of the base of the thumb
(D) an inability to oppose the thumb
(E) an inability to abduct the fingers

8. A patient is unable to flex the proximal interphalangeal joints, owing to paralysis of the

(A) palmar interossei
(B) flexor digitorum profundus
(C) dorsal interossei
(D) flexor digitorum superficialis
(E) lumbricals

9. A patient is incapable of adducting the arm because of paralysis of the

(A) teres minor
(B) supraspinatus
(C) latissimus dorsi
(D) infraspinatus
(E) subscapularis

10. The major contents of the proximal portion of the cubital fossa, in order from medial to lateral side, are

(A) biceps brachii tendon, median nerve, brachial artery, radial nerve
(B) median nerve, biceps brachii tendon, brachial artery, radial nerve
(C) brachial artery, median nerve, biceps brachii tendon, radial nerve
(D) brachial artery, biceps brachii tendon, median nerve, radial nerve
(E) median nerve, brachial artery, biceps brachii tendon, radial nerve

11. Injury to the radial nerve results in which of the following conditions?

(A) Waiter's tip hand
(B) Claw hand
(C) Wrist drop
(D) Ape hand
(E) Carpal tunnel syndrome

12. The extensor expansion on the middle digit is related to

(A) the extensor digitorum, one lumbrical muscle, one dorsal interosseous muscle, and one palmar interosseous muscle
(B) the extensor digitorum, one lumbrical muscle, one palmar interosseous muscle, and two dorsal interosseous muscles
(C) the extensor digitorum, one lumbrical muscle, and two dorsal interosseous muscles
(D) the extensor indicis, one lumbrical muscle, and one dorsal interosseous muscle
(E) the extensor digitorum, one lumbrical muscle, and one palmar interosseous muscle

13. Which of the following statements concerning the radial artery is true?

(A) It passes through the carpal tunnel
(B) It accompanies the posterior interosseous nerve in the forearm
(C) It is the principal source of blood to the superficial palmar arch
(D) It has the princeps pollicis artery as one of its branches
(E) It runs distally between the flexor digitorum superficialis and the flexor digitorum profundus

14. Which of the following group of nerves is intimately related to a portion of the humerus and can be affected by fractures of the humerus?

(A) Axillary, musculocutaneous, radial
(B) Axillary, median, ulnar
(C) Axillary, radial, ulnar
(D) Axillary, median, musculocutaneous
(E) Median, radial, ulnar

15. Which of the following statements concerning the position of the flexor retinaculum is true?

(A) It lies superficial to the ulnar and median nerves
(B) It lies deep to the ulnar and median nerves
(C) It lies superficial to the ulnar nerve and deep to the median nerve
(D) It lies deep to the ulnar nerve and superficial to the median nerve
(E) It lies deep to the ulnar nerve and superficial to the ulnar artery

16. Fracture of the first metacarpal bone may injure which of the following intrinsic muscles of the thumb?

(A) Abductor pollicis brevis
(B) Flexor pollicis brevis (superficial head)
(C) Opponens pollicis
(D) Adductor pollicis
(E) Flexor pollicis brevis (deep head)

17. A lesion of the ulnar nerve produces a paralysis of which of the following muscles?

(A) Palmar interossei and adductor pollicis
(B) Dorsal interossei and lateral two lumbricals
(C) Medial two lumbricals and opponens pollicis
(D) Abductor pollicis brevis and palmar interossei
(E) Medial and lateral lumbricals

18. An infection in the ulnar bursa could result in necrosis of which of the following tendons?

(A) Tendon of the flexor carpi ulnaris
(B) Tendon of the flexor pollicis longus
(C) Tendon of the flexor digitorum profundus
(D) Tendon of the flexor carpi radialis
(E) Tendon of the palmaris longus

19. Abductors of the arm are paralyzed resulting from a lesion of which of the following nerves?

(A) Axillary and musculocutaneous
(B) Thoracodorsal and upper subscapular
(C) Suprascapular and axillary
(D) Radial and lower subscapular
(E) Suprascapular and dorsal scapular

20. Which of the following statements concerning the muscles of the upper limb is true?

(A) All intrinsic muscles of the thumb insert into the base of the proximal phalanx
(B) All heads of the biceps brachii and the triceps brachii arise from the scapula
(C) The little finger does not have a named adductor
(D) The flexor digitorum profundus tendons attach to the middle phalanges of the fingers
(E) The flexor digitorum superficialis tendons attach to the terminal phalanges of the fingers

21. Which of the following pairs of nerves innervate the muscle that moves the metacarpophalangeal joint of the ring finger?

(A) Median and ulnar
(B) Radial and median
(C) Musculocutaneous and ulnar
(D) Ulnar and radial
(E) Radial and axillary

22. Which of the following pairs of muscles form the floor of the cubital fossa?

(A) Brachioradialis and supinator
(B) Brachialis and supinator
(C) Pronator teres and supinator
(D) Supinator and pronator quadratus
(E) Brachialis and pronator teres

23. Which of the following structures contain cell bodies of nerve fibers in the medial brachial cutaneous nerve?

(A) Dorsal root ganglia and anterior horn of the spinal cord
(B) Anterior and lateral horns of the spinal cord
(C) Sympathetic chain ganglia and dorsal root ganglia
(D) Lateral horn of the spinal cord and sympathetic chain ganglia
(E) Dorsal root ganglia and lateral horn of the spinal cord

24. Inability to supinate the forearm could result from an injury to which of the following pairs of nerves?

(A) Suprascapular and axillary
(B) Musculocutaneous and median
(C) Axillary and radial
(D) Radial and musculocutaneous
(E) Median and ulnar

25. A patient complains of sensory loss over the anterior and posterior surfaces of the medial third of the hand and the medial one and one-half fingers. Which of the following nerves is injured?

(A) Axillary
(B) Radial
(C) Median
(D) Ulnar
(E) Musculocutaneous

26. The radial nerve contains axons that have cell bodies in which of the following structures?

(A) Dorsal root ganglia and the posterior horn of the spinal cord
(B) Sympathetic chain ganglia and the lateral horn of the spinal cord
(C) The lateral horn of the spinal cord and dorsal root ganglia
(D) The anterior horn of the spinal cord and sympathetic chain ganglia
(E) The anterior and lateral horns of the spinal cord

Questions 27–30

A patient with a fracture of the clavicle at the junction of the inner and middle third of the bone exhibits overriding of the medial and lateral fragments. The arm is rotated medially.

27. The lateral portion of the fractured clavicle is displaced downward by

(A) the deltoid and trapezius muscles
(B) the pectoralis major and deltoid muscles
(C) the pectoralis minor muscle and gravity
(D) the trapezius and pectoralis minor muscles
(E) the deltoid muscle and gravity

28. Which of the following muscles causes upward displacement of the medial fragment?

(A) Pectoralis major
(B) Deltoid
(C) Trapezius
(D) Sternocleidomastoid
(E) Scalenus anterior

29. Each of the following conditions could occur secondary to the fractured clavicle EXCEPT

(A) a fatal hemorrhage from the subclavian vein
(B) thrombosis of the subclavian vein, causing a pulmonary embolism
(C) thrombosis of the subclavian artery, causing an embolism in the brachial artery
(D) damage to the lower trunk of the brachial plexus
(E) damage to the long thoracic nerve, causing the winged scapula

30. All of the following muscles are responsible for medial rotation of the patient's arm EXCEPT

(A) pectoralis major
(B) subscapularis
(C) teres major
(D) latissimus dorsi
(E) teres minor

Directions: Each of the numbered items or incomplete statements in this section is negatively phrased, as indicated by a capitalized word such as NOT, LEAST, or EXCEPT. Select the ONE lettered answer or completion that is BEST in each case.

31. The vascular anastomosis around the shoulder involves all of the following arteries EXCEPT

(A) the dorsal scapular artery
(B) the thoracoacromial artery
(C) the subscapular artery
(D) the posterior circumflex humeral artery
(E) the superior ulnar collateral artery

32. Each of the following statements concerning the breast and mammary gland is true EXCEPT

(A) the nipple usually lies at the level of the fourth intercostal space
(B) the breast receives arterial blood through branches of the lateral thoracic and internal thoracic arteries
(C) the mammary gland is connected to the skin by strong connective tissue strands called suspensory ligaments
(D) breast cancer causes dimpling of the overlying skin and nipple retraction
(E) the mammary gland lies in the deep fascia

33. Each of the following statements concerning the medial pectoral nerve is true EXCEPT

(A) it contains general somatic efferent (GSE) fibers
(B) it contains axons that have cell bodies in the sympathetic ganglion
(C) it innervates the pectoralis major muscle
(D) it is connected to the lateral pectoral nerve
(E) it contains axons that have cell bodies in the lateral horn of the spinal cord

34. Damage to the posterior cord of the brachial plexus results in paralysis of all of the following muscles EXCEPT

(A) subscapularis muscle
(B) teres major muscle
(C) latissimus dorsi muscle
(D) teres minor muscle
(E) infraspinatus muscle

35. Paralysis that impairs flexion of the distal interphalangeal joint of the index finger will also produce all of the following conditions EXCEPT

(A) similar paralysis of the third digit
(B) atrophy of the thenar eminence
(C) loss of sensation over the distal part of the second digit
(D) complete paralysis of the thumb
(E) loss of pronation

36. If a 24-year-old carpenter crushed his entire little finger, all of the following muscles would be damaged EXCEPT

(A) flexor digitorum profundus
(B) extensor digitorum
(C) palmar interossei
(D) dorsal interossei
(E) lumbricals

37. Each of the following statements concerning the axillary nerve is true EXCEPT

(A) it arises from the posterior cord of the brachial plexus
(B) it lies adjacent to the medial and posterior surface of the surgical neck of the humerus
(C) it supplies the deltoid and teres major muscles
(D) it can be damaged by inferior dislocation of the head of the humerus
(E) it passes through the quadrangular space formed by the teres minor, the teres major, the long head of the triceps, and the surgical neck of the humerus

38. Damage to the median nerve results in paralysis of all of the following muscles EXCEPT

(A) flexor digitorum superficialis
(B) opponens pollicis
(C) two medial lumbricals
(D) pronator teres
(E) flexor pollicis longus

39. Inflammation of the synovial sheath of the common flexor tendons damages the nerve passing through the carpal tunnel and can cause carpal tunnel syndrome. Each of the following statements concerning this syndrome is true EXCEPT

(A) the dorsal and palmar interosseous muscles are normal
(B) the flexor digitorum superficialis and profundus muscles are abnormal
(C) the thenar eminence is flattened
(D) the medial one and one-half fingers have diminished sensitivity
(E) the adductor pollicis muscle is not atrophied

40. Each of the following statements concerning the lymphatic drainage of the breast is true EXCEPT

(A) it may constitute a path for the spread of malignant tumors to axillary nodes
(B) it provides a rich drainage of cutaneous and glandular regions
(C) it has paths to the venous system
(D) it includes collecting vessels that follow the perforating blood vessels through the pectoralis major muscle
(E) its vessels have no valves

41. If the musculocutaneous nerve is severed, all of the following types of axons could be damaged EXCEPT

(A) postganglionic sympathetic axons
(B) somatic afferent axons
(C) preganglionic sympathetic axons
(D) somatic efferent axons
(E) visceral afferent axons

42. Each of the following statements concerning supination is true EXCEPT

(A) the palm faces forward
(B) it is partially impaired when a nerve that lies in the spiral groove of the humerus is severed
(C) it involves the active participation of the elbow and wrist joints
(D) it involves the participation of the proximal and distal radioulnar joints
(E) during its movement, the upper end of the radius nearly rotates within the annular ligament

Directions: Each set of matching questions in this section consists of a list of four to twenty-six lettered options (some of which may be in figures) followed by several numbered items. For each numbered item, select the ONE lettered option that is most closely associated with it. To avoid spending too much time on matching sets with large numbers of options, it is generally advisable to begin each set by reading the list of options. Then, for each item in the set, try to generate the correct answer and locate it in the option list, rather than evaluating each option individually. Each lettered option may be selected once, more than once, or not at all.

Questions 43–47

Match each characteristic below with the appropriate nerve.

(A) Axillary nerve
(B) Radial nerve
(C) Median nerve
(D) Musculocutaneous nerve
(E) Ulnar nerve

43. Passes through the quadrangular space of the shoulder

44. Innervates a muscle that originates from the third metacarpal and adducts the thumb

45. Innervates the flexor of the distal phalanx of the thumb; usually passes between the two heads of the pronator teres

46. Innervates muscles that abduct or adduct the fingers

47. Is usually accompanied by the profunda brachii artery along part of its course

Questions 48–52

Match each characteristic below with the appropriate bone.

(A) Scapula
(B) Clavicle
(C) Humerus
(D) Radius
(E) Ulna

48. The brachialis muscle inserts into a rough area on the anterior surface of its coronoid process

49. Its head is at the distal end

50. Its head is at the proximal end

51. It is the origination of the long head of the biceps brachii

52. The medial margin of its trochlea projects more than its lateral margin

Questions 53–57

Match each characteristic below with the appropriate muscle.

(A) Teres minor
(B) Latissimus dorsi
(C) Biceps brachii
(D) Supraspinatus
(E) Brachioradialis

53. Supinates the forearm; originates from the scapula

54. Helps stabilize the glenohumeral joint; is innervated by the axillary nerve

55. Forms part of the posterior axillary fold; is innervated by a nerve that branches from the posterior cord of the brachial plexus

56. Can flex the forearm; is innervated by the radial nerve

57. Abducts the arm; forms part of the rotator cuff; attaches on the greater tuberosity of the humerus

Questions 58–62

Match each characteristic below with the appropriate artery.

(A) Palmar metacarpal
(B) Anterior interosseous
(C) Posterior interosseous
(D) Radial
(E) Ulnar

58. Gives rise to the princeps pollicis artery

59. Lies superficial to the flexor retinaculum

60. Most superficial structure visible in the anatomical snuff-box

61. Gives rise to the interosseous recurrent artery

62. Descends between the flexor digitorum superficialis and the flexor digitorum profundus

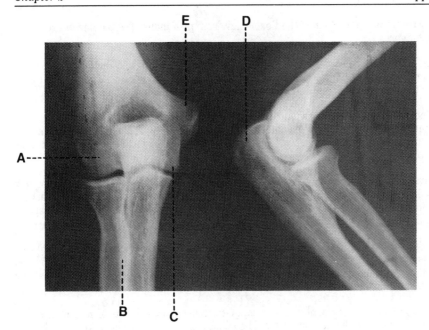

Questions 63–66

Match each description with the appropriate lettered site or structure in this radiograph of the elbow joint and its associated structures.

63. Site for tendinous attachment of the biceps brachii muscle

64. Site for origin of the common flexor tendon

65. The capitulum

66. The olecranon

Answers and Explanations

1–B. The brachial plexus is formed by the ventral primary rami of the lower four cervical nerves and first thoracic nerve. The spinal nerves are formed by the union of the dorsal and ventral roots. The roots of the brachial plexus are formed by the ventral primary rami of the spinal nerves. The ventral primary ramus of the seventh cervical nerve remains single as the middle trunk. The upper trunk of the brachial plexus is the only trunk that sends branches to innervate muscles such as the suprascapular nerve and subclavius nerve.

2–E. The superior thoracic artery, a branch of the axillary artery, would remain unobstructed if the thoracoacromial trunk were ligated. The thoracoacromial trunk has four branches: the pectoral, clavicular, acromial, and deltoid branches.

3–D. Winged scapula is caused by paralysis of the serratus anterior muscle as a result of damage to the long thoracic nerve that arises from the roots of the brachial plexus (C5–C7).

4–D. The dorsal and palmar interosseous and lumbrical muscles can flex the metacarpophalangeal joints and extend the interphalangeal joints. The adductor pollicis muscle is innervated by the ulnar nerve. The thenar muscles are innervated by the median nerve, which is formed by the union of the lateral and medial heads from the lateral and medial cords of the brachial plexus, respectively. The lumbrical muscles arise from the tendons of the flexor digitorum profundus. The palmar interosseous muscles adduct the fingers.

5–D. The pectoralis minor is invested by the clavipectoral fascia. The muscle divides the axillary artery into three parts and forms the anterior wall of the axilla. It originates from the second to the fifth ribs, inserts on the coracoid process, and is innervated by the medial and lateral pectoral nerves.

6–A. The intercostobrachial nerve arises from the lateral cutaneous branch of the second intercostal nerve and pierces the intercostal and serratus anterior muscles. It may communicate with the medial brachial cutaneous nerve, which is a branch of the medial cord of the brachial plexus. It contains sympathetic postganglionic fibers.

7–A. Injury to the radial nerve will result in loss of wrist extension, leading to wrist drop. The pronator teres, pronator quadratus, and opponens pollicis muscles, as well as the skin over the ventral aspect of the thumb, are innervated by the median nerve. The dorsal interosseous muscles, which act to abduct the fingers, are innervated by the ulnar nerve.

8–D. The flexor digitorum superficialis muscle flexes the proximal interphalangeal joints. The flexor digitorum profundus muscle flexes the distal interphalangeal joints. The palmar and dorsal interossei and lumbricals can flex metacarpophalangeal joints and extend the interphalangeal joints. The palmar interossei adduct the fingers; the dorsal interossei abduct the fingers.

9–C. The latissimus dorsi adducts the arm; the supraspinatus muscle abducts the arm. The infraspinatus and the teres minor rotate the arm laterally. The subscapularis rotates the arm medially.

10–E. The contents of the cubital fossa from medial to lateral side are the median nerve, the brachial artery, the biceps brachii tendon, and the radial nerve.

11–C. Injury to the radial nerve will result in loss of wrist extension, leading to wrist drop. The waiter's tip hand is produced by damage to the upper trunk of the brachial plexus, the claw hand by damage to the ulnar nerve or lower trunk of the brachial plexus, and the ape hand and carpal tunnel syndrome by damage to the median nerve.

12–C. The extensor expansion on the middle digit is related to the extensor digitorum, one lumbrical muscle, and two distal interosseous muscles, but no palmar interosseous muscles. The dorsal interossei abduct the fingers; the palmar interossei adduct them. Thus, the middle digit has an abductor on each side but no adductors.

13–D. The radial artery descends laterally beneath the brachioradialis with the superficial radial nerve, passes through the anatomical snuff-box, enters the palm by passing between the two heads of the first dorsal interosseous muscle, and divides into the princeps pollicis artery and the deep palmar arch.

14–C. To be injured in a fracture, the nerve must lie close to, or contact, the bone. The axillary nerve passes posteriorly around the surgical neck of the humerus; the radial nerve lies in the radial groove of the middle of the shaft of the humerus; and the ulnar nerve passes behind the medial epicondyle.

15–D. The flexor retinaculum lies deep to the ulnar artery and the ulnar nerve and superficial to the median nerve.

16–C. Fracture of the first metacarpal bone may result in injury to the opponens pollicis muscle. The opponens pollicis inserts on the first metacarpal; all other short muscles of the thumb insert on the proximal phalanges.

17–A. The ulnar nerve innervates all interossei (palmar and dorsal), the adductor pollicis, and two medial lumbricals. The abductor pollicis brevis, opponens pollicis, and two lateral lumbricals are innervated by the median nerve.

18–C. The ulnar bursa, or common synovial flexor sheath, contains the tendons of both the flexor digitorum superficialis and profundus muscles. The radial bursa envelops the tendon of the flexor pollicis longus.

19–C. The abductors of the arm are the deltoid and supraspinatus muscles, which are innervated by the axillary and suprascapular nerves, respectively.

20–C. There is no named adductor for the little finger. All intrinsic muscles of the thumb except the opponens pollicis insert into the base of the proximal phalanx. The opponens pollicis inserts on the first metacarpal bone. The short head of the biceps brachii arises from the coracoid process of the scapula; however, only the long head of the triceps brachii arises from the infraglenoid tubercle of the scapula. The flexor digitorum profundus tendons attach to the distal phalanges of the fingers; the flexor digitorum superficialis tendons attach to the middle phalanges.

21–D. The metacarpophalangeal joint of the ring finger is flexed by the lumbrical and palmar and dorsal interosseous muscles, which are innervated by the ulnar nerve. This joint is extended by the extensor digitorum, which is innervated by the radial nerve.

22–B. The floor of the cubital fossa is formed by the brachialis and supinator muscles. The brachioradialis and pronator teres muscles form the lateral and medial boundaries of the cubital fossa, respectively.

23–C. The medial brachial cutaneous nerve contains sensory fibers that have cell bodies in the dorsal root ganglia. It also contains sympathetic postganglionic fibers that have cell bodies in the sympathetic chain ganglia. The anterior horn contains cell bodies of skeletal motor fibers, and the lateral horn contains cell bodies of sympathetic preganglionic fibers.

24–D. Supination of the forearm is produced by the supinator and biceps brachii muscles, which are innervated by the radial and musculocutaneous nerves, respectively.

25–D. The ulnar nerve supplies sensory fibers to the skin over the palmar and dorsal surfaces of the medial third of the hand and the medial one and one-half fingers.

26–D. The radial nerve contains axons that have cell bodies in the dorsal root ganglia for general somatic and visceral afferent (GSA and GVA) nerve fibers, sympathetic chain ganglia for sympathetic postganglionic (GVE) fibers, and anterior horn of the spinal cord for general somatic efferent (GSE) fibers.

27–E. The lateral fragment of the clavicle is displaced downward by the pull of the deltoid muscle and gravity.

28–D. The sternocleidomastoid muscle is attached to the superior border of the medial third of the clavicle, and the medial fragment of a fractured clavicle will be displaced upward owing to the pull of the muscle.

29–E. The fractured clavicle may damage the subclavian vein, resulting in a pulmonary embolism; cause thrombosis of the subclavian artery, resulting in embolism of the brachial artery; or damage the lower trunk of the brachial plexus.

30–E. The pectoralis major, subscapularis, teres major, and latissimus dorsi muscles can rotate the arm medially. The teres minor muscle rotates the arm laterally.

31–E. The superior ulnar collateral artery enters the anastomosis around the elbow joint.

32–E. The mammary gland is a modified sweat gland located in the superficial fascia. The suspensory ligaments (Cooper's ligaments) are strong fibrous processes that run from the dermis of the skin to the deep layer of the superficial fascia through the breast.

33–E. The medial pectoral nerve communicates with the lateral pectoral nerve to form a loop in front of the axillary artery. It contains somatic motor fibers that supply the pectoralis major and minor muscles. It also contains sympathetic postganglionic fibers that supply blood vessels and have cell bodies in the sympathetic chain ganglia.

34–E. The infraspinatus is innervated by the suprascapular nerve, which originates in the upper trunk of the brachial plexus. The subscapularis muscle is innervated by the upper and lower subscapular nerves. The teres major muscle is innervated by the lower subscapular nerve. The latissimus dorsi muscle is innervated by the thoracodorsal nerve. The teres minor muscle is innervated by the axillary nerve. All of these nerves originate in the posterior cord of the brachial plexus.

35–D. The thumb would not be completely paralyzed in this case. Flexion of the distal interphalangeal joints of the index and middle fingers is accomplished by the flexor digitorum profundus muscle, which is innervated by the median nerve. This nerve also innervates the skin over the distal part of the second digit and the thenar muscles; however, the adductor pollicis and the deep head of the flexor pollicis brevis are innervated by the ulnar nerve. The median nerve also innervates the pronator teres and pronator quadratus muscles.

36–D. The dorsal interossei are abductors of the fingers. The little finger has no attachment for the dorsal interosseous muscle because it has its own abductor.

37–C. The axillary nerve innervates the deltoid and teres minor muscles. It arises from the posterior cord of the brachial plexus, accompanies the posterior humeral circumflex artery around the surgical neck of the humerus, and supplies the deltoid and teres minor muscles. Because it lies close to the surgical neck of the humerus, it can be damaged by inferior dislocation of the head of the humerus.

38–C. The median nerve innervates the two lateral lumbricals; however, the ulnar nerve innervates the two medial lumbricals.

39–D. Carpal tunnel syndrome results from injury to the median nerve. Diminished sensitivity in the medial one and one-half fingers would result from damage to the ulnar nerve.

40–E. The lymphatic vessels are constricted at the sites of valves, showing a beaded appearance. Most of the lymphatic drainage in the lateral quadrants of the breast goes to the axillary nodes, particularly the pectoral nodes. Lymphatic vessels in the medial quadrants follow the perforating vessels through the pectoralis major muscle and enter the parasternal nodes. The remaining lymphatic drainage goes to the apical nodes, nodes of the opposite breast, and nodes of the anterior abdominal wall.

41–C. The musculocutaneous nerve contains postganglionic sympathetic axons that innervate blood vessels, hair follicles, and sweat glands; afferent axons that innervate cutaneous tissues; and somatic efferent fibers that innervate skeletal muscles.

42–C. Supination involves participation of the radioulnar joints, but not of the wrist joint. In supination the palm faces forward (lateral rotation). This movement is carried out by the supinator and biceps brachii muscles, which are innervated by the radial and musculocutaneous nerves, respectively. Severing the radial nerve running in the spiral groove of the humerus partially impairs supination by affecting the action of the supinator muscle.

43–A. The axillary nerve and the posterior humeral circumflex artery pass through the quadrangular space.

44–E. The ulnar nerve innervates the adductor pollicis muscle, which arises from the second and third metacarpals, inserts on the proximal phalanx, and draws the thumb back into the plane of the palm (i.e., adducts the thumb).

45–C. The median nerve innervates the flexor pollicis longus muscle and usually passes between the two heads of the pronator teres.

46–E. The ulnar nerve supplies the dorsal interossei (abductors) and the palmar interossei (adductors).

47–B. The radial nerve occupies the spiral groove on the back of the humerus with the profunda brachii artery and separates the lateral and medial heads of the triceps brachii.

48–E. The brachialis muscle inserts on a rough area on the anterior surface of the coronoid process and tuberosity of the ulna.

49–E. The head of the ulna is at its distal end.

50–D. The head of the radius is at its proximal end.

51–A. The long head of the biceps brachii originates from the supraglenoid tubercle of the scapula.

52–C. The humerus has a trochlea with a medial edge that projects more inferiorly than its lateral edge. Consequently, the long axis of the humerus is oblique to the long axis of the ulna, forming an angulation called a carrying angle.

53–C. The long head of the biceps brachii muscle originates from the supraglenoid tubercle of the scapula; the short head arises from the coracoid process. This muscle supinates the forearm.

54–A. The teres minor muscle forms part of the rotator cuff, which helps stabilize the glenohumeral joint, and is innervated by the axillary nerve.

55–B. The latissimus dorsi muscle forms part of the posterior axillary fold. It is innervated by the thoracodorsal nerve, which arises from the posterior cord of the brachial plexus.

56–E. The brachioradialis muscle flexes the forearm and is innervated by the radial nerve.

57–D. The supraspinatus muscle arises from the supraspinous fossa and inserts on the superior facet of the greater tubercle of the humerus. This muscle is innervated by the suprascapular nerve and abducts the arm.

58–D. The radial artery divides into the princeps pollicis artery and the deep palmar arch.

59–E. The ulnar artery enters the hand superficial to the flexor retinaculum and lateral to the pisiform, and divides into the superficial and deep branches.

60–D. The radial artery runs through the anatomical snuff-box and then enters the palm by passing between the two heads of the first dorsal interosseous muscle.

61–C. The posterior interosseous artery gives rise to the interosseous recurrent artery, which anastomoses with a branch of the profunda brachii artery.

62–E. The ulnar artery descends between the flexor digitorum superficialis and the flexor digitorum profundus.

63–B. The tendon of the biceps brachii muscle inserts on the radial tuberosity.

64–E. The common flexor tendon of the arm attaches to the medial epicondyle.

65–A. The capitulum articulates with the head of the radius.

66–D. The olecranon is the curved projection on the back of the elbow.

3

Lower Limb

Bones and Joints

I. Coxal (Hip) Bone (Figures 3-1 and 3-2)
– is formed by the fusion of the **ilium, ischium,** and **pubis** on each side of the pelvis.
– articulates with the sacrum at the sacroiliac joint to form the **pelvic girdle.**

A. Ilium
– forms the upper part of the **acetabulum** and the lateral part of the hip bone.
– comprises the anterior–superior iliac spine, anterior–inferior iliac spine, posterior iliac spine, **greater sciatic notch, iliac fossa,** and gluteal lines.

B. Pubis
– forms the anterior part of the acetabulum and the anteromedial part of the hip bone.

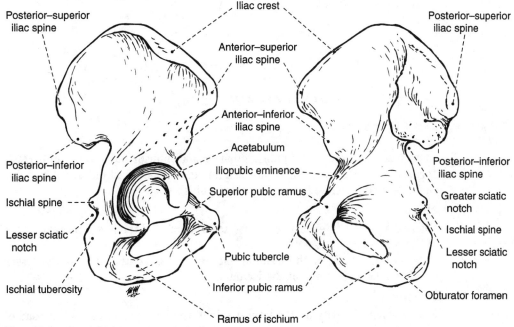

Figure 3-1. Coxal (hip) bone (lateral view).

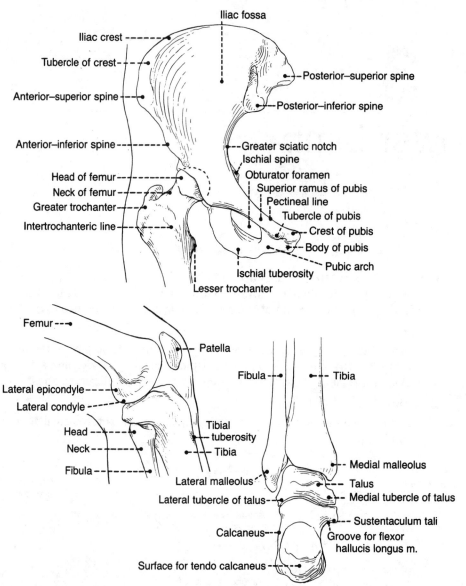

Figure 3-2. Bones of the lower limb.

- comprises the body of the pubis, superior and inferior pubic rami, **obturator foramen** (formed by fusion of the ischium and pubis), crest of the pubis, pubic tubercle, and **pectineal line** (or pecten pubis).

C. Ischium

- forms the posteroinferior part of the acetabulum and the lower posterior part of the hip bone.
- comprises the body of the ischium, **ischial spine, ischial tuberosity,** lesser sciatic notch, and ramus of the ischium.

D. Acetabulum

- is a cup-shaped cavity on the lateral side of the hip bone in which the head of the femur fits.

– has a deep notch, the **acetabular notch,** which is bridged by the transverse acetabular ligament.

– is formed by the **ilium** superiorly, the **ischium** posteroinferiorly, and the **pubis** anteromedially.

II. Bones of the Thigh and Leg (see Figure 3-2)

A. Femur

– is the long bone of the thigh.

1. Head of the femur

– forms about two-thirds of a sphere and is directed medially, upward, and slightly forward to fit into the acetabulum.

– has a depression in its articular surface, the **fovea capitis femoris,** to which the ligamentum capitis femoris is attached.

2. Neck of the femur

– connects the head to the body (shaft) and forms an angle of about 125°.

– is separated from the shaft in front by the **intertrochanteric line,** to which the iliofemoral ligament is attached.

3. Greater trochanter

– projects upward from the junction of the neck with the shaft.

– provides an insertion for the gluteus medius and minimus, piriformis, and obturator internus muscles.

– The **trochanteric fossa** on its medial aspect receives the obturator externus tendon.

4. Lesser trochanter

– lies in the angle between the neck and the shaft.

– projects at the inferior end of the **intertrochanteric crest.**

– provides an insertion for the iliopsoas tendon.

5. Linea aspera

– is the rough line or ridge on the body (shaft) of the femur.

– exhibits lateral and medial lips that provide attachments for many muscles and the three intermuscular septa.

6. Pectineal line

– runs from the lesser trochanter to the medial lip of the linea aspera.

– provides an insertion for the pectineus muscle.

7. Adductor tubercle

– is a small prominence at the uppermost part of the medial femoral condyle.

– provides an insertion for the adductor magnus muscle.

B. Patella

– is the largest sesamoid bone located within the tendon of the quadriceps, and articulates with the femur but not with the tibia.

– attaches to the tibial tuberosity by a continuation of the quadriceps tendon called the **patellar ligament.**

– functions to obviate wear and attrition on the quadriceps tendon as it passes across the trochlear groove and to increase the angle of pull of the quadriceps femoris, thereby magnifying its power.

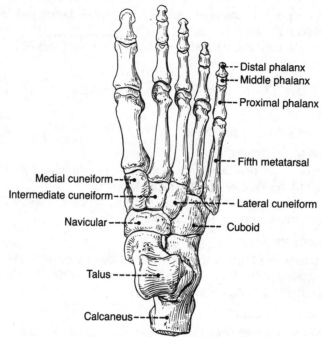

Figure 3-3. Bones of the foot.

C. **Tibia**
   – is the weight-bearing medial bone of the leg.
   – contains the **tibial tuberosity,** into which the patellar ligament inserts.
   – Its two condyles articulate with the condyles of the femur.
   – Its distal end projects medially and inferiorly as the **medial malleolus,** which
     has a **malleolar groove** for the tendons of the tibialis posterior and flexor
     digitorum longus muscles and another groove (lateral to the malleolus groove)
     for the tendon of the flexor hallucis longus muscle. It also provides attachment
     for the deltoid ligament.

D. **Fibula**
   – has little or no function in weight bearing but provides attachment for muscles.
   – attaches via its head (apex) to the fibular collateral ligament of the knee joint.
   – Its distal end projects laterally and inferiorly as the **lateral malleolus,** which
     lies more inferior and posterior than does the medial malleolus. The lateral
     malleolus attaches via its fossa to the transverse tibiofibular and posterior
     talofibular ligaments and provides attachment for the anterior talofibular, pos-
     terior talofibular, and calcaneofibular ligaments. It also presents the sulcus for
     the tendons of the peroneus longus and brevis muscles.

III. **Bones of the Ankle and Foot** (Figure 3-3)
   A. **Tarsus**
      – consists of seven tarsal bones: **talus, calcaneus, navicular bone, cuboid bone,**
        and **three cuneiform bones.**
      1. **Talus**
         – transmits the weight of the body from the tibia to other weight-bearing
           bones of the foot and is the only tarsal bone without muscle attachments.

- has a deep groove, the **sulcus tali,** for the interosseous ligaments between the talus and the calcaneus.
- has a groove on the posterior surface of its body for the flexor hallucis longus tendon.
- **Head** serves as **keystone of the medial longitudinal arch** of the foot.

2. Calcaneus

- is the largest and strongest bone of the foot and lies below the talus.
- forms the **heel** and articulates with the talus superiorly and the cuboid anteriorly.
- The **sustentaculum tali,** which supports the talus, is a shelf-like medial projection from the medial surface of the calcaneus; on its inferior surface, there is a groove for the flexor hallucis longus tendon.

3. Navicular bone

- is a boat-shaped tarsal bone lying between the head of the talus and the three cuneiform bones.

4. Cuboid bone

- is the most laterally placed tarsal bone and has a notch and groove for the tendon of the peroneus longus muscle.
- serves as the **keystone of the lateral longitudinal arch** of the foot.

5. Cuneiform bones

- are wedge-shaped bones that are related to the transverse arch.
- articulate with the navicular bone posteriorly and with three metatarsals anteriorly.

B. Metatarsus

- consists of **five metatarsals** and has prominent medial and lateral sesamoid bones on the first metatarsal.

C. Phalanges

- There are 14 (2 in the first digit and 3 in each of the others).

Joints and Ligaments of the Lower Limb

I. Hip (Coxal) Joint

- is a **multiaxial ball-and-socket synovial joint** between the acetabulum of the hip bone and the head of the femur, and allows abduction and adduction, flexion and extension, and circumduction and rotation.
- Its cavity is deepened by the **acetabular rim** and is completed below by the **transverse acetabular ligament,** which converts the **acetabular notch** into a foramen for passage of nutrient vessels and nerves.
- receives blood from branches of the medial and lateral femoral circumflex, superior and inferior gluteal, and obturator arteries. The posterior branch of the obturator artery gives rise to the artery of the ligamentum capitis femoris (ligamentum teres femoris).
- is innervated by branches of the femoral, obturator, and superior gluteal nerves and by the nerve to the quadratus femoris.

A. **Structures of the hip joint**

1. **Acetabular labrum**

    – is a complete fibrocartilage rim that deepens the articular socket for the head of the femur and consequently stabilizes the hip joint.

2. **Fibrous capsule**

    – is attached proximally to the margin of the acetabulum and to the transverse acetabular ligament.
    – is attached distally to the neck of the femur as follows: anteriorly to the intertrochanteric line and the root of the greater trochanter and posteriorly to the intertrochanteric crest.
    – encloses part of the head and most of the neck of the femur.
    – is reinforced anteriorly by the iliofemoral ligament, posteriorly by the ischio-femoral ligament, and inferiorly by the pubofemoral ligament.

B. **Ligaments of the hip joint**

1. **Iliofemoral ligament**

    – is the largest and most important ligament that reinforces the fibrous capsule anteriorly and is in the form of an inverted Y.
    – is attached proximally to the anterior–inferior iliac spine and the acetabular rim and distally to the intertrochanteric line and the front of the greater trochanter of the femur.
    – strongly resists hyperextension at the hip joint.

2. **Ischiofemoral ligament**

    – reinforces the fibrous capsule posteriorly and extends from the ischial portion of the **acetabular rim** to the neck of the femur, medial to the base of the greater trochanter.

3. **Pubofemoral ligament**

    – reinforces the fibrous capsule inferiorly and extends from the pubic portion of the acetabular rim and the superior pubic ramus to the lower part of the femoral neck.

4. **Ligamentum capitis femoris (ligamentum teres femoris)**

    – arises from the floor of the acetabular fossa (more specifically, from the margins of the acetabular notch and from the transverse ligament) and attaches to the **fovea capitis femoris.**

5. **Transverse acetabular ligament**

    – is a fibrous band that bridges the acetabular notch and converts it into a foramen.

II. Knee Joint (Figures 3-4 and 3-5)

– is a **hinge-type synovial joint** between two condyles of the femur and tibia, and also includes a saddle joint between the femur and the patella.
– permits flexion, extension, and some rotation in the flexed position of the knee; full extension is accompanied by medial rotation of the femur on the tibia, pulling all ligaments taut.
– is encompassed by **a fibrous capsule** that is rather thin, weak, and in many cases incomplete.

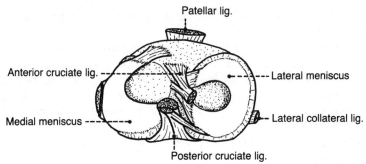

Figure 3-4. Ligaments of the knee.

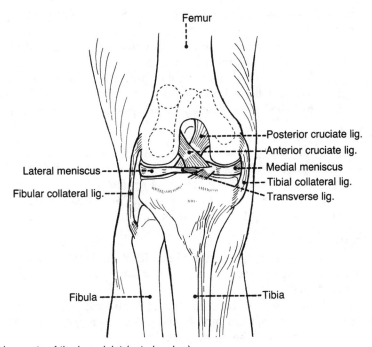

Figure 3-5. Ligaments of the knee joint (anterior view).

- is attached via this capsule to the margins of the femoral condyles, to the patella and the patellar ligament, and to the tibia at the margins of the tibial condyles.
- is stabilized on the lateral side by the biceps and gastrocnemius (lateral head) tendons, the **iliotibial tract,** and the fibular collateral ligaments and on the medial side by the sartorius, gracilis, gastrocnemius (medial head), semitendinosus, and semimembranosus muscles, and the tibial collateral ligament.
- receives blood from the genicular branches (superior medial and lateral, inferior medial and lateral, and middle) of the popliteal artery, a descending branch of the lateral femoral circumflex artery, an articular branch of the descending genicular artery, and the anterior tibial recurrent artery.
- is innervated by branches of the sciatic, femoral, and obturator nerves.

A. Ligaments of the knee joint

1. Anterior cruciate ligament

- lies inside the knee joint capsule but outside the synovial cavity of the joint.

– attaches to the anterior intercondylar area of the tibia, hence the "anterior" part of its name.

– passes upward, backward, and laterally to be inserted into the posterior aspect of the medial surface of the lateral femoral condyle.

– prevents backward slipping (dislocation) of the femur on the tibia (or forward slipping of the tibia on the femur) and hyperextension of the knee joint.

– limits excessive anterior mobility of the tibia on the femur when the knee is extended.

– is lax when the knee is flexed and becomes taut when the knee is fully extended.

2. Posterior cruciate ligament

– lies outside the synovial cavity but within the fibrous joint capsule.

– attaches to a depression of the posterior intercondylar area.

– passes upward, forward, and medially to insert into the anterior part of the lateral surface of the medial femoral condyle.

– limits hyperflexion of the knee and prevents forward displacement of the femur on the tibia (or backward slipping of the tibia on the femur).

– is lax when the knee is extended and becomes taut when the knee is flexed.

3. Medial meniscus

– lies outside the synovial cavity but within the joint capsule.

– is **C-shaped** (i.e., forms a semicircle) and is attached to the interarticular area of the tibia or medial collateral ligament.

– acts as a cushion or shock absorber and facilitates lubrication.

– is more frequently torn in injuries than the lateral meniscus.

4. Lateral meniscus

– lies outside the synovial cavity but within the joint capsule.

– is approximately circular and is incompletely attached to the upper aspect of the tibia.

– is separated laterally from the fibular (or lateral) collateral ligament by the tendon of the popliteal muscle and aids in forming a more stable base for the articulation of the femoral condyle.

5. Transverse ligament

– binds the anterior horns (ends) of the lateral and medial semilunar cartilages (menisci).

6. Medial (tibial) collateral ligament

– is a broad band separated from the capsule anteriorly by a bursa.

– is attached to the medial meniscus as well as to the medial aspects of the articular capsule and tibial condyle.

– prevents medial displacement of the two long bones and thus abduction of the leg at the knee.

– Its **firm attachment to the medial meniscus** is of clinical significance because injury to the ligament results in concomitant damage to the medial meniscus.

7. Lateral (fibular) collateral ligament

– is a rounded cord that stands well away from the capsule of the joint.

– extends between the lateral femoral epicondyle and the head of the fibula.

8. **Patellar ligament**

- is a strong flattened fibrous band that is the continuation of the **quadriceps femoris tendon** and extends from the apex of the patella to the tuberosity of the tibia.

9. **Arcuate popliteal ligament**

- arises from the head of the fibula and passes upward and medially over the tendon of the popliteus muscle on the back of the knee joint.

10. **Oblique popliteal ligament**

- is an oblique expansion of the **semimembranosus tendon** and passes upward obliquely across the posterior surface of the knee joint from the medial condyle of the tibia.
- resists hyperextension of the leg and lateral rotation during the final phase of extension.

B. **Bursae of the knee joint**

1. **Suprapatellar bursa**

- lies deep to the quadriceps femoris muscle and extends about 8 cm superior to the patella.
- is the major bursa communicating with the knee joint cavity (the semimembranosus bursa also may communicate with it).

2. **Prepatellar bursa**

- lies over the superficial surface of the patella.

3. **Infrapatellar bursa**

- consists of a **subcutaneous infrapatellar bursa,** which lies over the patellar ligament, and a **deep infrapatellar bursa,** which lies deep to the patellar ligament.

III. Tibiofibular Joints

A. **Proximal tibiofibular joint**

- is a plane synovial joint between the head of the fibula and the tibia.

B. **Distal tibiofibular joint**

- is a fibrous joint between the tibia and the fibula.

IV. Ankle (Talocrural) Joint (Figure 3-6; see Figure 3-2)

- is a **hinge-type (ginglymus) synovial joint** between the inferior ends of the tibia and fibula and the superior surface of the talus, permitting dorsiflexion and plantar flexion.

A. **Articular capsule**

- is a thin fibrous capsule that lies both anteriorly and posteriorly, allowing movement.
- is reinforced medially by the medial (or deltoid) ligament and laterally by the lateral ligament, which prevents anterior and posterior slipping of the tibia and fibula on the talus.

B. **Ligaments of the ankle joint**

1. **Medial (deltoid) ligament**

- has four parts: the tibionavicular, tibiocalcaneal, anterior tibiotalar, and posterior tibiotalar ligaments.

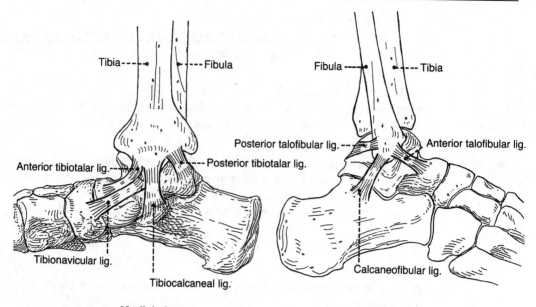

Figure 3-6. Ligaments of the ankle joint.

 – extends from the medial malleolus to the navicular bone, calcaneus, and talus.
 – prevents overeversion of the foot and helps to maintain the medial longitudinal arch.

2. **Lateral ligament**
 – consists of the anterior talofibular, posterior talofibular, and calcaneofibular (cord-like) ligaments.
 – resists inversion of the foot and may be torn during an **ankle sprain** (inversion injury).

V. Tarsal Joints

A. Intertarsal joints

1. **Talocalcaneal (subtalar) joint**
 – is part of the talocalcaneonavicular joint formed between the talus and calcaneus bones.
 – allows inversion and eversion of the foot.

2. **Talocalcaneonavicular joint**
 – is part of the transverse tarsal joint.
 – resembles a ball-and-socket joint in which the head of the talus (ball) is in contact with a socket formed by the calcaneus and navicular bones.
 – is supported by the spring (plantar calcaneonavicular) ligament.

3. **Calcaneocuboid joint**
 – is part of the transverse tarsal joint.
 – resembles a saddle joint between the calcaneus and the cuboid bones.
 – is supported by the short plantar (plantar calcaneocuboid) and long plantar ligaments and by the tendon of the peroneus longus muscle.

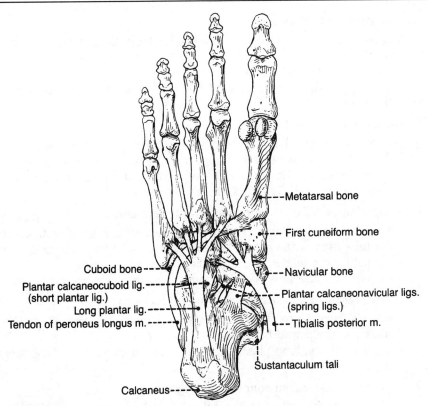

Figure 3-7. Plantar ligaments.

> **4. Transverse tarsal (midtarsal) joint**
> – is a collective term for the **talonavicular part** of the talocalcaneonavicular joint and the calcaneocuboid joint. The two joints are separated anatomically but act together functionally.
> – is important in inversion and eversion of the foot.

B. Tarsometatarsal joints
> – are **plane joints** that strengthen the transverse arch.
> – are united by articular capsules and are reinforced by the plantar, dorsal, and interosseous ligaments.

C. Metatarsophalangeal joints
> – are **ellipsoid (condyloid) joints** that are joined by articular capsules and are reinforced by the plantar and collateral ligaments.

D. Interphalangeal joints
> – are **hinge (ginglymus) joints** that are enclosed by articular capsules and are reinforced by the plantar and collateral ligaments.

VI. Ligaments and Arches of the Foot (Figure 3-7)

A. Transverse (metatarsal) arch
> – is formed by the navicular bone, the three cuneiform bones, the cuboid bone, and the five metatarsal bones of the foot.
> – is maintained anteriorly by the transverse head of the adductor hallucis.

B. Lateral longitudinal arch

- is formed by the calcaneus, the cuboid bone, and the lateral two metatarsal bones.
- **Keystone** is the **cuboid bone.**
- is supported by the peroneus longus tendon and the long and short plantar ligaments.

C. Medial longitudinal arch

- is formed by the talus, calcaneus, navicular bone, cuneiform bones, and medial three metatarsal bones and is supported by the spring ligament.
- **keystone** is the **head of the talus,** which is located at the summit between the sustentaculum tali and the navicular bone.

D. Long plantar (plantar calcaneocuboid) ligament

- extends from the plantar aspect of the calcaneus in front of its tuberosity to the tuberosity of the cuboid bone and the base of the metatarsals and forms a canal for the tendon of the peroneus longus.
- supports the lateral side of the longitudinal arch of the foot.

E. Short plantar (plantar calcaneocuboid) ligament

- extends from the front of the plantar surface of the calcaneus to the plantar surface of the cuboid bone.
- lies deep to the long plantar ligament and supports the lateral longitudinal arch.

F. Spring (plantar calcaneonavicular) ligament

- passes from the sustentaculum tali of the calcaneus to the navicular bone.
- supports the **head of the talus** and the medial longitudinal arch.
- is called the spring ligament because of its **elasticity** under the pressure of the head of the talus.

Cutaneous Nerves, Superficial Veins, and Lymphatics

I. Cutaneous Nerves of the Lower Limb (Figure 3-8)

A. Lateral femoral cutaneous nerve

- arises from the lumbar plexus (L2–L3), emerges from the lateral border of the psoas major, crosses the iliacus, and passes under the inguinal ligament near the anterior–superior iliac spine.
- innervates the **skin on the anterior and lateral aspects of the thigh** as far as the knee.

B. Clunial (buttock) nerves

- innervate the **skin of the gluteal region.**
- consist of **superior** (lateral branches of the dorsal rami of the upper three lumbar nerves), **middle** (lateral branches of the dorsal rami of the upper three sacral nerves), and **inferior** (gluteal branches of the posterior femoral cutaneous nerve) nerves.

C. Posterior femoral cutaneous nerve

- arises from the **sacral plexus** (S1–S3), passes through the greater sciatic foramen below the piriformis muscle, runs deep to the gluteus maximus muscle, and emerges from the inferior border of this muscle.

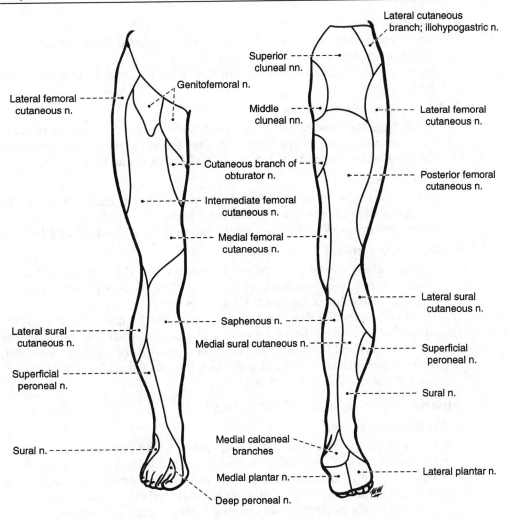

Figure 3-8. Cutaneous nerves of the lower limb.

- descends in the posterior midline of the thigh deep to the fascia lata and pierces the fascia lata near the popliteal fossa.
- innervates the **skin of the buttock, thigh, and calf.**

D. Saphenous nerve

- arises from the **femoral nerve** in the **femoral triangle** and descends with the femoral vessels through the femoral triangle and the adductor canal.
- pierces the fascial covering of the adductor canal at its distal end in company with the saphenous branch of the descending genicular artery.
- becomes cutaneous between the sartorius and the gracilis, descends behind the condyles of the femur and tibia and medial aspect of the leg in company with the great saphenous vein, and innervates the **skin on the medial side of the leg and foot.**

E. Lateral sural cutaneous nerve

– arises from the **common peroneal nerve** in the **popliteal fossa** and innervates the **skin on the posterolateral side of the leg.**

– may have a **communicating branch** that joins the medial sural cutaneous nerve.

F. Medial sural cutaneous nerve

– arises from the tibial nerve in the popliteal fossa, may join the lateral sural nerve or its communicating branch to form the **sural nerve,** and innervates the **back of the leg** and the **lateral side of the heel and foot.**

G. Sural nerve

– is formed by the union of the medial sural and lateral sural nerves (or the communicating branch of the lateral sural nerve).

– innervates the **back of the leg** and the **lateral side of the heel and foot.**

H. Superficial peroneal nerve

– passes distally between the peroneus muscles and the extensor digitorum longus, pierces the deep fascia in the lower third of the leg, and innervates the **skin of the lower part of the leg.**

– divides into a **medial dorsal cutaneous nerve,** which supplies the medial side of the foot and ankle, the medial sides of the great toe, and the adjacent sides of the second and third toes, and an **intermediate dorsal cutaneous nerve,** which supplies the skin of the lateral side of the foot and ankle and the adjacent sides of the third, fourth, and little toes.

II. Superficial Veins of the Lower Limb

A. Great saphenous vein

– begins at the medial end of the **dorsal venous arch** of the foot.

– ascends in front of the medial malleolus and along the medial aspect of the tibia along with the saphenous nerve, passes behind the medial condyles of the tibia and femur, and then ascends along the medial side of the femur.

– passes through the **saphenous opening (fossa ovalis)** in the fascia lata and pierces the femoral sheath to join the femoral vein.

– is a suitable vessel for use in arterial bypass surgery.

B. Small (short) saphenous vein

– begins at the lateral end of the **dorsal venous arch** and passes upward along the lateral side of the foot with the sural nerve, behind the lateral malleolus.

– passes to the popliteal fossa, where it perforates the deep fascia and terminates in the **popliteal vein.**

III. Lymphatics of the Lower Limb

A. Vessels

1. Superficial lymph vessels

– are divided into a **medial group,** which follows the great saphenous vein, and a **lateral group,** which follows the small saphenous vein.

2. Deep lymph vessels

– consist of the **anterior tibial, posterior tibial, and peroneal vessels,** which follow the course of the corresponding blood vessels and enter the **popliteal lymph nodes.**

B. Lymph nodes

   1. Superficial inguinal group of lymph nodes

     – is located subcutaneously near the **saphenofemoral junction** and drains the superficial thigh region.

     – receives lymph from the anterolateral abdominal wall below the umbilicus, gluteal region, lower parts of the vagina and anus, and external genitalia except the glans, and drains into the **external iliac nodes.**

   2. Deep inguinal group of lymph nodes

     – lies deep to the fascia lata on the medial side of the femoral vein.

     – receives lymph from deep lymph vessels (i.e., efferents of the popliteal nodes) that accompany the femoral vessels and from the glans penis or glans clitoris, and drains into the external iliac nodes through the femoral canal.

Muscles and Fasciae of the Lower Limb

I. Fibrous Structures in the Gluteal Region and Posterior Thigh

A. Sacrotuberous ligament

   – extends from the ischial tuberosity to the posterior iliac spines, lower sacrum, and coccyx.

   – converts, with the sacrospinous ligament, the lesser sciatic notch into the lesser sciatic foramen.

B. Sacrospinous ligament

   – extends from the ischial spine to the lower sacrum and the coccyx and converts the greater sciatic notch into the greater sciatic foramen.

C. Sciatic foramina

   1. Greater sciatic foramen

     – transmits the piriformis muscle, superior and inferior gluteal vessels and nerves, internal pudendal vessels and pudendal nerve, sciatic nerve, posterior femoral cutaneous nerve, and the nerves to the obturator internus and quadratus femoris muscles.

   2. Lesser sciatic foramen

     – transmits the tendon of the obturator internus, the nerve to the obturator internus, and the internal pudendal vessels and pudendal nerve.

   3. Structures that pass through both the greater and lesser sciatic foramina

     – include the pudendal nerve, the internal pudendal vessels, and the nerve to the obturator internus.

D. Iliotibial tract

   – is a thick lateral portion of the **fascia lata.**

   – provides insertion for the gluteus maximus and tensor fasciae latae muscles.

   – helps to form the **fibrous capsule of the knee joint.**

E. Fascia lata

   – is a membranous deep fascia covering muscles of the thigh and forms the **lateral and medial intermuscular septa** by its inward extension to the femur.

– is attached to the pubic symphysis, pubic crest, pubic rami, ischial tuberosity, inguinal and sacrotuberous ligaments, and the sacrum and coccyx.

II. Muscles of the Gluteal Region

Muscle	Origin	Insertion	Nerve	Action
Gluteus maximus	Ilium; sacrum; coccyx; sacrotuberous ligament	Gluteal tuberosity; iliotibial tract	Inferior gluteal	Extends and rotates thigh laterally
Gluteus medius	Ilium between iliac crest, and anterior and posterior gluteal lines	Greater trochanter	Superior gluteal	Abducts and rotates thigh medially
Gluteus minimus	Ilium between anterior and inferior gluteal lines	Greater trochanter	Superior gluteal	Abducts and rotates thigh medially
Tensor fasciae latae	Iliac crest; anterior–superior iliac spine	Iliotibial tract	Superior gluteal	Flexes, abducts, and rotates thigh medially
Piriformis	Pelvic surface of sacrum; sacrotuberous ligament	Upper end of greater trochanter	Sacral (S1–S2)	Rotates thigh laterally
Obturator internus	Ischiopubic rami; obturator membrane	Greater trochanter	Nerve to obturator internus	Abducts and rotates thigh laterally
Superior gemellus	Ischial spine	Obturator internus tendon	Nerve to obturator internus	Rotates thigh laterally
Inferior gemellus	Ischial tuberosity	Obturator internus tendon	Nerve to quadratus femoris	Rotates thigh laterally
Quadratus femoris	Ischial tuberosity	Intertrochanteric crest	Nerve to quadratus femoris	Rotates thigh laterally

III. Posterior Muscles of the Thigh*

Muscle	Origin	Insertion	Nerve	Action
Semitendinosus	Ischial tuberosity	Medial surface of upper part of tibia	Tibial portion of sciatic nerve	Extends thigh; flexes and rotates leg medially
Semimembranosus	Ischial tuberosity	Medial condyle of tibia	Tibial portion of sciatic nerve	Extends thigh; flexes and rotates leg medially

Muscle	Origin	Insertion	Nerve	Action
Biceps femoris	Long head from ischial tuberosity; short head from linea aspera and upper supracondylar line	Head of fibula	Tibial (long head) and common peroneal (short head) divisions of sciatic nerve	Extends thigh; flexes and rotates leg laterally

* These three muscles collectively are called hamstrings.

IV. Fibrous Structures of the Anterior Thigh and Popliteal Fossa

A. Femoral triangle
- is bounded by the inguinal ligament superiorly, the sartorius muscle laterally, and the adductor longus muscle medially.
- contains the femoral nerve and vessels.

B. Femoral ring
- is the abdominal opening of the femoral canal.
- is bounded by the inguinal ligament anteriorly, the femoral vein laterally, the lacunar ligament medially, and the pectineal ligament posteriorly.

C. Femoral canal
- lies medial to the femoral vein in the femoral sheath.
- contains fat, areolar connective tissue, and lymph nodes.
- transmits **lymphatics** from the lower limb and perineum to the peritoneal cavity.
- is a potential weak area and a site of **femoral herniation.**

D. Femoral sheath
- is formed by a prolongation of the **transversalis and iliac fasciae** in the thigh.
- contains the femoral artery and vein, the femoral branch of the genitofemoral nerve, and the femoral canal. (The femoral nerve lies outside the femoral sheath, lateral to the femoral artery.)
- Its distal end reaches the level of the proximal end of the saphenous opening.

E. Adductor canal
- begins at the apex of the femoral triangle and ends at the **adductor hiatus** (hiatus tendineus).
- lies between the adductor magnus and longus muscles and the vastus medialis muscle and is covered by the sartorius muscle and fascia.
- contains the femoral vessels, the saphenous nerve, and the nerve to the vastus medialis.

F. Adductor hiatus (hiatus tendineus)
- is the aperture in the tendon of insertion of the adductor magnus.
- allows the passage of the femoral vessels into the popliteal fossa.

G. Popliteal fossa
- is bounded superomedially by the semitendinosus and semimembranosus muscles and superolaterally by the biceps muscle.

– is bounded inferolaterally by the lateral head of the gastrocnemius muscle and inferomedially by the medial head of the gastrocnemius muscle.

– contains the popliteal vessels, the common peroneal and tibial nerves, and the small saphenous vein.

– floor is composed of the femur, the oblique popliteal ligament, and the popliteus muscle.

V. Anterior Muscles of the Thigh

Muscle	Origin	Insertion	Nerve	Action
Iliacus	Iliac fossa; ala of sacrum	Lesser trochanter	Femoral	Flexes and rotates thigh medially (with psoas major)
Sartorius	Anterior–superior iliac spine	Upper medial side of tibia	Femoral	Flexes and rotates thigh laterally; flexes and rotates leg medially
Rectus femoris	Anterior–inferior iliac spine; posterior–superior rim of acetabulum	Base of patella; tibial tuberosity	Femoral	Flexes thigh; extends leg
Vastus medialis	Intertrochanteric line; linea aspera; medial intermuscular septum	Medial side of patella; tibial tuberosity	Femoral	Extends leg
Vastus lateralis	Intertrochanteric line; greater trochanter; linea aspera; gluteal tuberosity; lateral intermuscular septum	Lateral side of patella; tibial tuberosity	Femoral	Extends leg
Vastus intermedius	Upper shaft of femur; lower lateral intermuscular septum	Upper border of patella; tibial tuberosity	Femoral	Extends leg

VI. Medial Muscles of the Thigh

Muscle	Origin	Insertion	Nerve	Action
Adductor longus	Body of pubis below its crest	Middle third of linea aspera	Obturator	Adducts and flexes thigh
Adductor brevis	Body and inferior pubic ramus	Pectineal line; upper part of linea aspera	Obturator	Adducts and flexes thigh
Adductor magnus	Ischiopubic ramus; ischial tuberosity	Linea aspera; medial supracondylar line; adductor tubercle	Obturator and sciatic	Adducts, flexes, and extends thigh
Pectineus	Pectineal line of pubis	Pectineal line of femur	Obturator and femoral	Adducts and flexes thigh

Muscle	Origin	Insertion	Nerve	Action
Gracilis	Body and inferior pubic ramus	Medial surface of upper quarter of tibia	Obturator	Adducts and flexes thigh; flexes and rotates leg medially
Obturator externus	Margin of obturator foramen and obturator membrane	Intertrochanteric fossa of femur	Obturator	Rotates thigh laterally

VII. Anterior and Lateral Muscles of the Leg

Muscle	Origin	Insertion	Nerve	Action
Anterior Tibialis anterior	Lateral tibial condyle; interosseous membrane	First cuneiform; first metatarsal	Deep peroneal	Dorsiflexes and inverts foot
Extensor hallucis longus	Middle half of anterior surface of fibula; interosseous membrane	Base of distal phalanx of big toe	Deep peroneal	Extends big toe; dorsiflexes and inverts foot
Extensor digitorum longus	Lateral tibial condyle; upper two-thirds of fibula; interosseous membrane	Bases of middle and distal phalanges	Deep peroneal	Extends toes; dorsiflexes foot
Peroneus tertius	Distal one-third of fibula; interosseous membrane	Base of fifth metatarsal	Deep peroneal	Dorsiflexes and everts foot
Lateral Peroneus longus	Lateral tibial condyle; head and upper lateral side of fibula	Base of first metatarsal; medial cuneiform	Superficial peroneal	Everts and plantar flexes foot
Peroneus brevis	Lower lateral side of fibula; intermuscular septa	Base of fifth metatarsal	Superficial peroneal	Everts and plantar flexes foot

VIII. Posterior Muscles of the Leg

Muscle	Origin	Insertion	Nerve	Action
Superficial group Gastrocnemius	Lateral (lateral head) and medial (medial head) femoral condyles	Posterior aspect of calcaneus via tendo calcaneus	Tibial	Flexes knee; plantar flexes foot

Muscle	Origin	Insertion	Nerve	Action
Soleus	Upper fibula head; soleal line on tibia	Posterior aspect of calcaneus via tendo calcaneus	Tibial	Plantar flexes foot
Plantaris	Lower lateral supra-condylar line	Posterior surface of calcaneus	Tibial	Flexes and rotates leg medially
Deep group Popliteus	Lateral condyle of femur; popliteal ligament	Upper posterior side of tibia	Tibial	Flexes and rotates leg medially
Flexor hallucis longus	Lower two-thirds of fibula; interosseous membrane; intermuscular septa	Base of distal phalanx of big toe	Tibial	Flexes distal phalanx of big toe
Flexor digitorum longus	Middle posterior aspect of tibia	Distal phalanges of lateral four toes	Tibial	Flexes lateral four toes; plantar flexes foot
Tibialis posterior	Interosseous membrane; upper parts of tibia and fibula	Tuberosity of navicular; sustentacula tali; three cuneiforms; cuboid; bases of metatarsals 2–4	Tibial	Plantar flexes and inverts foot

IX. Muscles of the Foot

Muscle	Origin	Insertion	Nerve	Action
Dorsum of foot Extensor digitorum brevis	Dorsal surface of calcaneus	Tendons of extensor digitorum longus	Deep peroneal	Extends toes
Extensor hallucis brevis	Dorsal surface of calcaneus	Base of proximal phalanx of big toe	Deep peroneal	Extends big toe
Sole of foot Abductor hallucis	Medial tubercle of calcaneus	Base of proximal phalanx of big toe	Medial plantar	Abducts big toe
Flexor digitorum brevis	Medial tubercle of calcaneus	Middle phalanges of lateral four toes	Medial plantar	Flexes middle phalanges of lateral four toes
Abductor digiti minimi	Medial and lateral tubercles of calcaneus	Proximal phalanx of little toe	Lateral plantar	Abducts little toe
Quadratus plantae	Medial and lateral side of calcaneus	Tendons of flexor digitorum longus	Lateral plantar	Aids in flexing toes

Muscle	Origin	Insertion	Nerve	Action
Lumbricals (4)	Tendons of flexor digitorum longus	Proximal phalanges; extensor expansion	First by medial plantar; lateral three by lateral plantar	Flex metatarsophalangeal joints and extend interphalangeal joints
Flexor hallucis brevis	Cuboid; third cuneiform	Proximal phalanx of big toe	Medial plantar	Flexes big toe
Adductor hallucis:				
Oblique head	Bases of metatarsals 2–4	Proximal phalanx of big toe	Lateral plantar	Adducts big toe
Transverse head	Capsule of lateral four metatarsophalangeal joints			
Flexor digiti minimi brevis	Base of metatarsal 5	Proximal phalanx of little toe	Lateral plantar	Flexes little toe
Plantar interossei (3)	Medial sides of metatarsals 3–5	Medial sides of base of proximal phalanges 3–5	Lateral plantar	Adduct toes; flex proximal and extend distal phalanges
Dorsal interossei (4)	Adjacent shafts of metatarsals	Proximal phalanges of second toes (medial and lateral sides), and third and fourth toes (lateral sides)	Lateral plantar	Abduct toes; flex proximal and extend distal phalanges

X. Fascial Structures of the Foot

A. Superior extensor retinaculum

– is a **broad band of deep fascia** extending between the tibia and fibula, above the ankle.

B. Inferior extensor retinaculum

– is a **Y-shaped band of deep fascia,** which forms a loop for the tendons of the extensor digitorum longus and the peroneus tertius and then divides into an upper band, which attaches to the medial malleolus, and a **lower band,** which attaches to the deep fascia of the foot and the plantar aponeurosis.

C. Flexor retinaculum

– is a deep fascial band that passes between the medial malleolus and the medial surface of the calcaneus.
– holds three tendons in place beneath it: the tibialis posterior, flexor digitorum longus, and flexor hallucis longus.
– transmits the tibial nerve and posterior tibial artery beneath it.

D. Tendo calcaneus (Achilles tendon)

– is the tendon of insertion of the **triceps surae** (gastrocnemius and soleus) into the tuberosity of the calcaneus.

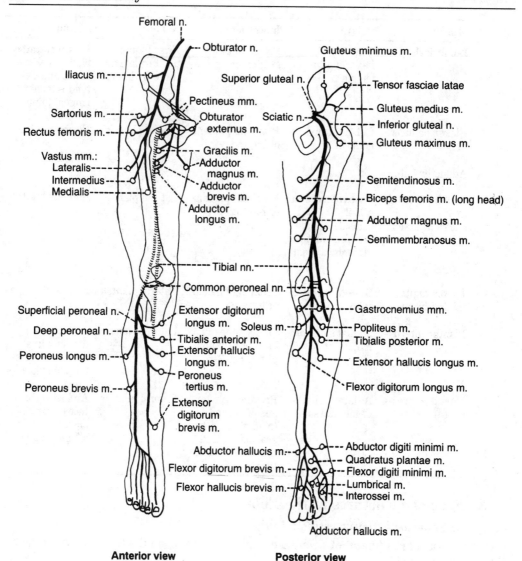

Figure 3-9. Innervation of the lower limb.

E. **Plantar aponeurosis**

 – is a thick fascia investing the plantar muscles.
 – radiates from the **calcaneal tuberosity** (tuber calcanei) toward the toes and gives attachment to the short flexor muscles of the toes.

Nerves and Vasculature of the Lower Limb

I. Nerves of the Lower Limb (Figure 3-9)

A. Obturator nerve

 – arises from the **lumbar plexus** (L3–L4) and enters the thigh through the obturator foramen.
 – divides into anterior and posterior branches.

1. Anterior branch

– descends between the adductor longus and adductor brevis muscles and innervates the adductor longus, adductor brevis, gracilis, and pectineus muscles.

2. Posterior branch

– descends between the adductor brevis and adductor magnus muscles and innervates the obturator externus and adductor magnus muscles.

B. Femoral nerve

– arises from the **lumbar plexus** (L2–L4) within the substance of the psoas major, emerges between the iliacus and psoas major muscles, and enters the thigh by passing deep to the inguinal ligament and lateral to the femoral sheath.
– gives rise to **muscular branches; articular branches** to the hip and knee joints; and **cutaneous branches,** including the anterior femoral cutaneous nerve and the saphenous nerve, which descends through the femoral triangle and accompanies the femoral vessels in the adductor canal.

C. Superior gluteal nerve

– arises from the **sacral plexus** (L4–S1) and enters the buttock through the greater sciatic foramen above the piriformis.
– passes between the gluteus medius and minimus muscles and divides into numerous branches to innervate the gluteus medius and minimus, the tensor fasciae latae, and the hip joint.

D. Inferior gluteal nerve

– arises from the **sacral plexus** (L5–S2) and enters the buttock through the greater sciatic foramen below the piriformis.
– divides into numerous branches to innervate the overlying gluteus maximus.

E. Posterior femoral cutaneous nerve

– arises from the **sacral plexus** (S1–S3) and enters the buttock through the greater sciatic foramen below the piriformis.
– runs deep to the gluteus maximus and emerges from the inferior border of this muscle.
– descends on the posterior thigh and innervates the skin of the buttock, thigh, and calf.

F. Sciatic nerve

– arises from the **sacral plexus** (L4–S3) and is the **largest nerve in the body,** consisting of the tibial and common peroneal components.
– enters the buttock through the greater sciatic foramen below the piriformis.
– descends over the obturator internus gemelli and quadratus femoris muscles between the ischial tuberosity and the greater trochanter.
– **innervates the hamstring muscles** by its tibial division, except for the short head of the biceps femoris, which is innervated by its common peroneal division.
– provides articular branches to the hip and knee joints.

1. Common peroneal (fibular) nerve

– is separated from the tibial portion at the apex of the popliteal fossa and descends through the fossa.
– superficially crosses the lateral head of the gastrocnemius muscle, then turns laterally around the neck of the fibula deep to the peroneus longus, where it divides into the deep peroneal and superficial peroneal nerves.

– gives rise to the **lateral sural cutaneous nerve,** which supplies the skin on the lateral part of the back of the leg, and the **recurrent articular branch** to the knee joint.

a. Superficial peroneal nerve

– arises from the common peroneal nerve between the peroneus longus and the neck of the fibula, descends in the lateral compartment, and innervates the peroneus longus and brevis muscles.

– emerges between the peroneus longus and brevis muscles by piercing the deep fascia at the lower third of the leg to become subcutaneous and innervates the skin of the lower leg and foot.

b. Deep peroneal nerve

– arises from the common peroneal nerve between the peroneus longus and the neck of the fibula.

– gives rise to a recurrent branch to the knee joint.

– passes around the neck of the fibula and through the extensor digitorum longus muscle.

– descends on the **interosseous membrane** between the extensor digitorum longus and the tibialis anterior and then between the extensor digitorum longus and the extensor hallucis longus muscles.

– innervates the anterior muscles of the leg and divides into a **lateral branch,** which supplies the extensor digitorum brevis, and a **medial branch,** which accompanies the dorsalis pedis artery to supply adjacent sides of the first and second toes.

2. Tibial nerve

– descends through the popliteal fossa and then lies on the popliteus muscle.

– gives rise to **three articular branches,** which accompany the medial superior genicular, middle genicular, and medial inferior genicular arteries to the knee joint.

– gives rise to **muscular branches** to the posterior muscles of the leg.

– gives rise to the medial sural cutaneous nerve, the medial calcaneal branch to the skin of the heel and sole, and the articular branches to the ankle joint.

– terminates beneath the flexor retinaculum by dividing into the medial and lateral plantar nerves.

a. Medial plantar nerve

– arises beneath the flexor retinaculum, deep to the posterior portion of the abductor hallucis muscle, as the larger terminal branch from the tibial nerve.

– passes distally between the abductor hallucis and flexor digitorum brevis muscles and innervates them.

– gives rise to **common digital branches** that divide into proper digital branches, which supply the flexor hallucis brevis and the first lumbrical and the skin of the medial three and one-half toes.

b. Lateral plantar nerve

– is the smaller terminal branch of the tibial nerve.

– runs distally and laterally between the quadratus plantae and the flexor digitorum brevis, innervating the quadratus plantae and the abductor digiti minimi muscles.

– divides into a **superficial branch,** which innervates the flexor digiti minimi brevis, and a **deep branch,** which innervates the plantar and dorsal interossei, the lateral three lumbricals, and the adductor hallucis.

II. Arteries of the Lower Limb (Figure 3-10)

A. Superior gluteal artery

- arises from the **internal iliac artery,** passes between the lumbosacral trunk and the first sacral nerve, and enters the buttock through the greater sciatic foramen above the piriformis muscle.
- runs deep to the gluteus maximus muscle and divides into a **superficial branch,** which forms numerous branches to supply the gluteus maximus, and a **deep branch,** which runs between the gluteus medius and minimus muscles and supplies these muscles and the tensor fasciae latae.

B. Inferior gluteal artery

- arises from the **internal iliac artery,** usually passes between the first and second sacral nerves, and enters the buttock through the greater sciatic foramen below the piriformis.
- enters the deep surface of the gluteus maximus and descends on the medial side of the sciatic nerve, in company with the posterior femoral cutaneous nerve.

C. Obturator artery

- arises from the **internal iliac artery** in the pelvis and passes through the obturator foramen, where it divides into anterior and posterior branches.
- **Anterior branch** descends in front of the adductor brevis muscle and gives rise to muscular branches.
- **Posterior branch** descends behind the adductor brevis muscle to supply the adductor muscles, and gives rise to the **acetabular branch,** which passes through the acetabular notch and provides an **artery to the head of the femur,** which accompanies the ligament of the head of the femur.

D. Femoral artery

- begins as the continuation of the **external iliac artery** distal to the inguinal ligament, descends through the femoral triangle, and enters the adductor canal.
- Its branches include the following:

  #### 1. Superficial epigastric artery
  - runs subcutaneously upward toward the umbilicus.

  #### 2. Superficial circumflex iliac artery
  - runs laterally almost parallel with the inguinal ligament.

  #### 3. External pudendal artery
  - emerges through the saphenous ring, runs medially over the spermatic cord (or the round ligament of the uterus), and sends inguinal branches and anterior scrotal (or labial) branches.

  #### 4. Profunda femoris (deep femoral) artery
  - arises from the **femoral artery** within the femoral triangle.
  - descends in front of the pectineus, adductor brevis, and adductor magnus muscles, but behind the adductor longus muscle.
  - gives rise to the medial and lateral femoral circumflex and muscular branches.

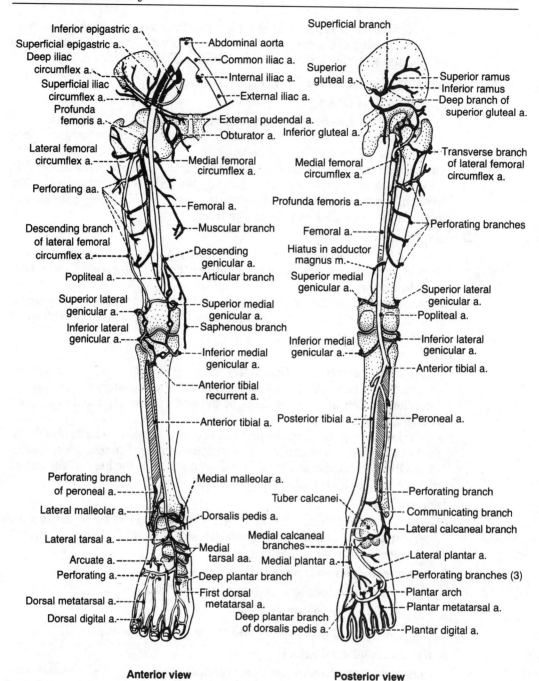

Anterior view   **Posterior view**

Figure 3-10. Blood supply to the lower limb.

– provides, in the adductor canal, **four perforating arteries** that perforate
  and supply the adductor magnus and hamstring muscles.
– Its first perforating branch anastomoses with the inferior gluteal artery and
  the transverse branches of the medial and lateral femoral circumflex arteries.
  This is known as the **cruciate anastomosis** of the buttock.

5. Medial femoral circumflex artery

– arises from the **femoral or profunda femoris artery** in the femoral triangle.
– runs between the pectineus and iliopsoas muscles, continues between the obturator externus and adductor brevis muscles, and enters the gluteal region between the adductor magnus and quadratus femoris muscles.
– gives rise to **muscular branches** and an **acetabular branch** to the hip joint and then divides into an **ascending branch,** which anastomoses with the gluteal arteries, and a **transverse branch,** which joins the cruciate anastomosis.
– is clinically important because it **supplies most of the blood to the head and neck of the femur.**

6. Lateral femoral circumflex artery

– arises from the **femoral or profunda femoris artery.**
– passes laterally deep to the sartorius and rectus femoris muscles.
– divides into an **ascending branch,** which gives a branch to the hip joint and anastomoses with the superior gluteal artery; a **transverse branch,** which joins the cruciate anastomosis; and a **descending branch,** which anastomoses with the superior lateral genicular branch of the popliteal artery.

7. Descending genicular artery

– arises from the **femoral artery** just before it passes through the adductor canal.
– divides into the **articular branch,** which enters the anastomosis around the knee, and the **saphenous branch,** which supplies the superficial tissue and skin on the medial side of the knee.

E. Popliteal artery

– is a continuation of the **femoral artery** at the adductor hiatus and runs through the popliteal fossa.
– terminates at the lower border of the popliteus muscle by dividing into the anterior and posterior tibial arteries.
– gives rise to five genicular arteries:

1. Superior lateral genicular artery, which passes deep to the biceps femoris tendon

2. Superior medial genicular artery, which passes deep to the semimembranosus and semitendinosus muscles and enters the substance of the vastus medialis

3. Inferior lateral genicular artery, which passes laterally above the head of the fibula and then deep to the fibular collateral ligament

4. Inferior medial genicular artery, which passes medially along the upper border of the popliteus muscle, deep to the popliteus fascia

5. Middle genicular artery, which pierces the oblique popliteal ligament and enters the knee joint

F. Posterior tibial artery

– arises from the popliteal artery at the lower border of the popliteus, between the tibia and the fibula.
– is accompanied by two venae comitantes and the tibial nerve.

- gives rise to the **peroneal (fibular) artery,** which descends between the tibialis posterior and the flexor hallucis longus muscles and supplies the lateral muscles in the posterior compartment. The peroneal artery passes behind the lateral malleolus, gives rise to the **posterior lateral malleolar branch,** and ends in branches to the ankle and heel.
- also gives rise to the posterior medial malleolar, perforating, and muscular branches and terminates by dividing into the medial and lateral plantar arteries.

   1. **Medial plantar artery**
      - is the smaller terminal branch of the posterior tibial artery.
      - runs between the abductor hallucis and the flexor digitorum brevis muscles.
      - gives rise to a **superficial branch,** which supplies the big toe, and a **deep branch,** which forms three superficial digital branches.

   2. **Lateral plantar artery**
      - is the larger terminal branch of the posterior tibial artery.
      - runs forward laterally in company with the lateral plantar nerve between the quadratus plantae and the flexor digitorum brevis muscles and then between the flexor digitorum brevis and the adductor digiti minimi muscles.
      - forms the **plantar arch** by joining the deep plantar branch of the dorsalis pedis artery. The plantar arch gives rise to four plantar metatarsal arteries.

G. Anterior tibial artery

- arises from the **popliteal artery** and enters the anterior compartment by passing through the gap at the upper end of the interosseous membrane.
- descends on the interosseous membrane between the tibialis anterior and extensor digitorum longus muscles.
- gives rise to the **anterior tibial recurrent artery,** which ascends to the knee joint, and the **anterior medial and lateral malleolar arteries** at the ankle.
- runs distally across the ankle midway between the lateral and medial malleoli and continues onto the dorsum of the foot as the dorsalis pedis artery.

H. Dorsalis pedis artery

- begins anterior to the ankle joint midway between the two malleoli as the continuation of the **anterior tibial artery.**
- gives rise to the **medial tarsal, lateral tarsal, arcuate, and first dorsal metatarsal arteries** and terminates as the **deep plantar artery,** which enters the sole of the foot by passing between the two heads of the first dorsal interosseous muscle and joins the lateral plantar artery to form the **plantar arch.** The **arcuate artery** gives rise to the second, third, and fourth dorsal metatarsal arteries.

III. Deep Veins of the Lower Limb

A. Popliteal vein

- ascends through the popliteal fossa behind the popliteal artery.
- receives the small saphenous vein and those veins corresponding to the branches of the popliteal artery.

B. Femoral vein

- accompanies the femoral artery as a continuation of the popliteal vein through the upper two-thirds of the thigh.

– has three valves, receives tributaries corresponding to branches of the femoral artery, and is joined by the great saphenous vein, which passes through the saphenous opening.

Clinical Considerations

I. Reflexes

A. Knee-jerk (patellar) reflex

– occurs when the patellar ligament is tapped, resulting in a sudden contraction of the quadriceps femoris.

B. Ankle-jerk reflex

– is a reflex twitch of the triceps surae (i.e., the medial and lateral heads of the gastrocnemius and the soleus muscles).
– is induced by tapping the tendo calcaneus.
– Its reflex center is in the fifth lumbar and first sacral segments of the spinal cord.

II. Syndromes and Abnormal Signs

A. Femoral hernia

– lies lateral to the pubic tubercle and deep to the inguinal ligament.
– sac is formed by the parietal peritoneum.
– passes through the femoral ring and canal.
– is more common in women than in men.

B. Gluteal gait (gluteus medius limp)

– is a **waddling gait,** characterized by the pelvis falling toward the unaffected side at each step due to **paralysis of the gluteus medius muscle.** This muscle normally functions to stabilize the pelvis when the opposite foot is off the ground.

C. Trendelenburg's sign

– is seen in a fracture of the femoral neck, dislocated hip joint, or weakness and paralysis of the gluteus medius muscle. When the patient stands on the affected limb, the pelvis on the sound side will sag; if normal, the pelvis will rise.

D. Congenital dislocation of the hip joint

– occurs because of **faulty development of the upper lip of the acetabulum;** the head of the femur moves out of the acetabulum through the ruptured capsule onto the gluteal surface of the ilium.
– The affected limb is **shortened, adducted, and medially rotated.**

E. Traumatic dislocation of the hip joint

– In **anterior dislocation,** the joint capsule is torn anteriorly and the femoral head moves out from the acetabulum; the femoral head lies inferior to the pubic bone.
– In **posterior dislocation,** the joint capsule is torn posteriorly and the femoral head moves out from the acetabulum; the femoral head rests on the gluteal surface of the ischium, and both the posterior acetabulum and the ligamentum capitis femoris are likely to be ruptured.

– In **medial or intrapelvic dislocation,** the joint capsule is torn medially and the femoral head is dislocated; this may be accompanied by acetabular fracture and rupture of the bladder.

F. The unhappy triad of the knee joint

– can occur when a football player's cleated shoe is planted firmly in the turf and the knee is struck from the lateral side.
– is indicated by a knee that is markedly swollen, particularly in the suprapatellar region.
– results in tenderness on application of pressure along the extent of the tibial collateral ligament.
– is characterized by:

1. **Rupture of the tibial collateral ligament,** as a result of excessive abduction

2. **Tearing of the anterior cruciate ligament,** as a result of forward displacement of the tibia

3. **Injury to the medial meniscus,** as a result of the tibial collateral ligament attachment

G. Prepatellar bursitis (housemaid's knee)

– is inflammation and swelling of the prepatellar bursa.

H. Popliteal (Baker's) cyst

– is a **swelling behind the knee,** caused by escape of synovial fluid posteriorly through the joint capsule.
– impairs flexion and extension of the knee joint.

I. Anterior tibial compartment syndrome

– is characterized by **ischemic necrosis** of the muscles of the anterior tibial compartment of the leg. Presumably this happens as a result of compression of arteries by swollen muscles, following excessive exertion.
– is accompanied by extreme tenderness and pain on the anterolateral aspect of the leg.

J. Talipes planus (flat foot)

– is characterized by a **waddling gait** with the feet turned out.
– results in disappearance of the medial portion of the longitudinal arch, which appears completely flattened.
– causes greater wear on the inner border of the soles and heels of shoes than on the outer border.
– causes pain as a result of stretching of the spring ligament and the long and short plantar ligaments.

K. Talipes equinovarus (clubfoot)

– is characterized by plantar flexion, inversion, and adduction of the foot.
– is a congenitally deformed foot that is twisted from its natural position.

III. Lesions of Peripheral Nerves

A. Damage to the femoral nerve

– causes impaired flexion of the hip and impaired extension of the leg due to paralysis of the quadriceps femoris.

B. **Damage to the obturator nerve**

   – causes a weakness of adduction and a **lateral swinging of the limb during walking** because of the unopposed abductors.

C. **Damage to the sciatic nerve**

   – causes impaired extension at the hip and impaired flexion at the knee, loss of dorsiflexion at the ankle and of eversion of the foot, and **peculiar gait** because of increased flexion at the hip in order to lift the dropped foot off the ground.

D. **Damage to the common peroneal nerve**

   – results in **foot drop** and loss of sensation on the dorsum of the foot and lateral aspect of the leg.
   – causes **paralysis** of all of the dorsiflexor and evertor muscles of the foot.

E. **Damage to the tibial nerve**

   – causes loss of plantar flexion of the foot and impaired inversion due to paralysis of the tibialis posterior.
   – causes a difficulty in getting the heel off the ground and a **shuffling of the gait.**
   – results in a **characteristic clawing of the toes** and secondary loss on the sole of the foot, affecting posture and locomotion.

F. **Damage to the deep peroneal nerve**

   – results in **foot drop** and hence a characteristic **high-stepping gait.**

G. **Damage to the superficial peroneal nerve**

   – causes no foot drop but loss of eversion of the foot.

Summary of Muscle Actions of the Lower Limb

Movements at the Hip Joint (Ball-and-Socket Joint)
   Flexion—iliopsoas, tensor fasciae latae, rectus femoris, adductors, sartorius, pectineus
   Extension—hamstrings, gluteus maximus, adductor magnus
   Adduction—adductor magnus, adductor longus, adductor brevis, pectineus, gracilis
   Abduction—gluteus medius, gluteus minimus
   Medial rotation—tensor fasciae latae, gluteus medius, gluteus minimus
   Lateral rotation—obturator internus, obturator externus, gemelli, piriformis, quadratus femoris, gluteus maximus

Movements at the Knee Joint (Hinge Joint)
   Flexion—hamstrings, gracilis, sartorius, gastrocnemius, popliteus
   Extension—quadriceps femoris
   Medial rotation—semitendinosus, popliteus
   Lateral rotation—biceps femoris

Movements at the Ankle Joint (Hinge Joint)
   Dorsiflexion—anterior tibialis, extensor digitorum longus, extensor hallucis longus, peroneus tertius
   Plantar flexion—triceps surae, plantaris, posterior tibialis, peroneus longus, flexor digitorum longus, flexor hallucis longus (when the knee is fully flexed)

Movements at the Intertarsal Joint (Talocalcaneal, Transverse Tarsal Joint)
Inversion—tibialis posterior, tibialis anterior, triceps surae
Eversion—peroneus longus, peroneus brevis, peroneus tertius

Movements at the Metatarsophalangeal Joint (Ellipsoid Joint)
Flexion—lumbricals, interossei, flexor hallucis brevis, flexor digiti minimi brevis
Extension—extensor digitorum longus and brevis, extensor hallucis longus

Movements at the Interphalangeal Joint (Hinge Joint)
Flexion—flexor digitorum longus and brevis, flexor hallucis longus
Extension—extensor digitorum longus and brevis, extensor hallucis longus

---

Summary of Muscle Innervations of the Lower Limb

Muscles of the Thigh

Muscles of the Anterior Compartment: Femoral Nerve
Sartorius
Quadriceps femoris–rectus femoris; vastus medialis; vastus intermedius; and vastus lateralis

Muscles of the Medial Compartment: Obturator Nerve
Adductor longus; adductor brevis; adductor magnus (obturator and tibial nerves)*; gracilis; obturator externus; pectineus (femoral and obturator nerves)*

Muscles of the Posterior Compartment: Tibial Part of Sciatic Nerve
Semitendinosus; semimembranosus; biceps femoris, long head; biceps femoris, short head (common peroneal part of sciatic nerve)*; adductor magnus (tibial part of sciatic and obturator nerve)*

Muscles of the Lateral Compartment
Gluteus maximus (inferior gluteal nerve)
Gluteus medius (superior gluteal nerve)
Gluteus minimus (superior gluteal nerve)
Tensor fasciae latae (superior gluteal nerve)
Piriformis (nerve to piriformis)
Obturator internus (nerve to obturator internus)
Superior gemellus (nerve to obturator internus)
Inferior gemellus (nerve to quadratus femoris)
Quadratus femoris (nerve to quadratus femoris)

Muscles of the Leg

Muscles of the Anterior Compartment: Deep Peroneal Nerve
Tibialis anterior; extensor digitorum longus; extensor hallucis longus; peroneus tertius

Muscles of the Lateral Compartment: Superficial Peroneal Nerve
Peroneus longus; peroneus brevis

Muscles of the Posterior Compartment: Tibial Nerve
Superficial layer—gastrocnemius; soleus; plantaris
Deep layer—popliteus; tibialis posterior; flexor digitorum longus; flexor hallucis longus

---

* Indicates exception.

Review Test

Directions: Each of the numbered items or incomplete statements in this section is followed by answers or by completions of the statement. Select the ONE lettered answer or completion that is BEST in each case.

1. If there is a loss of skin sensation and paralysis of muscles on the plantar aspect of the medial side of the foot, which of the following nerves is damaged?

(A) Common peroneal nerve
(B) Tibial nerve
(C) Superficial peroneal nerve
(D) Deep peroneal nerve
(E) Anterior tibial nerve

2. A patient walks with a waddling gait that is characterized by the pelvis falling toward one side at each step. Which of the following nerves is damaged?

(A) Obturator nerve
(B) Nerve to obturator internus
(C) Superior gluteal nerve
(D) Inferior gluteal nerve
(E) Femoral nerve

3. Which of the following ligaments prevents forward displacement of the femur on the tibia when the knee is flexed?

(A) Anterior cruciate ligament
(B) Fibular collateral ligament
(C) Patellar ligament
(D) Posterior cruciate ligament
(E) Tibial collateral ligament

4. Lesion of the femoral nerve results in

(A) paralysis of the psoas major muscle
(B) loss of skin sensation on the lateral side of the foot
(C) loss of skin sensation over the greater trochanter
(D) paralysis of the sartorius muscle
(E) paralysis of the tensor fasciae latae

5. A patient is unable to invert his foot, indicating lesions of which of the following pairs of nerves?

(A) Superficial and deep peroneal nerves
(B) Deep peroneal and tibial nerves
(C) Superficial peroneal and tibial nerves
(D) Medial and medial plantar nerves
(E) Obturator and tibial nerves

6. Which of the following muscles is responsible for unlocking of the knee joint to permit flexion of the leg?

(A) Rectus femoris
(B) Semimembranosus
(C) Popliteus
(D) Gastrocnemius
(E) Biceps femoris

7. A patient presents with sensory loss on adjacent sides of the great and second toes and cannot dorsiflex the foot. These signs probably indicate damage to the

(A) superficial peroneal nerve
(B) lateral plantar nerve
(C) deep peroneal nerve
(D) sural nerve
(E) tibial nerve

8. When the common peroneal nerve is severed in the popliteal fossa but the tibial nerve is spared, the foot will be

(A) plantar flexed and inverted
(B) dorsiflexed and everted
(C) dorsiflexed and inverted
(D) plantar flexed and everted
(E) dorsiflexed only

9. To avoid damaging the sciatic nerve during an intramuscular injection in the right gluteal region, the clinician should insert the needle

(A) in the area over the sacrospinous ligament
(B) midway between the ischial tuberosity and the lesser trochanter
(C) at the midpoint of the gemelli muscles
(D) in the upper right quadrant of the gluteal region
(E) in the lower right quadrant of the gluteal region

10. Which of the following muscles can flex the thigh and extend the leg?

(A) Semimembranosus
(B) Sartorius
(C) Rectus femoris
(D) Vastus medialis
(E) Gastrocnemius

11. Which of the following muscles has a tendon that occupies the groove in the lower surface of the cuboid bone?

(A) Peroneus tertius
(B) Peroneus brevis
(C) Peroneus longus
(D) Tibialis anterior
(E) Tibialis posterior

12. Which of the following muscles has a tendon that occupies the groove on the undersurface of the sustentaculum tali of the calcaneus?

(A) Flexor digitorum brevis
(B) Flexor digitorum longus
(C) Flexor hallucis brevis
(D) Flexor hallucis longus
(E) Tibialis posterior

13. Which of the following statements concerning the great saphenous vein is true?

(A) It ascends posterior to the medial malleolus
(B) It empties into the popliteal vein
(C) It courses anterior to the medial condyles of the tibia and femur
(D) It passes superficial to the fascia lata of the thigh
(E) It runs along with the femoral vessels

14. The inability to extend the leg at the knee joint indicates paralysis of the

(A) semitendinosus muscle
(B) sartorius muscle
(C) gracilis muscle
(D) quadriceps femoris muscle
(E) biceps femoris muscle

15. Which of the following muscles can dorsiflex and invert the foot?

(A) Peroneus longus
(B) Peroneus brevis
(C) Extensor hallucis longus
(D) Extensor digitorum longus
(E) Peroneus tertius

Questions 16–20

A 62-year-old woman slipped and fell on the bathroom floor, which resulted in a posterior dislocation of the hip joint and a fracture of the neck of the femur.

16. Rupture of the ligamentum capitis femoris results in damage to a branch of which of the following arteries?

(A) Medial circumflex femoral artery
(B) Lateral circumflex femoral artery
(C) Obturator artery
(D) Superior gluteal artery
(E) Inferior gluteal artery

17. Fracture of the neck of the femur would result in avascular necrosis of the femoral head, probably owing to lack of blood supply from the

(A) obturator and inferior gluteal arteries
(B) superior gluteal and femoral arteries
(C) inferior gluteal and superior gluteal arteries
(D) lateral and medial femoral circumflex arteries
(E) medial femoral circumflex and deep femoral arteries

18. If the acetabulum is fractured at its posterosuperior margin, which of the following bones could be involved?

(A) Ilium and pubis
(B) Ischium and sacrum
(C) Ilium and ischium
(D) Pubis and pelvic bone
(E) Pelvic bone and femur

19. The patient exhibits a flexed, adducted, and medially rotated thigh. Which of the following muscles is a medial rotator of the hip joint?

(A) Piriformis
(B) Obturator internus
(C) Quadratus femoris
(D) Gluteus maximus
(E) Gluteus minimus

20. All of the following arteries participate in the cruciate anastomosis of the thigh EXCEPT

(A) medial femoral circumflex artery
(B) lateral femoral circumflex artery
(C) superior gluteal artery
(D) inferior gluteal artery
(E) first perforating artery

Directions: Each of the numbered items or incomplete statements in this section is negatively phrased, as indicated by a capitalized word such as NOT, LEAST, or EXCEPT. Select the ONE lettered answer or completion that is BEST in each case.

21. Each of the following statements concerning the dorsalis pedis artery is true EXCEPT

(A) it begins anterior to the ankle joint
(B) it is a continuation of the anterior tibial artery
(C) it gives rise to the medial malleolar artery
(D) it terminates as the deep plantar artery, which enters the sole of the foot by passing between the two heads of the first dorsal interosseous muscle
(E) it gives rise to the arcuate artery

22. Each of the following muscles contributes directly to the stability of the knee joint EXCEPT

(A) soleus
(B) semimembranosus
(C) sartorius
(D) biceps femoris
(E) gastrocnemius

23. Each of the following structures passes deep to the inferior or superior extensor retinaculum of the ankle EXCEPT

(A) anterior tibial nerve
(B) extensor digitorum longus muscle
(C) dorsalis pedis artery
(D) peroneus tertius muscle
(E) superficial peroneal nerve

24. Loss of ability to flex the leg could result from lesion of all of the following nerves EXCEPT

(A) tibial portion of the sciatic nerve
(B) common peroneal portion of the sciatic nerve
(C) deep peroneal nerve
(D) femoral nerve
(E) obturator nerve

25. Each of the following statements concerning the longitudinal arches of the foot is true EXCEPT

(A) they include the proximal phalanges distally
(B) they are supported by ligaments and muscle tendons
(C) the keystone of the lateral longitudinal arch is the cuboid bone
(D) the talus transmits weight from the tibia to the longitudinal arches
(E) the medial longitudinal arch is supported by the spring ligaments

26. Each of the following statements concerning the femoral triangle is true EXCEPT

(A) it is covered superficially by fascia lata
(B) it contains the femoral artery, femoral vein, and femoral nerve
(C) it is bounded medially by the adductor longus muscle
(D) the femoral nerve is covered by the femoral sheath within it
(E) it is bounded superiorly by the inguinal ligament

27. The following muscles bound the adductor canal in the lower extremity EXCEPT

(A) sartorius
(B) vastus medialis
(C) adductor longus
(D) gracilis
(E) adductor magnus

28. The adductor canal contains all of the following structures EXCEPT

(A) femoral artery
(B) femoral vein
(C) saphenous nerve
(D) great saphenous vein
(E) nerve to the vastus medialis

29. While playing football, a 19-year-old college student received a twisting injury to his knee when he was tackled from the lateral side. Each of the following statements concerning the injury is true EXCEPT

(A) tear of the medial meniscus occurs as a result of lateral rotation of the partially flexed leg
(B) the medial (tibial) collateral ligament is ruptured because the medial meniscus is firmly attached to this ligament
(C) there is tenderness on pressure along the tibial collateral ligament and over the torn medial meniscus
(D) the posterior cruciate ligament is usually ruptured
(E) swelling is observed on the front of the joint

30. All of the following muscles are attached to the greater trochanter of the femur EXCEPT

(A) piriformis
(B) obturator internus
(C) gluteus maximus
(D) gluteus medius
(E) gluteus minimus

Directions: Each set of matching questions in this section consists of a list of four to twenty-six lettered options (some of which may be in figures) followed by several numbered items. For each numbered item, select the ONE lettered option that is most closely associated with it. To avoid spending too much time on matching sets with large numbers of options, it is generally advisable to begin each set by reading the list of options. Then, for each item in the set, try to generate the correct answer and locate it in the option list, rather than evaluating each option individually. Each lettered option may be selected once, more than once, or not at all.

Questions 31–35

Match each description below with the most appropriate ligament.

(A) Calcaneofibular ligament
(B) Long plantar ligament
(C) Plantar calcaneonavicular ligament
(D) Short plantar ligament
(E) Deltoid ligament

31. The thickened medial part of the ankle joint capsule

32. Forms a canal for the tendon of the peroneus longus muscle

33. The cord-like ligament that reinforces the lateral side of the ankle joint

34. Extends from the front of the inferior surface of the calcaneus to the plantar surface of the cuboid bone; supports the longitudinal arch

35. Supports the head of the talus in maintaining the medial longitudinal arch

Questions 36–40

Match each description below with the most appropriate muscle.

(A) Biceps femoris
(B) Gluteus medius
(C) Iliopsoas
(D) Rectus femoris
(E) Tensor fasciae latae

36. Rotates the leg laterally

37. The chief flexor of the thigh

38. Flexes the thigh and extends the leg

39. Can flex and medially rotate the thigh during running and climbing

40. Its paralysis causes the pelvis to tilt away from the paralyzed side when the body is supported by the leg on the same side

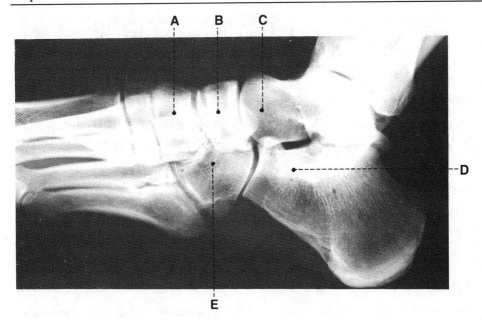

Questions 41–45

Match each bone below with the appropriate lettered structure in this radiograph of the foot.

41. Navicular bone

42. Cuboid bone

43. Cuneiform bones

44. Calcaneus

45. Talus

Questions 46–50

Match each description below with the most appropriate structure.

(A) Femoral sheath
(B) Fascia lata
(C) Iliotibial tract
(D) Tibionavicular ligament
(E) Calcaneofibular ligament

46. Provides attachment for the gluteus maximus muscle

47. Could be torn during an ankle sprain (inversion injury)

48. Is formed by a prolongation of the transversalis and iliac fasciae in the thigh

49. Contains the femoral canal but not the femoral nerve

50. Has a saphenous opening for the passage of the great saphenous vein

Answers and Explanations

1–B. The tibial nerve divides into medial and lateral plantar nerves, which innervate the plantar aspect of the foot.

2–C. Gluteal gait, a waddling gait characterized by a falling of the pelvis toward the unaffected side at each step, is caused by paralysis of the gluteus medius muscle, which is innervated by the superior gluteal nerve. The gluteus medius muscle normally functions to stabilize the pelvis when the opposite foot is off the ground.

3–D. The posterior cruciate ligament is important because it prevents forward displacement of the femur on the tibia when the knee is flexed.

4–D. The femoral nerve innervates the quadratus femoris and sartorius muscles, and so damage to the nerve will result in paralysis of these muscles. The second and third lumbar nerves innervate the psoas major muscle. The sural nerve innervates the skin on the lateral side of the foot. The iliohypogastric nerve and superior clunial nerves supply the skin over the greater trochanter. The superior gluteal nerve innervates the tensor fasciae latae.

5–B. Inversion of the foot is produced by the action of the tibialis anterior muscle, which is innervated by the deep peroneal nerve, and by the action of the tibialis posterior and the triceps surae, which are innervated by the tibial nerve.

6–C. The popliteus muscle acts to rotate the tibia medially or to rotate the femur laterally, depending on which bone is fixed. This action results in unlocking of the knee joint to initiate flexion of the leg at the joint.

7–C. The deep peroneal nerve supplies the anterior muscles of the leg, including the tibialis anterior, extensor hallucis longus, extensor digitorum longus, and peroneus tertius muscles, which dorsiflex the foot. The medial branch of the deep peroneal nerve supplies adjacent sides of the great and second toes.

8–A. Severance of the common peroneal nerve paralyzes the muscles for dorsiflexion and eversion of the foot; thus, the foot is plantar flexed and inverted. The common peroneal nerve supplies the anterior and lateral muscles of the leg; the tibial nerve supplies the posterior muscles of the leg.

9–D. To avoid damaging the sciatic nerve during an intramuscular injection, the clinician should insert the needle in the upper right quadrant of the gluteal region.

10–C. The rectus femoris can flex the thigh and extend the leg. The hamstring muscles—the semitendinosus, semimembranosus, and biceps femoris—can extend the thigh and flex the leg. The sartorius can flex the thigh and the leg. The vastus medialis can extend the leg. The gastrocnemius can flex the leg and plantar flex the foot.

11–C. The groove in the lower surface of the cuboid bone is occupied by the tendon of the peroneus longus muscle.

12–D. The groove on the undersurface of the sustentaculum tali is occupied by the tendon of the flexor hallucis longus muscle.

13–D. The greater saphenous vein courses anterior to the medial malleolus, medial to the tibia, and posterior to the medial condyles of the tibia and femur; ascends superficial to the fascia lata; and terminates in the femoral vein by passing through the saphenous opening.

14–D. The quadriceps femoris muscle includes the rectus femoris muscle and the vastus medialis, intermedialis, and lateralis muscles. They extend the leg at the knee joint. The semitendinosus, semimembranosus, and biceps femoris muscles (the hamstrings) extend the thigh and flex the leg. The sartorius and gracilis muscles can flex the thigh and the leg.

15–C. The peroneus longus and brevis muscles can plantar flex and evert the foot. The peroneus tertius can dorsiflex and evert the foot. The extensor digitorum longus can dorsiflex the foot and extend the toes.

16–C. The obturator artery gives rise to a branch that follows the ligament of the head of the femur (ligamentum capitis femoris).

17–D. In the adult, the chief arterial supply to the head of the femur is from the branches of the medial and lateral femoral circumflex arteries. The posterior branch of the obturator artery gives rise to the artery of the head of the femur, which is usually insufficient to supply the head of the femur in the adult.

18–C. The acetabulum is a cup-shaped cavity on the lateral side of the hip bone and is formed superiorly by the ilium, posteroinferiorly by the ischium, and anteromedially by the pubis.

19–E. The gluteus minimis abducts and rotates the thigh medially. The piriformis, obturator internus, quadratus femoris, and gluteus maximus muscles can rotate the thigh laterally.

20–C. The superior gluteal artery is not part of the cruciate anastomosis of the thigh. The cruciate anastomosis of the thigh is formed by the inferior gluteal artery, transverse branches of the medial and lateral femoral circumflex arteries, and an ascending branch of the first perforating artery.

21–C. The medial and lateral malleolar arteries are branches of the anterior tibial artery, not of the dorsalis pedis artery. The dorsalis pedis artery begins anterior to the ankle joint midway between the two malleoli as the continuation of the anterior tibial artery.

22–A. The soleus muscle arises from the fibula and tibia below the knee and therefore does not contribute to the stability of the knee joint.

23–E. The superficial peroneal nerve emerges between the peroneus longus and peroneus brevis muscles and descends superficial to the extensor retinaculum of the ankle, innervating the skin of the lower leg and foot.

24–C. The deep peroneal nerve supplies the muscles that extend, dorsiflex, evert, and invert the foot. Flexion of the knee joint is performed by the hamstring, gracilis, sartorius, gastrocnemius, and popliteus muscles. The tibial portion of the sciatic nerve supplies the hamstring muscles with the exception of the short head of the biceps femoris, which can extend the thigh and flex the knee. The femoral nerve supplies the sartorius muscle, which can flex the thigh and the leg. The obturator nerve supplies the gracilis muscle, which can flex the thigh and the leg.

25–A. Bones that form the longitudinal arch of the foot include metatarsal bones but not the proximal phalanges distally. The talus transmits weight from the tibia to the longitudinal arch. The lateral longitudinal arch is supported by the long and short plantar ligaments, and its keystone is the cuboid bone. The medial longitudinal arch is supported by the spring ligament, and its keystone is the head of the talus. The arches are also supported by tendons of the posterior tibialis and peroneus longus muscles.

26–D. The femoral nerve lies outside the femoral sheath, lateral to the femoral artery. From lateral to medial, the femoral sheath contains the femoral artery, femoral vein, and femoral canal.

27–D. The adductor canal is bounded by the sartorius, vastus medialis, adductor longus, and adductor magnus muscles.

28–D. The adductor canal contains the femoral vessels, the saphenous nerve, and the nerve to the vastus medialis.

29–D. The posterior cruciate ligament is not ruptured in this injury. The unhappy triad of the knee joint is characterized by rupture of the tibial collateral ligament, tear of the medial meniscus, and rupture of the anterior cruciate ligament. This injury can occur when a football player's cleated shoe is planted firmly in the turf and the knee is struck from the lateral side. Tenderness along the medial collateral ligament and over the medial meniscus and swelling on the front of the joint are due to excessive production of synovial fluid, which fills the joint cavity and the suprapatellar bursa.

30–D. The gluteus maximus is inserted into the gluteal tuberosity of the femur and the iliotibial tract.

31–E. The deltoid ligament is the thickened medial part of the ankle joint capsule. It consists of the anterior tibiotalar, tibionavicular, tibiocalcaneal, and posterior tibiotalar ligaments.

32–B. The long plantar ligament forms a canal for the tendon of the peroneus longus muscle.

33–A. The calcaneofibular ligament is the cord-like ligament that reinforces the lateral side of the ankle joint.

34–D. The short plantar ligament extends from the front of the inferior surface of the calcaneus to the plantar surface of the cuboid bone and supports the longitudinal arch.

35–C. The plantar calcaneonavicular (spring) ligament supports the head of the talus in maintaining the medial longitudinal arch.

36–A. The biceps femoris muscle rotates the leg laterally when the knee is flexed.

37–C. The iliopsoas is the primary flexor of the thigh.

38–D. The rectus femoris muscle crosses the hip and knee joints; thus, it can flex the thigh and extend the leg.

39–E. The tensor fasciae latae can flex and medially rotate the thigh during running and climbing.

40–B. The gluteus medius muscle supports the pelvis. When the opposing leg is raised—for example, during walking—the gluteus medius muscle swings the pelvis forward and prevents it from tilting to the opposite side.

41–B. The navicular bone is a boat-shaped tarsal bone between the head of the talus and the three cuneiform bones.

42–E. The cuboid bone is the most lateral tarsal bone. It has a notch and groove for the tendon of the peroneus longus muscle.

43–A. The cuneiform bones are wedge-shaped bones that articulate posteriorly with the navicular bone and anteriorly with three metatarsals.

44–D. The calcaneus forms the heel. It articulates superiorly with the talus and anteriorly with the cuboid bone.

45–C. The talus transmits the weight of the body from the tibia to other weight-bearing bones of the foot and is the only tarsal bone without muscle attachments. It articulates anteriorly with the navicular bone and posteriorly with the calcaneus. It has a groove on the posterior surface for the tendon of the flexor hallucis longus muscle.

46–C. The iliotibial tract functions as the tendon of insertion for the tensor fasciae latae muscle and the gluteus maximus muscle.

47–E. The calcaneofibular ligament is a part of the lateral ligament of the ankle joint and could be torn during an ankle sprain (inversion injury).

48–A. The femoral sheath is a funnel-shaped extension of the transversalis and iliac fasciae into the thigh, deep to the inguinal ligament.

49–A. The femoral sheath contains the femoral artery and vein, the femoral branch of the genitofemoral nerve, and the femoral canal. However, the femoral nerve lies outside the femoral sheath.

50–B. The fascia lata has a saphenous opening (fossa ovalis) through which the saphenous vein passes to enter the femoral vein.

4

Thorax

Thoracic Wall

I. Skeleton of the Thorax (Figure 4-1)

A. Sternum

1. Manubrium
- has a superior margin, the **jugular notch,** which can be readily palpated at the root of the neck.
- has a **clavicular notch** on each side for articulation with the clavicle.
- also articulates with the cartilage of the first rib, the upper half of the second rib, and the body of the sternum at the **manubriosternal joint,** or sternal angle.

2. Sternal angle (angle of Louis)
- is the junction between the manubrium and the body of the sternum.
- is located at the level where:

a. The second ribs articulate with the sternum.

b. The aortic arch begins and ends.

c. The trachea bifurcates into the right and left bronchi.

d. The inferior border of the superior mediastinum is demarcated.

e. A transverse plane can pass through the vertebral column between T4 and T5.

3. Body of the sternum
- articulates with the second to seventh costal cartilages.
- also articulates with the xiphoid process at the **xiphosternal joint,** which is level with the ninth thoracic vertebra.

4. Xiphoid process
- is a flat, cartilaginous process at birth that ossifies slowly from the central core and unites with the body of the sternum after middle age.
- can be palpated in the epigastrium.
- is attached via its pointed caudal end to the **linea alba.**

111

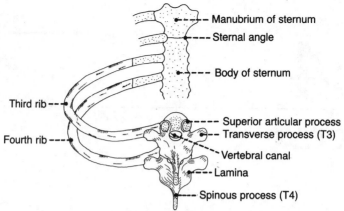

Manubrium of sternum
Sternal angle
Body of sternum
Third rib
Fourth rib
Superior articular process
Transverse process (T3)
Vertebral canal
Lamina
Spinous process (T4)

Figure 4-1. Articulations of the ribs with the vertebrae and the sternum.

B. Ribs

- consist of 12 pairs of bones that form the main part of the **thoracic cage,** extending from the vertebrae to or toward the sternum.
- **increase the anteroposterior and transverse diameters of the thorax** by their movements.

1. Parts of the ribs

- Each rib is divided into head, neck, tubercle, and body (shaft).
- The **head** articulates with the corresponding vertebral bodies and intervertebral disks and suprajacent vertebral bodies.
- The **tubercle** articulates with the transverse processes of the corresponding vertebrae, with the exception of ribs 11 and 12.

2. Classification of ribs

a. True ribs

- are the first seven ribs (**ribs 1 to 7**), which are attached to the sternum by their costal cartilages.

b. False ribs

- are the lower five ribs (**ribs 8 to 12**); ribs 8 to 10 are connected to the costal cartilages immediately above them to form the **anterior costal margin.**

c. Floating ribs

- are the last two ribs (**ribs 11 and 12**), which are connected only to the vertebrae.

3. First rib

- is the broadest and shortest of the true ribs.
- has a single articular facet on its head, which articulates with the first thoracic vertebra.
- has a **scalene tubercle** for the insertion of the anterior scalene muscle.

4. Second rib

- has two articular facets on its head, which articulate with the bodies of the first and second thoracic vertebrae.
- is about twice as long as the first rib.

5. **Tenth rib**
   – has a single articular facet on its head, which articulates with the tenth thoracic vertebra.

6. **Eleventh and twelfth ribs**
   – have a single articular facet on their heads.
   – have no neck or tubercle.

II. Articulations of the Thorax (see Figure 4-1)

A. Sternoclavicular joint
   – provides the only bony attachment between the appendicular and axial skeletons.
   – is a saddle-type synovial joint but has the movements of a ball-and-socket joint.
   – has a fibrocartilaginous articular surface and contains two separate synovial cavities.

B. Sternocostal (sternochondral) joints
   – are synchondroses in which the sternum articulates with the first seven costal cartilages.

C. Costochondral joints
   – are synchondroses in which the ribs articulate with their respective costal cartilages.

III. Muscles of the Thoracic Wall

Muscle	Origin	Insertion	Nerve	Action
External intercostals	Lower border of ribs	Upper border of rib below	Intercostal	Elevate ribs in inspiration
Internal intercostals	Lower border of ribs	Upper border of rib below	Intercostal	Depress ribs; interchondral part elevates ribs
Innermost intercostals	Lower border of ribs	Upper border of rib below	Intercostal	Elevate ribs
Transverse thoracic	Posterior surface of lower sternum and xiphoid	Inner surface of costal cartilages 2–6	Intercostal	Depresses ribs
Subcostalis	Inner surface of lower ribs near their angles	Upper borders of ribs 2 or 3 below	Intercostal	Elevates ribs
Levator costarum	Transverse processes of T7–T11	Subjacent ribs between tubercle and angle	Dorsal primary rami of C8–T11	Elevates ribs

IV. Nerves and Blood Vessels of the Thoracic Wall

A. Intercostal nerves
   – are the **anterior primary rami** of the first 11 thoracic spinal nerves. The anterior primary ramus of the 12th thoracic spinal nerve is the **subcostal nerve**, which runs beneath rib 12.

– are called **typical intercostal nerves** (for the third to sixth nerves).

– run between the internal and innermost layers of muscles, with the intercostal veins and arteries above (**van: veins-arteries-nerves**).

– are lodged in the **costal grooves** on the inferior surface of the ribs.

– give rise to lateral and anterior cutaneous branches and muscular branches.

B. Internal thoracic artery

– usually arises from the **first part of the subclavian artery** and descends directly behind the first six costal cartilages, just lateral to the sternum.

– gives rise to two anterior intercostal arteries in each of the upper six intercostal spaces and terminates at the sixth intercostal space by dividing into the superior epigastric and musculophrenic arteries.

1. Pericardiophrenic artery

– accompanies the phrenic nerve between the pleura and the pericardium to the diaphragm.

– supplies the pleura, pericardium, and diaphragm (upper surface).

2. Anterior intercostal arteries

– are **twelve small arteries,** two in each of the upper six intercostal spaces that run laterally, one each at the upper and lower borders of each space.

– The upper one in each intercostal space anastomoses with the **posterior intercostal artery,** and the lower one joins the **collateral branch** of the posterior intercostal artery.

– supply the upper six intercostal spaces.

– provide muscular branches to the intercostal, serratus anterior, and pectoral muscles.

3. Anterior perforating branches

– perforate the internal intercostal muscles in the upper six intercostal spaces, course with the anterior cutaneous branches of the intercostal nerves, and supply the pectoralis major muscle and the skin and subcutaneous tissue over it.

– The second, third, and fourth branches provide the **medial mammary branches.**

4. Musculophrenic artery

– follows the **costal arch** on the inner surface of the costal cartilages.

– gives rise to two anterior arteries in the seventh, eighth, and ninth spaces; perforates the diaphragm; and ends in the tenth intercostal space, where it anastomoses with the **deep circumflex iliac artery.**

– supplies the pericardium, diaphragm, and muscles of the abdominal wall.

5. Superior epigastric artery

– descends on the deep surface of the rectus abdominis muscle within the rectus sheath; supplies this muscle and anastomoses with the **inferior epigastric artery.**

– supplies the diaphragm, peritoneum, and anterior abdominal wall.

C. Thoracoepigastric vein

– is a venous connection between the lateral thoracic vein and the superficial epigastric vein.

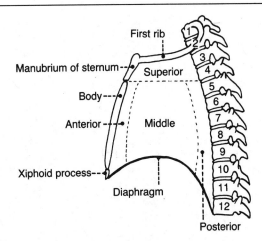

Figure 4-2. Mediastinum.

V. Lymphatic Drainage of the Thorax

A. Sternal or parasternal (internal thoracic) nodes

– are placed along the **internal thoracic artery.**
– receive lymph from the medial portion of the breast, intercostal spaces, diaphragm, and supraumbilical region of the abdominal wall.
– drain into the junction of the internal jugular and subclavian veins.

B. Intercostal nodes

– lie near the heads of the ribs.
– receive lymph from the intercostal spaces and the pleura.
– drain into the **cisterna chyli** or the **thoracic duct.**

C. Phrenic nodes

– lie on the thoracic surface of the diaphragm.
– receive lymph from the pericardium, diaphragm, and liver.
– drain into the sternal and posterior mediastinal nodes.

Mediastinum, Pleura, and Organs of Respiration

I. Mediastinum (Figure 4-2)

– is an **interpleural space** (area between the pleural cavities) in the thorax and is bounded laterally by the pleural cavities, anteriorly by the sternum, and posteriorly by the vertebral column (does not contain the lungs).
– consists of the superior mediastinum above the pericardium and the three lower divisions: anterior, middle, and posterior.

A. Superior mediastinum

– is bounded superiorly by the oblique plane of the first rib and inferiorly by the imaginary line running from the sternal angle to the intervertebral disk between the fourth and fifth thoracic vertebrae.
– contains the superior vena cava, brachiocephalic veins, arch of the aorta, thoracic duct, trachea, esophagus, thymus, vagus nerve, left recurrent laryngeal nerve, and phrenic nerve.

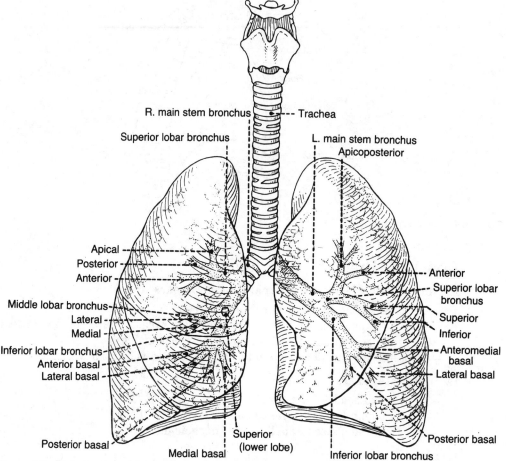

Figure 4-3. Anterior view of the trachea, bronchi, and lungs.

B. Anterior mediastinum

– lies anterior to the pericardium and contains the remnants of the thymus gland, lymph nodes, fat, and connective tissue.

C. Middle mediastinum

– lies between the right and left pleural cavities and contains the heart, pericardium, phrenic nerves, roots of the great vessels, arch of the azygos vein, and main bronchi.

D. Posterior mediastinum

– lies posterior to the pericardium between the mediastinal pleurae and contains the esophagus, thoracic aorta, azygos and hemiazygos veins, thoracic duct, vagus nerves, sympathetic trunk, and splanchnic nerves.

II. Trachea and Bronchi (Figure 4-3)

A. Trachea

– begins at the inferior border of the **cricoid cartilage** (C6).
– has **16 to 20 incomplete hyaline cartilaginous rings** that prevent the trachea from collapsing and that open posteriorly.

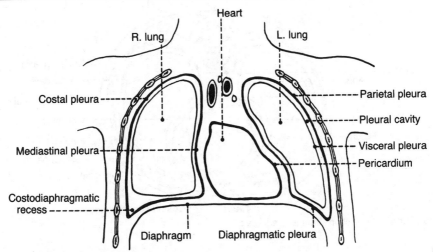

Figure 4-4. Frontal section of the thorax.

- is about 9–15 cm in length and bifurcates into the right and left main stem bronchi at the level of the **sternal angle** (junction of T4 and T5).
- The **carina,** a downward and backward projection of the last tracheal cartilage, forms a keel-like ridge separating the openings of the right and left main stem bronchi.

B. Right main stem (primary) bronchus

- is shorter, wider, and more vertical than the left main stem bronchus; therefore, more foreign bodies that enter through the trachea are lodged in this bronchus.
- runs under the arch of the azygos vein and divides into the **superior, middle, and inferior lobar (secondary) bronchi.** The right superior lobar (secondary) bronchus is known as the **eparterial bronchus** because it passes above the level of the pulmonary artery. All others are the **hyparterial bronchi.**

C. Left main stem (primary) bronchus

- crosses anterior to the esophagus and divides into **two lobar (secondary) bronchi,** the upper and lower.

III. Pleurae and Pleural Cavities (Figures 4-4 and 4-5)

A. Pleura

- is a thin serous membrane that consists of a parietal pleura and a visceral pleura.

1. Parietal pleura

- lines the inner surface of the thoracic wall and the mediastinum, and has costal, diaphragmatic, mediastinal, and cervical parts. The cervical pleura **(cupula)** is the dome of the pleura, projecting into the neck above the neck of the first rib. It is reinforced by **Sibson's fascia,** which is a thickening of the endothoracic fascia.
- is separated from the thoracic wall by the **endothoracic fascia,** which is an extrapleural fascial sheet lining the thoracic wall.
- The costal pleura and the peripheral portion of the diaphragmatic pleura are innervated by the **intercostal nerves,** and the central portion of the

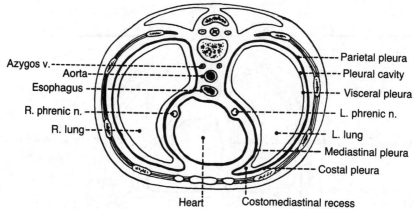

Figure 4-5. Horizontal section through the thorax.

diaphragmatic pleura and the mediastinal pleura are innnervated by **phrenic nerves.** The pleura is **very sensitive to pain.**
– is supplied by branches of the internal thoracic, superior phrenic, posterior intercostal, and superior intercostal arteries. However, the visceral pleura is supplied by the bronchial arteries.
– The **pulmonary ligament** is a two-layered vertical fold of mediastinal pleura, extending along the mediastinal surface of each lung from the **hilus** to the **base** (diaphragmatic surface). It supports the lungs in the **pleural sac** by retaining the lower parts of the lungs in position.

2. Visceral pleura (pulmonary pleura)

– intimately invests the lungs and dips into all of the fissures.
– is supplied by bronchial arteries, but its venous blood is drained by pulmonary veins.
– is insensitive to pain but contains vasomotor fibers and sensory endings of vagal origin, which may be involved in respiratory reflexes.

B. Pleural cavity

– is a **potential space** between the parietal and visceral pleurae, and represents a closed sac with no communication between right and left parts.
– contains a film of fluid that lubricates the surface of the pleurae and facilitates the movement of the lungs.

1. Costodiaphragmatic recesses

– are the **pleural recesses** formed by the reflection of the costal and diaphragmatic pleurae.
– can accumulate fluid when in the erect position.
– allow the lungs to be pulled in and expanded during inspiration.

2. Costomediastinal recesses

– are part of the pleural cavity where the costal and mediastinal pleurae meet.

IV. Lungs (see Figure 4-3)

– are the **essential organs of respiration** and are attached to the heart and trachea by their roots and the pulmonary ligaments.

- contain nonrespiratory tissues, which are nourished by the **bronchial arteries** and drained by the **bronchial veins** for the larger subdivisions of the bronchi and by the **pulmonary veins** for the smaller subdivisions of the bronchial tree.
- Their **bases** rest on the convex surface of the diaphragm, descend during inspiration, and ascend during expiration.
- receive parasympathetic fibers that innervate the smooth muscle and glands of the bronchial tree and probably are excitatory to these structures (bronchoconstrictor and secretomotor).
- receive sympathetic fibers that innervate blood vessels, smooth muscle, and glands of the bronchial tree and probably are inhibitory to these structures (bronchodilator and vasoconstrictor).
- have some sensory endings of vagal origin, which are stimulated by the stretching of the lung during inspiration and are concerned in the reflex control of respiration.

A. Right lung

- Its **apex** projects into the root of the neck and is smaller than that of the left lung.
- is heavier and shorter than the left lung (because of the higher right dome of the diaphragm) and is wider because the heart bulges more to the left.
- is divided into upper, middle, and lower lobes by the **oblique and horizontal fissures,** but usually receives a single bronchial artery.
- has 3 lobar (secondary) bronchi and 10 segmental (tertiary) bronchi.

B. Left lung

- has two lobes, upper and lower, and usually receives two bronchial arteries.
- has an **oblique fissure** that follows the line of the sixth rib.
- contains the **lingula,** a tongue-shaped portion of its upper lobe that corresponds to the middle lobe of the right lung.
- contains a **cardiac notch,** a deep indentation of the anterior border of the superior lobe of the left lung.
- has 2 lobar (secondary) bronchi and 8 (or possibly 9 or 10) segmental bronchi.

C. Bronchopulmonary segment

- is the anatomical, functional, and surgical unit (subdivision) of the lungs.
- consists of a segmental (tertiary or lobular) bronchus, a segmental branch of the pulmonary artery, and a segment of the lung tissue, surrounded by a delicate connective-tissue septum.
- refers to the **portion of the lung supplied by each segmental bronchus and segmental artery.** The pulmonary veins are intersegmental.
- is clinically important because the surgeon can remove a segment of the lung without seriously disrupting the surrounding lung tissue.

V. Lymphatic Vessels of the Lung

- drain the bronchial tree, pulmonary vessels, and connective-tissue septa.
- run along the bronchiole and bronchi toward the hilus, where they end in the **pulmonary and bronchopulmonary nodes,** which in turn drain into the **tracheobronchial nodes.**
- are not present in the walls of the pulmonary alveoli.

VI. Blood Vessels of the Lung (Figure 4-6)

A. Pulmonary trunk

- extends upward from the **conus arteriosus** of the right ventricle of the heart.
- passes superiorly and posteriorly from the front of the ascending aorta to its left side for about 5 cm and bifurcates into the right and left pulmonary arteries.

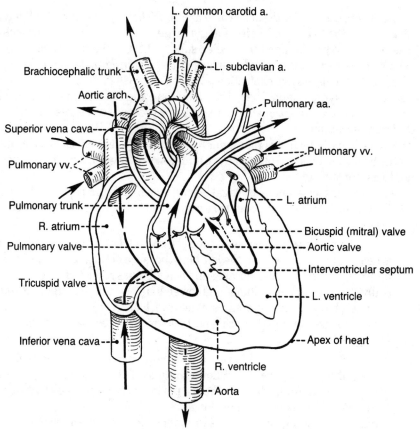

Figure 4-6. Pulmonary circulation and circulation through the heart chambers.

- has much lower blood pressure than that in the aorta, and is partially invested with fibrous pericardium.

1. **Left pulmonary artery**

    - carries unoxygenated blood to the left lung and is shorter and narrower than the right pulmonary artery.
    - is connected to the arch of the aorta by the **ligamentum arteriosum,** the fibrous remains of the ductus arteriosus.

2. **Right pulmonary artery**

    - runs horizontally toward the hilus of the right lung under the arch of the aorta behind the ascending aorta and superior vena cava and anterior to the right bronchus.

B. **Pulmonary veins**

    - are **intersegmental** in drainage (do not accompany the bronchi or the segmental artery within the parenchyma of the lungs).
    - Five pulmonary veins leave the lung, one from each lobe of the lungs, but the right upper and middle veins usually join so that only four veins enter the **left atrium.**
    - collect oxygenated blood from the respiratory part of the lung and deoxygenated blood from the visceral pleura and from a part of the bronchi.

C. Bronchial arteries

- arise from the **thoracic aorta;** usually there is one artery for the right lung and two for the left lung.
- supply oxygenated blood to the **nonrespiratory conducting tissues of the lungs.** Anastomoses occur between the capillary beds of the bronchial and the pulmonary arterial systems.

D. Bronchial veins

- receive blood from the larger subdivisions of the bronchi and empty into the **azygos vein** on the right and into the **accessory hemiazygos vein** on the left. (Other venous blood is drained by the pulmonary veins.)
- may receive twigs from the tracheobronchial lymph nodes.

VII. Respiration

- is the vital exchange of oxygen and carbon dioxide that occurs in the lungs.

A. During inspiration

- The **diaphragm contracts,** pulling the dome inferiorly into the abdomen, thereby increasing the vertical diameter of the thorax and decreasing intrathoracic and intrapulmonary pressures.
- The **pleural cavities and the lungs enlarge,** the intra-alveolar pressure is reduced, and the air passively rushes into the lungs as a result of atmospheric pressure.
- The **ribs are elevated** and the **abdominal pressure is increased** with decreased abdominal volume.
- **Muscles of inspiration** include the diaphragm, external intercostal muscles, interchondral part of internal intercostal muscles, innermost intercostal muscle, sternocleidomastoid, levator scapulae, serratus anterior, scalenus, pectoralis major and minor, erector spinae, and serratus posterosuperior muscles.

B. During expiration

- The diaphragm, the intercostal muscles, and other muscles relax; the thoracic volume is decreased; and the intrathoracic pressure is increased.
- The stretched elastic tissue of the lungs recoils, and much of the air is expelled.
- The **abdominal pressure is decreased** and the **ribs are depressed.**
- **Muscles of expiration** include the muscles of the abdominal wall, internal intercostal muscles, and serratus posteroinferior muscles.

VIII. Nerve Supply to the Lung

A. Pulmonary plexus

- receives afferent and efferent (parasympathetic preganglionic) fibers from the vagus nerve, joined by branches (sympathetic postganglionic fibers) from the sympathetic trunk and cardiac plexus.
- is divided into the **anterior pulmonary plexus,** which lies in front of the root of the lung, and the **posterior pulmonary plexus,** which lies behind the root of the lung.
- Its branches accompany the blood vessels and bronchi into the lung.
- Its sympathetic nerve fibers dilate the lumina of the bronchi, and parasympathetic fibers constrict the lumina and increase glandular secretion.

B. Phrenic nerve

– arises from the third through fifth cervical nerves (C3–C5) and lies in front of the anterior scalene muscle.
– enters the thorax by passing deep to the subclavian vein and superficial to the subclavian arteries.
– runs anterior to the root of the lung, whereas the vagus nerve runs posterior to the root of the lung.
– is accompanied by the pericardiophrenic vessels of the internal thoracic vessels and descends between the mediastinal pleura and the pericardium.
– innervates the pericardium, the mediastinal and diaphragmatic pleurae, and the diaphragm.

IX. Clinical Considerations

A. Disorders of the lungs

1. Pneumothorax

– is the presence of air or gas in the pleural cavity.

2. Emphysema

– is the accumulation of air in the terminal bronchioles and alveolar sacs.
– reduces the surface area available for gas exchange and thereby reduces oxygen absorption.

3. Pneumonia (pneumonitis)

– is an inflammation of the lungs.

B. Pleural tap (thoracentesis or pleuracentesis)

– is a surgical puncture of the thoracic wall into the pleural cavity for aspiration of fluid.
– is performed posterior to the midaxillary line one or two intercostal spaces below the fluid level but not below the ninth intercostal space.

C. Lesion of the phrenic nerve

– may not produce complete paralysis of the corresponding half of the diaphragm because the **accessory phrenic nerve,** derived from the fifth cervical nerve as a branch of the nerve to the subclavius, usually joins the phrenic nerve in the root of the neck or in the upper part of the thorax.

D. Hiccup

– is an **involuntary spasmodic sharp contraction of the diaphragm,** accompanied by the approximation of the vocal folds and closure of the glottis of the larynx.
– may occur as a result of the stimulation of nerve endings in the digestive tract or the diaphragm.
– when chronic, can be stopped by **sectioning or crushing the phrenic nerve.**

Pericardium and Heart

I. Pericardium

– is a fibroserous sac that encloses the heart and the roots of the great vessels and occupies the **middle mediastinum.**
– is composed of the fibrous pericardium and serous pericardium.

– receives blood from the pericardiophrenic, bronchial, and esophageal arteries.
– is innervated by vasomotor and sensory fibers from the phrenic and vagus nerves and the sympathetic trunks.

A. Fibrous pericardium

– is a strong, dense, fibrous layer that blends with the adventitia of the roots of the great vessels and the central tendon of the diaphragm.

B. Serous pericardium

– consists of the **parietal layer,** which lines the inner surface of the fibrous pericardium, and the **visceral layer,** which forms the outer layer (epicardium) of the heart wall and the roots of the great vessels.

C. Pericardial cavity

– is a **potential space** between the visceral layer of the serous pericardium (epicardium) and the parietal layer of the serous pericardium lining the inner surfaces of the fibrous pericardium.

D. Pericardial sinuses

1. Transverse sinus

– is a subdivision of the **pericardial sac,** lying posterior to the ascending aorta and pulmonary trunk and anterosuperior to the left atrium and the pulmonary veins.

2. Oblique sinus

– is a subdivision of the **pericardial sac** behind the heart, surrounded by the reflection of the serous pericardium around the right and left pulmonary veins and the inferior vena cava.

II. Heart (Figure 4-7)

A. General characteristics

– The **apex of the heart** lies in the left fifth intercostal space slightly medial to the midclavicular (or nipple) line, about 9 cm from the midline; this location is useful clinically for determining the left border of the heart and for **auscultation of the mitral valve.**
– Its posterior aspect, called the **base,** is formed mainly by the left atrium and only partly by the posterior right atrium.
– Its **right (acute) border** is formed by the superior vena cava, right atrium, and inferior vena cava, and its **left (obtuse) border** is formed by the left ventricle.
– The heart wall consists of three layers: inner **endocardium,** middle **myocardium,** and outer **epicardium.**
– The **sulcus terminalis,** a groove on the external surface of the right atrium, marks the junction of the primitive sinus venosus with the atrium in the embryo and corresponds to a ridge on the internal heart surface, the **crista terminalis.**
– The **coronary sulcus,** a groove on the external surface of the heart, marks the division between the atria and the ventricles.
– The **cardiovascular silhouette,** or cardiac shadow, is the contour of the heart and great vessels seen on PA chest radiographs. Its right border is formed by the superior vena cava, the right atrium, and the inferior vena cava. Its left

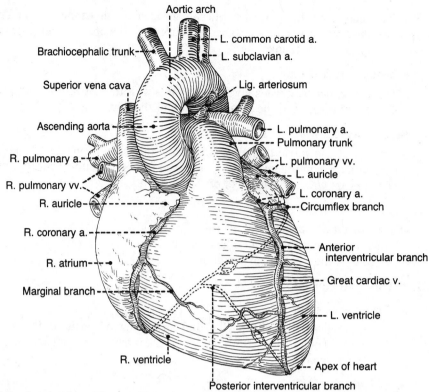

Figure 4-7. Anterior view of the heart with coronary arteries.

border is formed by the aortic arch (which produces the **aortic knob**), the pulmonary trunk, the left auricle, and the left ventricle.

B. Internal anatomy of the heart (Figure 4-8)

1. Right atrium

– Its walls are relatively smooth except for the presence of the pectinate muscles.

– is larger than the left atrium, but has a thinner wall.

– **Right atrial pressure** is normally slightly lower than left atrial pressure.

a. Right auricle

– is the conical muscular pouch of the upper anterior portion of the right atrium, which covers the first part of the right coronary artery.

b. Sinus venarum (sinus venarum cavarum)

– is a posteriorly situated, smooth-walled area that is separated from the more muscular atrium proper by the **crista terminalis.**

– develops from the embryonic **sinus venosus** and receives the superior vena cava, inferior vena cava, coronary sinus, and anterior cardiac veins.

c. Pectinate muscles

– are **prominent ridges of atrial myocardium** located in the interior of both auricles and the right atrium.

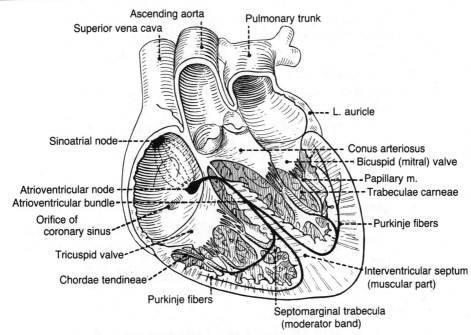

Figure 4-8. Internal anatomy and conducting system of the heart.

   d. Crista terminalis
- is a **vertical muscular ridge** running anteriorly along the right atrial wall from the opening of the superior vena cava to the inferior vena cava, providing the **origin of the pectinate muscles**.
- represents the junction between the primitive sinus venosus and the right atrium proper and is indicated externally by the **sulcus terminalis.**

   e. Venae cordis minimae
- are the smallest cardiac veins, which begin in the substance of the heart (endocardium and innermost layer of the myocardium) and end chiefly in the atria at the **foramina venarum minimarum cordis.**

   f. Fossa ovalis
- represents the position of the **foramen ovale,** through which blood runs from the right to the left atrium before birth.

  2. Left atrium
- is smaller and has thicker walls than the right atrium.
- is the most posterior of the four chambers, and its walls are smooth, except for a few pectinate muscles in the auricle.
- receives oxygenated blood through four pulmonary veins.

  3. Right ventricle
- makes up the major portion of the anterior surface of the heart.
- contains the following structures:

   a. Trabeculae carneae cordis
- are anastomosing muscular ridges of myocardium in the ventricles.

   b. Papillary muscles
- are cone-shaped muscles enveloped by endocardium.
- extend from the anterior and posterior ventricular walls and the septum, and their apices are attached to the **chordae tendineae.**

– contract to tighten the chordae tendineae, preventing the cusps of the **tricuspid valve** (right atrioventricular valve) from being everted into the atrium by the pressure developed by the pumping action of the heart.

c. **Chordae tendineae**

– extend from one papillary muscle to more than one cusp of the tricuspid valve.
– prevent eversion of the valve during ventricular contractions.

d. **Conus arteriosus (infundibulum)**

– is the upper end of the right ventricle.
– has smooth walls and leads into the pulmonary orifice.

e. **Septomarginal trabecula (moderator band)**

– is an isolated trabecula of the bridge type, extending from the **interventricular septum** to the base of the anterior papillary muscle of the right ventricle.
– carries the right limb of the **atrioventricular bundle** from the septum to the sternocostal wall of the ventricle.

f. **Interventricular septum**

– gives an attachment to the septal cusp of the tricuspid valve.
– is mostly muscular but has a small membranous upper part, which is a common site of ventricular septal defects.

4. **Left ventricle**

– is divided into the left ventricle proper and the **aortic vestibule,** which is the upper anterior part of the left ventricle and leads into the aorta.
– contains two **papillary muscles** (anterior and posterior) with their **chordae tendineae** and a meshwork of muscular ridges, the **trabeculae carneae cordis.**
– performs more work than the right ventricle; its wall is usually twice as thick.
– is longer, narrower, and more conical-shaped than the right ventricle.

C. **Heart valves** (Figure 4-9)

1. **Pulmonary valve**

– lies behind the medial end of the left third costal cartilage and adjoining part of the sternum.
– is most audible over the **left second intercostal space.**

2. **Aortic valve**

– lies behind the left half of the sternum opposite the third intercostal space.
– is most audible over the **right second intercostal space.**

3. **Tricuspid (right atrioventricular) valve**

– lies between the right atrium and ventricle, behind the right half of the sternum opposite the fourth intercostal space and is covered by endocardium.
– is most audible over the **right lower part of the body of the sternum.**
– has **anterior, posterior, and septal cusps,** which are attached by the chordae tendineae to three papillary muscles that keep the valves against the pressure developed by the pumping action of the heart.

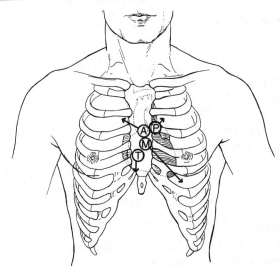

Figure 4-9. Positions of the valves of the heart and heart sounds. *P* = pulmonary valve; *A* = aortic valve; *M* = mitral valve; T = tricuspid valve. *Arrows* indicate positions of the heart sounds.

4. Mitral (left atrioventricular) valve

 – lies between the left atrium and ventricle, behind the left half of the sternum opposite the fourth costal cartilage, and has **two cusps: anterior and posterior.**
 – is most audible over the **left fifth intercostal space at the midclavicular line.**

D. Conducting system of the heart (see Figure 4-8)

1. Sinoatrial (SA) node

 – is a small mass of specialized cardiac muscle fibers that lies in the myocardium near the upper end of the crista terminalis, near the opening of the superior vena cava in the right atrium.
 – is the **"pacemaker" of the heart** and initiates the heartbeat, which can be altered by autonomic nervous stimulation (sympathetic stimulation speeds it up and vagal stimulation slows it down).
 – is supplied by the **sinus node artery,** which is usually a branch of the right coronary artery.

2. Atrioventricular (AV) node

 – lies beneath the endocardium in the lower part of the atrial septum, above the opening of the coronary sinus in the right atrium.
 – is supplied by the posterior interventricular branch of the right coronary artery.
 – is innervated by autonomic nerve fibers, although the cardiac muscle fibers lack motor endings.

3. Atrioventricular bundle (bundle of His)

 – begins at the **atrioventricular node** and runs along the membranous part of the interventricular septum.
 – splits into right and left branches, which descend into the muscular part of the interventricular septum, and breaks up into terminal conducting fibers **(Purkinje fibers)** to spread out into the ventricular walls.

E. Coronary arteries (see Figure 4-7)

  1. Right coronary artery

    – arises from the anterior (right) aortic sinus, runs between the root of the pulmonary trunk and the right auricle, and supplies the right atrium and ventricle.

    – gives rise to the following:

    a. The sinus node artery, which passes between the right atrium and the opening of the superior vena cava and supplies the SA node

    b. The marginal artery to the right ventricular wall

    c. The posterior ventricular (right posterior descending) artery

  2. Left coronary artery

    – arises from the left aortic sinus, just above the **aortic semilunar valve.**

    – is shorter than the right coronary artery and usually is distributed to more of the myocardium.

    – gives rise to the circumflex branch and the **anterior interventricular (left anterior descending; LAD) branch.**

  3. Blood flow in the coronary arteries

    – is maximal during diastole and minimal during systole, owing to compression of the arterial branches in the myocardium during systole.

F. Cardiac veins and coronary sinus (Figure 4-10)

  1. Coronary sinus

    – is the **largest vein draining the heart** and lies in the **coronary sulcus,** which separates the atria from the ventricles.

    – opens into the right atrium between the opening of the inferior vena cava and the atrioventricular opening.

    – has a one-cusp valve at the right margin of its aperture.

    – receives the great, middle, and small cardiac veins; the oblique vein of the left atrium; and the posterior vein of the left ventricle.

  2. Great cardiac vein

    – begins at the apex of the heart, ascends in the anterior interventricular groove, and drains upward alongside the anterior interventricular branch of the left coronary artery.

    – turns to the left to lie in the coronary sinus, and continues as the **coronary sinus.**

  3. Middle cardiac vein

    – begins at the apex of the heart and ascends in the **posterior interventricular groove,** accompanying the posterior interventricular branch of the right coronary artery.

    – drains into the right end of the coronary sinus.

  4. Small cardiac vein

    – runs along the right margin of the heart in company with the marginal artery and then runs posteriorly in the coronary sulcus to end in the right end of the coronary sinus.

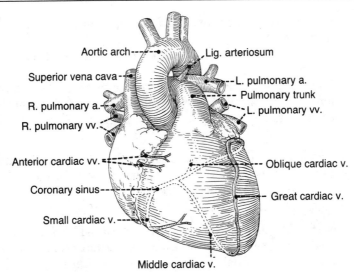

Aortic arch

Lig. arteriosum

Superior vena cava

L. pulmonary a.

Pulmonary trunk

R. pulmonary a.

L. pulmonary vv.

R. pulmonary vv.

Anterior cardiac vv.

Oblique cardiac v.

Coronary sinus

Great cardiac v.

Small cardiac v.

Middle cardiac v.

Figure 4-10. Anterior view of the heart.

5. Oblique vein of the left atrium

– descends to empty into the coronary sinus, near its left end.

6. Anterior cardiac vein

– drains the anterior right ventricle, crosses the coronary groove, and ends directly in the right atrium.

7. Smallest cardiac veins (venae cordis minimae)

– begin in the wall of the heart and empty directly into its chambers.

G. Lymphatic vessels of the heart

– receive lymph from the myocardium and epicardium.
– follow the right coronary artery to empty into the **anterior mediastinal nodes** and follow the left coronary artery to empty into a tracheobronchial node.

H. Cardiac plexus

– receives the superior, middle, and inferior cervical and thoracic cardiac nerves from the sympathetic trunks and vagus nerves.
– is divisible into the **superficial cardiac plexus**, which lies beneath the arch of the aorta, in front of the pulmonary artery, and the **deep cardiac plexus**, which lies posterior to the arch of the aorta, in front of the bifurcation of the trachea.
– richly innervates the conducting system of the heart: the right sympathetic and parasympathetic branches terminate chiefly in the region of the **SA node,** and the left branches end chiefly in the region of the **AV node.** The cardiac muscle fibers are devoid of motor endings and are activated by the conducting system.
– supplies the heart with **sympathetic fibers,** which increase the heart rate and the force of the heartbeat, causing dilation of the coronary arteries; and **parasympathetic fibers,** which decrease the heart rate.

I. Clinical considerations

1. Disorders of the heart and pericardium

a. Angina pectoris

– is characterized by attacks of **chest pain originating** in the heart and

felt beneath the sternum, in many cases radiating to the left shoulder and down the arm.
- is caused by an **insufficient supply of oxygen to the heart muscle,** owing to coronary artery disease.
- Its pain impulses travel in visceral afferent fibers through the middle and inferior cervical and thoracic cardiac branches of the sympathetic nervous system before entering the thoracic segment of the spinal cord.

b. Coronary atherosclerosis
- is characterized by the presence of **sclerotic plaques** containing cholesterol and lipoid material that impairs myocardial blood flow, leading to ischemia and myocardial infarction.

c. Pericardial tamponade
- is an **acute compression of the heart** caused by a rapid accumulation of fluid in the pericardial sac from **pericardial effusion** (passage of fluid from the pericardial capillaries into the pericardial sac).
- results in decreased diastolic capacity, reduced cardiac output with an increased heart rate, increased venous pressure with jugular vein distension, hepatic enlargement, and peripheral edema.

d. Pericarditis
- is an **inflammation of the parietal serous pericardium,** which may result in cardiac tamponade and precordial and epigastric pain.
- causes the surfaces of the pericardium to become rough; the resulting friction, which sounds like the rustle of silk, can be heard on auscultation.

2. Pericardiocentesis
- is a **surgical puncture of the pericardial cavity** for the aspiration of fluid, which is necessary to relieve the pressure of accumulated fluid on the heart. A needle is inserted into the pericardial sac through the fifth or sixth intercostal spaces adjacent to the sternum.

Structures in the Posterior Mediastinum

I. Esophagus

- is a muscular tube that is continuous with the pharynx in the neck and enters the thorax behind the trachea.
- has **three constrictions:** one at the level of the sixth cervical vertebra, where it begins; one at the crossing of the left main stem bronchus; and one at the tenth thoracic vertebra, where it pierces the diaphragm. The left atrium also presses against the anterior surface of the esophagus.
- receives blood from three branches of the aorta (the **inferior thyroid, bronchial, and esophageal arteries**) and from the left gastric and inferior phrenic arteries.

II. Blood Vessels and Lymphatic Vessels

A. Thoracic aorta

- begins at the level of the fourth thoracic vertebra.
- descends on the left side of the vertebral column and then approaches the median plane to end in front of the vertebral column by passing through the **aortic hiatus** of the diaphragm.

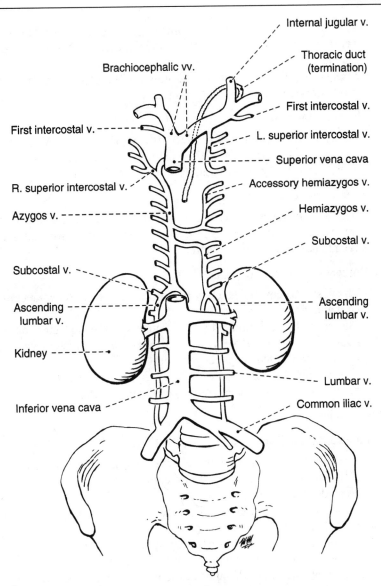

Figure 4-11. Azygos venous system.

- gives rise to nine pairs of **posterior intercostal arteries** and one pair of **subcostal arteries.** The first two intercostal arteries arise from the highest intercostal arteries of the costocervical trunk. The posterior intercostal artery gives rise to a collateral branch, which runs along the upper border of the rib below the space.
- also gives rise to pericardial, bronchial (one right and two left), esophageal, mediastinal, and superior phrenic branches.

B. Azygos venous system (Figure 4-11)

1. Azygos vein

- is formed by the union of the **right ascending lumbar and right subcostal veins.**

– enters the thorax through the aortic opening of the diaphragm.

– receives the right intercostal veins and the right posterior–superior intercostal vein.

– arches over the root of the right lung and **empties into the superior vena cava,** of which it is the first tributary.

2. Hemiazygos vein

– is formed by the union of the **left subcostal and ascending lumbar veins.**

– ascends on the left side of the vertebral bodies behind the thoracic aorta, receiving the lower four posterior intercostal veins.

3. Accessory hemiazygos vein

– begins at the fourth intercostal space, receives the fourth to seventh or eighth intercostal veins, descends in front of the posterior intercostal arteries, and terminates in the azygos vein.

4. Posterior intercostal veins

– The first intercostal vein on each side drains into the corresponding brachiocephalic vein.

– The second, third, and often the fourth intercostal veins join to form the **superior intercostal vein,** which drains into the azygos vein on the right and into the brachiocephalic vein on the left.

– The rest of the veins drain into the azygos vein on the right and into the hemiazygos or accessory hemiazygos veins on the left.

C. Lymphatics

1. Thoracic duct (Figure 4-12; see also Figure 4-11)

– begins in the abdomen at the **cisterna chyli,** which is the dilated junction of the intestinal, lumbar, and descending intercostal trunks.

– is usually beaded because of its numerous valves.

– drains the lower limbs, pelvis, abdomen, left thorax, left upper limb, and left side of the head and neck.

– passes through the aortic opening of the diaphragm and ascends through the posterior mediastinum between the aorta and the azygos vein.

– arches laterally over the apex of the left pleura and between the left carotid sheath in front and the vertebral artery behind, runs behind the left internal jugular vein, and then usually empties into the junction of the left internal jugular and subclavian veins.

2. Right lymphatic duct

– drains the right sides of the thorax, upper limb, head and neck, and empties into the junction of the right internal jugular and subclavian veins.

III. Autonomic Nervous System in the Thorax (Figure 4-13)

– is composed of motor, or efferent, nerves through which **cardiac muscle, smooth muscle, and glands** are innervated.

– involves two neurons: **preganglionic** and **postganglionic.**

– also includes **general visceral afferent (GVA) fibers** that run along with **general visceral efferent (GVE) fibers.**

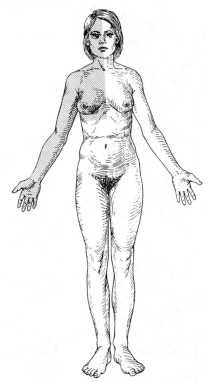

Figure 4-12. All areas except the *shaded area* (upper right quadrant) are drained by the thoracic duct.

- consists of sympathetic (or thoracolumbar outflow) and parasympathetic (or cranio-sacral outflow) systems.
- consists of **cholinergic** fibers (sympathetic preganglionic, parasympathetic preganglionic and postganglionic) that use acetylcholine as the neurotransmitter, and **adrenergic** fibers (sympathetic postganglionic) that use norepinephrine as the neurotransmitter except those to sweat glands (cholinergic).

A. Sympathetic nervous system

- enables the body to cope with crises or emergencies and thus often is referred to as the **fight-or-flight division.**
- Preganglionic cell bodies are located in the lateral horn or intermediolateral cell column of the spinal cord segments between T1 and L2.
- Preganglionic fibers pass through the white rami communicantes and enter the sympathetic chain ganglion, where they synapse.
- Postganglionic fibers join each spinal nerve by way of the gray rami communicantes and supply the blood vessels, hair follicles (arrector pili muscles), and sweat glands.

1. Sympathetic trunk

- is composed primarily of ascending and descending preganglionic sympathetic fibers and visceral afferent fibers, and contains the cell bodies of the postganglionic sympathetic (GVE) fibers.
- descends in front of the neck of the ribs and the posterior intercostal vessels.
- The **cervicothoracic (or stellate) ganglion** is formed by fusion of the inferior cervical ganglion with the first thoracic ganglion.

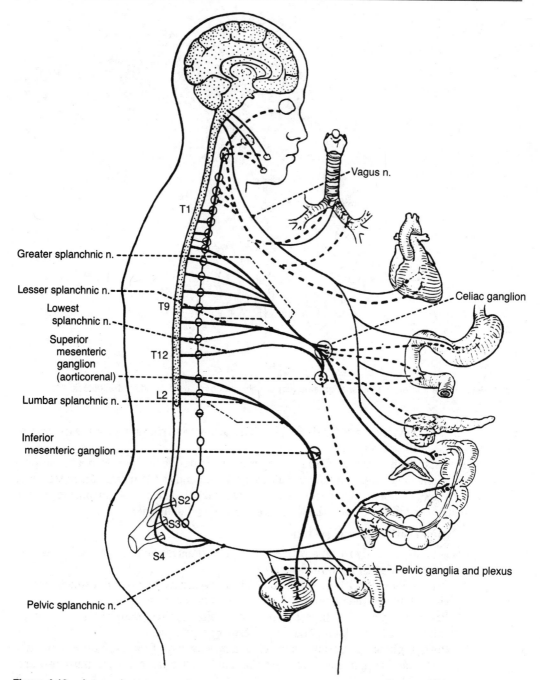

Figure 4-13. Autonomic nervous system.

- enters the abdomen through the crus of the diaphragm or behind the medial lumbocostal arch.
- gives rise to cardiac, pulmonary, mediastinal, and splanchnic branches.
- is connected to the thoracic spinal nerves by gray and white rami communicantes.

2. **Rami communicantes**
   a. **White rami communicantes**
      – contain preganglionic sympathetic (GVE; myelinated) fibers with cell bodies located in the lateral horn (intermediolateral cell column) of the spinal cord, and GVA fibers with cell bodies located in the dorsal root ganglia.
      – are connected to the spinal nerves, limited to the spinal cord segments between T1 and L2.
   b. **Gray rami communicantes**
      – contain postganglionic sympathetic (GVE; unmyelinated) fibers that supply the blood vessels, sweat glands, and arrector pili muscles of hair follicles.
      – are connected to every spinal nerve and contain fibers with cell bodies located in the sympathetic trunk.

3. **Thoracic splanchnic nerves**
   – contain sympathetic preganglionic fibers with cell bodies located in the lateral horn (intermediolateral cell column) of the spinal cord, and visceral afferent fibers with cell bodies located in the dorsal root ganglia.
   a. **Greater splanchnic nerve**
      – arises usually from the fifth through ninth thoracic sympathetic ganglia, perforates the crus of the diaphragm or occasionally passes through the aortic hiatus, and ends in the **celiac ganglion.**
   b. **Lesser splanchnic nerve**
      – is derived usually from the tenth and eleventh thoracic ganglia, pierces the crus of the diaphragm, and ends in the **aorticorenal ganglion.**
   c. **Least splanchnic nerve**
      – is derived usually from the twelfth thoracic ganglion, pierces the crus of the diaphragm, and ends in the ganglia of the **renal plexus.**

B. **Parasympathetic nervous system**
   – promotes quiet and orderly processes of the body, thereby conserving energy.
   – is not as widely distributed over the entire body as sympathetic fibers; the body wall and extremities have no parasympathetic nerve supply.
   – Preganglionic fibers running in cranial nerves (CN) III, VII, and IX pass to cranial autonomic ganglia (i.e., the ciliary, submandibular, pterygopalatine, and otic ganglia), where they synapse with postganglionic neurons.
   – Preganglionic fibers in CN X and in pelvic splanchnic nerves (originating from S2–S4) pass to terminal ganglia, where they synapse.
   – Parasympathetic fibers in the **vagus nerve** supply all of the thoracic and abdominal viscerae except the descending and sigmoid colons and other pelvic viscerae, which are innervated by the pelvic splanchnic nerves (S2–S4).
   – The vagus nerve contains the parasympathetic preganglionic fibers with cell bodies located in the medulla oblongata, and the GVA fibers with cell bodies located in the inferior (nodose) ganglion.
   1. **Right vagus nerve**
      – gives rise to the **right recurrent laryngeal nerve,** which hooks around the right subclavian artery and ascends into the neck between the trachea and the esophagus.

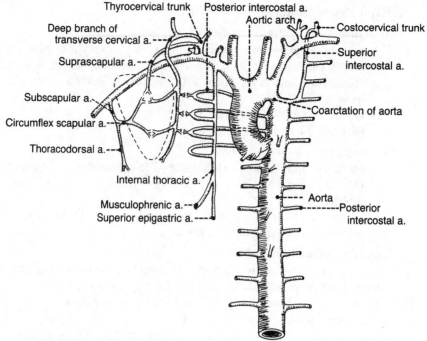

Figure 4-14. Coarctation of the aorta.

   – crosses anterior to the right subclavian artery, runs posterior to the superior
     vena cava, and descends at the right surface of the trachea and then posterior
     to the right main bronchus.
   – contributes to the cardiac, pulmonary, and esophageal plexuses.
   – forms the vagal trunks (or gastric nerves) at the lower part of the esophagus
     and enters the abdomen through the esophageal hiatus.

   2. **Left vagus nerve**
   – enters the thorax between the left common carotid and subclavian arteries
     and behind the left brachiocephalic vein and descends on the arch of the
     aorta.
   – gives rise to the left recurrent laryngeal nerve, which hooks around the arch
     of the aorta to the left of the ligamentum arteriosum. It ascends through the
     superior mediastinum in the groove between the trachea and the esophagus.
   – gives rise to the thoracic cardiac branches, breaks up into the pulmonary
     plexuses, and then continues into the esophageal plexus.

IV. Clinical Considerations

A. Conditions

   1. **Aneurysm of the aortic arch**
   – is a sac formed by dilation of the aortic arch that compresses the left recurrent
     laryngeal nerve, leading to coughing, hoarseness, and paralysis of the ipsilat-
     eral vocal cord because it innervates almost all of the muscles of the larynx.

   2. **Coarctation of the aorta** (Figure 4-14)
   – usually occurs distal to the point of entrance of the ductus arteriosus into
     the aorta, in which case an adequate collateral circulation develops before

birth. If this condition occurs proximal to the origin of the left subclavian artery of the ductus arteriosus, adequate collateral circulation does not develop.

– results in tortuous and enlarged blood vessels, especially the internal thoracic, intercostal, epigastric, and scapular arteries.

– results in elevated blood pressure in the radial artery and decreased pressure in the femoral artery.

– causes the femoral pulse to occur after the radial pulse. (Normally, the femoral pulse occurs slightly before the radial pulse and is under about the same pressure.)

– leads to the development of the important **collateral circulation** over the thorax. It occurs between the:

a. **Anterior intercostal** branches of the internal thoracic artery and the **posterior intercostal** arteries

b. **Superior epigastric** branch of the internal thoracic artery and the **inferior epigastric** artery

c. **Posterosuperior intercostal** branch of the costocervical trunk and the **third posterior intercostal** artery

d. **Posterior intercostal** arteries and the **descending scapular** (or dorsal scapular) artery, which anastomoses with the suprascapular and circumflex scapular arteries around the scapula

3. **Rupture of the thoracic duct**

– can be caused by fracture of the thoracic vertebrae, leading to **chylothorax,** an accumulation of chyle in the pleural sac.

B. **Vagotomy**

– is transection of the vagus nerves at the lower portion of the esophagus in an attempt to reduce gastric secretion in the treatment of peptic ulcer.

Review Test

Directions: Each of the numbered items or incomplete statements in this section is followed by answers or by completions of the statement. Select the ONE lettered answer or completion that is BEST in each case.

1. On the surface of the chest, the apex of the heart is located

(A) at the level of the sternal angle
(B) in the left fourth intercostal space
(C) in the left fifth intercostal space
(D) in the right fifth intercostal space
(E) at the level of the xiphoid process of the sternum

2. Normal, quiet expiration is achieved by contraction of the

(A) elastic tissue in the thoracic wall and lungs
(B) serratus posterior–superior muscles
(C) pectoralis minor muscles
(D) serratus anterior muscles
(E) diaphragm

3. Which of the following groups of fibers would be injured if the greater splanchnic nerves were severed?

(A) General somatic afferent (GSA) and preganglionic sympathetic fibers
(B) General visceral afferent (GVA) and postganglionic sympathetic fibers
(C) GVA and preganglionic sympathetic fibers
(D) General somatic efferent (GSE) and postganglionic sympathetic fibers
(E) GVA and GSE fibers

4. A stab wound that severed the white rami communicantes at the level of the sixth thoracic vertebra would result in damage to which of the following nerve fibers?

(A) Preganglionic sympathetic efferent and GSE fibers
(B) Postganglionic sympathetic efferent and GSA fibers
(C) Preganglionic sympathetic and postganglionic parasympathetic efferent fibers
(D) Preganglionic sympathetic efferent and GVA fibers
(E) Postganglionic sympathetic efferent and GVA fibers

5. Where should the physician place the stethoscope to listen to the sound of the mitral valve?

(A) Over the medial end of the second left intercostal space
(B) Over the medial end of the second right intercostal space
(C) In the left fourth intercostal space at the midclavicular line
(D) In the left fifth intercostal space at the midclavicular line
(E) Over the right half of the lower end of the body of the sternum

6. Which of the following statements concerning the cardiac veins is true?

(A) The great cardiac vein accompanies the posterior descending interventricular artery
(B) The middle cardiac vein accompanies the anterior descending interventricular artery
(C) The anterior cardiac vein ends in the right atrium
(D) The small cardiac vein accompanies the circumflex branch of the left coronary artery
(E) The oblique veins of the left atrium end in the left atrium

7. The largest portion of the sternocostal surface of the heart seen on a PA chest radiograph is composed of the

(A) left atrium
(B) right atrium
(C) left ventricle
(D) right ventricle
(E) base of the heart

8. Which of the following valves guards the opening between the right atrium and right ventricle?

(A) Pulmonary semilunar valve
(B) Mitral valve
(C) Valve of the coronary sinus
(D) Tricuspid valve
(E) Aortic semilunar valve

9. An artificial cardiac pacemaker is implanted in a 54-year-old patient. Which of the following conductive tissues of the heart has not been properly functioning?

(A) Atrioventricular bundle
(B) Atrioventricular node
(C) Sinoatrial node
(D) Purkinje fiber
(E) Moderator band

10. Which of the following bronchopulmonary segments connects to the right middle lobar bronchus?

(A) Medial and lateral
(B) Anterior and posterior
(C) Anterior basal and medial basal
(D) Anterior basal and posterior basal
(E) Lateral basal and posterior basal

11. Which of the following bronchi is called the eparterial bronchus?

(A) Left superior bronchus
(B) Left inferior bronchus
(C) Right superior bronchus
(D) Right middle bronchus
(E) Right inferior bronchus

12. Which portion of the heart musculature is most likely to be ischemic if a blood clot blocks the circumflex branch of the left coronary artery?

(A) Anterior portion of the left ventricle
(B) Anterior interventricular region
(C) Posterior interventricular region
(D) Posterior portion of the left ventricle
(E) Anterior portion of the right ventricle

13. Which of the following statements concerning the phrenic nerve is true?

(A) It contains only GSE fibers
(B) Its injury could cause difficulty in expiration
(C) It supplies the pericardium, mediastinal pleura, and diaphragm
(D) It enters the thorax by passing in front of the subclavian vein
(E) It passes posterior to the root of the lung as it courses through the thorax

14. Which of the following statements concerning the left recurrent laryngeal nerve is true?

(A) It hooks below the aortic arch, medial to the ligamentum arteriosum
(B) It ascends into the neck, passing in front of the subclavian artery
(C) It ascends through the superior mediastinum behind the esophagus
(D) It can be damaged by an aortic aneurysm, causing hoarseness
(E) It forms the majority of the esophageal plexus

15. Which of the following conditions could result from myocardial infarction limited to the interventricular septum?

(A) Tricuspid valve insufficiency
(B) Pulmonary valve insufficiency
(C) Inferior vena caval valve insufficiency
(D) Mitral valve insufficiency
(E) Aortic valve insufficiency

16. The cardiac notch is a deep indentation on the

(A) superior lobe of the right lung
(B) middle lobe of the right lung
(C) inferior lobe of the right lung
(D) superior lobe of the left lung
(E) inferior lobe of the left lung

Directions: Each of the numbered items or incomplete statements in this section is negatively phrased, as indicated by a capitalized word such as NOT, LEAST, or EXCEPT. Select the ONE lettered answer or completion that is BEST in each case.

17. All of the following conditions represent the normal changes that occur in the circulation at, or soon after, birth EXCEPT

(A) increased blood flow through the lungs
(B) closure of the ductus arteriosus
(C) increased left atrial pressure
(D) closure of the foramen ovale
(E) closure of the right umbilical vein

18. The right ventricle contains all of the following structures EXCEPT

(A) the papillary muscles
(B) the chordae tendineae
(C) the septomarginal trabecula
(D) the pectinate muscles
(E) the trabeculae carneae cordis

19. All of the following statements concerning the structures involved in fetal circulation are true EXCEPT

(A) the ductus venosus shunts blood from the umbilical vein to the inferior vena cava before birth, bypassing the liver
(B) the umbilical arteries carry blood to the placenta for reoxygenation before birth
(C) the ductus arteriosus closes functionally soon after birth
(D) the right umbilical vein is obliterated to form the ligamentum teres hepatis after birth
(E) the umbilical veins carry highly oxygenated blood from the placenta to the fetus

20. All of the following statements concerning the intercostal arteries are true EXCEPT

(A) the upper six pairs of anterior intercostal arteries arise from the internal thoracic artery
(B) all posterior intercostal arteries branch directly from the thoracic aorta
(C) the upper two posterior intercostal arteries are branches of the superior intercostal artery of the costocervical trunk
(D) the musculophrenic artery provides anterior intercostal arteries in the seventh to ninth intercostal spaces
(E) the intercostal arteries are located in the costal groove between the vein above and the nerve below

21. All of the following statements concerning the heart and its associated structures are true EXCEPT

(A) it is situated in the middle mediastinum
(B) the upper end of the right ventricle is known as the conus arteriosus
(C) the coronary sulcus lies between the atria and the ventricles
(D) blood flow in the coronary arteries is maximal during diastole and minimal during systole
(E) the coronary arteries are branches of the aortic arch

22. All of the following statements concerning the thoracic duct are true EXCEPT

(A) it originates from the cisterna chyli in the abdomen
(B) it passes upward through the aortic opening in the diaphragm
(C) it drains into the junction of the left internal jugular vein and the left subclavian vein
(D) it receives drainage from the right cervical lymph nodes
(E) it contains valves and ascends between the aorta and the azygos vein in the thorax

23. All of the following statements concerning the right lung are true EXCEPT

(A) it has ten bronchopulmonary segments
(B) it usually receives one bronchial artery
(C) it has three lobar (secondary) bronchi
(D) its upper lobe has a tongue-shaped portion called a lingula
(E) it has a slightly larger capacity than the left lung

24. All of the following muscles function to elevate the ribs EXCEPT

(A) levator costarum
(B) subcostalis
(C) external intercostal muscle
(D) innermost intercostal muscle
(E) transversus thoracis

25. All of the following structures are found in the superior mediastinum EXCEPT

(A) the brachiocephalic veins
(B) the trachea
(C) part of the left common carotid artery
(D) the arch of the aorta
(E) the hemiazygos vein

26. All of the following veins receive drainage from the posterior intercostal veins EXCEPT

(A) the hemiazygos vein
(B) the azygos vein
(C) the subclavian veins
(D) the brachiocephalic veins
(E) the superior intercostal vein

27. All of the following statements concerning the left vagus nerve are true EXCEPT

(A) it passes in front of the left subclavian artery as it enters the thorax
(B) it contributes to the anterior esophageal plexus
(C) it forms the anterior vagal trunk at the lower part of the esophagus
(D) it can be cut on the lower part of the esophagus to reduce gastric secretion
(E) it contains parasympathetic postganglionic fibers

28. All of the following anatomic features occur at the level of the sternal angle EXCEPT

(A) the trachea bifurcates into the right and left bronchi
(B) the aortic arch begins
(C) the aortic arch ends
(D) the third rib articulates with the sternum
(E) the inferior border of the superior mediastinum is demarcated

29. All of the following statements concerning the ductus arteriosus are true EXCEPT

(A) it is anatomically closed several weeks after birth
(B) it carries poorly oxygenated blood during prenatal life
(C) it is derived from the sixth aortic arch
(D) it connects the left pulmonary vein to the aorta
(E) it shunts blood from the pulmonary trunk to the aorta before birth, bypassing the pulmonary circulation

30. All of the following conditions can cause right ventricular hypertrophy EXCEPT

(A) a constricted pulmonary artery
(B) an abnormally small left atrioventricular opening
(C) improper closing of the pulmonary valves
(D) an abnormally large right atrioventricular opening
(E) stenosis of the aorta

31. The third rib articulates with all of the following structures EXCEPT

(A) the body of the sternum
(B) the body of the second thoracic vertebra
(C) the transverse process of the second thoracic vertebra
(D) the body of the third thoracic vertebra
(E) the transverse process of the third thoracic vertebra

32. All of the following statements concerning the right primary bronchus are true EXCEPT

(A) it has a larger diameter than the left primary bronchus
(B) more foreign bodies enter it via the trachea than enter the left primary bronchus
(C) it gives rise to the eparterial bronchus
(D) it is longer than the left primary bronchus
(E) it runs under the arch of the azygos vein

33. All of the following statements concerning the azygos vein are true EXCEPT

(A) it receives the left superior intercostal vein
(B) it receives the right superior intercostal vein
(C) it empties into the superior vena cava
(D) it arches over the root of the right lung
(E) it enters the thorax through the aortic opening

34. All of the following structures form the left border of the cardiovascular silhouette on PA chest radiographs EXCEPT

(A) the arch of the aorta
(B) the pulmonary trunk
(C) the left atrium
(D) the left auricle
(E) the left ventricle

35. All of the following statements concerning the nerve supply of the lungs and pleurae are true EXCEPT

(A) the mediastinal pleura is innervated by the phrenic nerve
(B) the peripheral part of the diaphragmatic pleura is innervated by the intercostal nerves
(C) the anterior pulmonary plexus contains preganglionic sympathetic fibers
(D) constriction of the bronchial lumen is controlled by the vagus nerve
(E) the smooth muscle in the wall of the pulmonary artery is controlled by sympathetic fibers

Directions: Each set of matching questions in this section consists of a list of four to twenty-six lettered options (some of which may be in figures) followed by several numbered items. For each numbered item, select the ONE lettered option that is most closely associated with it. To avoid spending too much time on matching sets with large numbers of options, it is generally advisable to begin each set by reading the list of options. Then, for each item in the set, try to generate the correct answer and locate it in the option list, rather than evaluating each option individually. Each lettered option may be selected once, more than once, or not at all.

Questions 36–40

Match each of the following statements with the nerve that is associated with it.

(A) Vagus nerve
(B) Phrenic nerve
(C) Sympathetic trunk
(D) Left recurrent laryngeal nerve
(E) Greater splanchnic nerve

36. Contributes to the formation of the esophageal plexus superior to the diaphragm

37. Loops around the arch of the aorta near the ligamentum arteriosum

38. Arises in the posterior mediastinum and enters the celiac ganglion

39. Lies between the fibrous pericardium and parietal mediastinal pleura along part of its course

40. Is connected to the intercostal nerves by white and gray rami

Questions 41–45

Match each of the following descriptions with the appropriate lettered structure in the superior cross-sectional view of the heart.

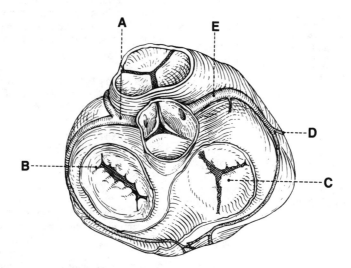

41. Runs toward the apex of the heart; supplies atrial blood to the anterior right ventricle

42. Supplies arterial blood to the sinoatrial node

43. Arises from the left aortic sinus

44. Gives rise to the anterior interventricular artery

45. Its sound is best heard through a stethoscope at the left fifth intercostal space at the midclavicular line

Questions 46–50

Match each statement below with the most appropriate chamber of the heart.

(A) Right atrium
(B) Left atrium
(C) Right ventricle
(D) Left ventricle
(E) Right auricle

46. Blood leaves this chamber through a valve that can be auscultated over the second left intercostal space, just lateral to the sternum

47. Receives blood from the anterior cardiac veins

48. Associated with the apex of the heart

49. Enlarges briefly in response to coarctation of the aorta

50. The sinoatrial node is in its wall

Questions 51–55

Match each of the following statements with the appropriate cardiac structure.

(A) Crista terminalis
(B) Septomarginal trabecula
(C) Chordae tendineae
(D) Pectinate muscle
(E) Anulus fibrosus

51. Maintains constant tension on the cusps of the atrioventricular valve

52. Attachment site of the cusps of the atrioventricular valve

53. Extends from the interventricular septum to the base of the anterior papillary muscle of the right ventricle

54. A vertical muscular ridge representing the junction between the sinus venarum and the atrium proper

55. Contains Purkinje fibers from the right limb of the atrioventricular bundle

Answers and Explanations

1–C. On the surface of the chest, the apex of the heart can be located in the left fifth intercostal space slightly medial to the midclavicular (or nipple) line.

2–A. Normal, quiet expiration is achieved by contraction of extensible tissue in the lungs and the thoracic wall. The serratus posterior–superior muscles, diaphragm, pectoralis major, and serratus anterior are muscles of inspiration.

3–C. The greater splanchnic nerves contain general visceral afferent (GVA) and preganglionic sympathetic (GVE) fibers.

4–D. The white rami communicantes contain preganglionic sympathetic (GVE) fibers and GVA fibers.

5–C. The mitral valve (left atrioventricular valve) produces the apex beat of the heart, which is most audible over the left fifth intercostal space at the midclavicular line.

6–C. The anterior cardiac vein drains the anterior right ventricle, crosses the coronary groove, and ends directly in the right atrium. The great cardiac vein ascends in the anterior interventricular groove, accompanied by the anterior interventricular branch of the left coronary artery. The middle cardiac vein ascends in the posterior interventricular groove, accompanied by the posterior interventricular branch of the right coronary artery. The anterior cardiac veins drain into the right atrium. The small cardiac vein accompanies the right marginal artery. The oblique veins of the left atrium empty into the coronary sinus, near its beginning.

7–D. The right ventricle forms a large part of the sternocostal surface of the heart.

8–D. The tricuspid valve guards the opening between the right atrium and right ventricle.

9–C. The sinoatrial node initiates the impulse of contraction and is known as the pacemaker of the heart.

10–A. The right middle lobar (secondary) bronchus leads to the medial and lateral bronchopulmonary segments.

11–C. The eparterial bronchus is the right superior lobar (secondary) bronchus; all of the other bronchi are hyparterial bronchi.

12–D. The circumflex branch of the left coronary artery supplies the posterior portion of the left ventricle. The right coronary artery gives rise to the marginal branch, which supplies the anterior portion of the right ventricle, and the posterior interventricular artery, which supplies the septum and the adjacent ventricles.

13–C. The phrenic nerve descends behind the subclavian vein, passes anterior to the root of the lung, and supplies the pericardium and mediastinal and diaphragmatic (central part) pleurae as well as the diaphragm, which is an important muscle of inspiration. It contains GSE, GSA, and GVE (postganglionic sympathetic) fibers.

14–D. The left recurrent laryngeal nerve hooks around the aortic arch lateral to the ligamentum arteriosum; therefore, it can be damaged by an aneurysm of the aortic arch, causing paralysis of the laryngeal muscles. This nerve ascends behind the subclavian artery and lies in a groove between the trachea and the esophagus. The majority of the esophageal plexus is formed by the vagus nerves.

15–A. The three cusps of the tricuspid valve are the anterior, posterior, and septal cusps. The septal cusp lies on the margin of the orifice adjacent to the interventricular septum. There is defective conduction of cardiac impulses because the atrioventricular bundle descends into the interventricular septum. The pulmonary, aortic, and mitral valves are not closely associated with the interventricular septum.

16–D. The cardiac notch is a deep indentation of the anterior border of the superior lobe of the left lung.

17–E. The right umbilical vein is obliterated during the embryonic period. The changes in circulation that occur at, or soon after, birth include obliteration of the umbilical arteries, left umbilical vein, and ductus venosus; functional closure of the ductus arteriosus and the foramen ovale (anatomic closure requires weeks or months); increased blood flow through the lungs; and increased left atrial pressure.

18–D. The pectinate muscles are prominent ridges of atrial myocardium located in the interior of both auricles and the right atrium.

19–D. The right umbilical vein is obliterated during the embryonic period; the left umbilical vein is obliterated after birth to form the ligamentum teres hepatis, which is located between the quadrate lobe and the left lobe of the liver.

20–C. The first two posterior intercostal arteries are branches of the highest (superior) intercostal artery of the costocervical trunk; the remaining nine branches are from the thoracic aorta. The musculophrenic artery gives off anterior intercostal arteries in the seventh, eighth, and ninth intercostal spaces, and ends in the tenth intercostal space where it anastomoses with the deep circumflex iliac artery.

21–E. The coronary arteries arise from the ascending aorta immediately above the aortic valve. Blood flow in the coronary arteries is maximal during diastole and minimal during systole, owing to the compression of the arterial branches in the myocardium during systole (ventricular contraction). The diastole is the dilation of the heart chambers, during which they fill with blood.

22–D. The thoracic duct drains most of the lymph in the body; however, it does not drain lymph from the right side of the head and neck, right upper limb, or right thorax. The lower, dilated end of the thoracic duct is called the cisterna chyli.

23–D. The lingula is the tongue-shaped portion of the upper lobe of the left lung. The oblique and horizontal fissures divide the right lung into upper, middle, and lower lobes; thus, the right lung has

3 lobar (secondary) bronchi and 10 lobar bronchopulmonary segments. The right lung receives one bronchial artery and has a slightly larger capacity than the left.

24–E. The transversus thoracis inserts into the inner surfaces of the costal cartilages and functions to depress the ribs.

25–E. The brachiocephalic veins, trachea, part of the left common carotid artery, and the arch of the aorta are located in the superior mediastinum. The hemiazygos vein is located in the inferior mediastinum.

26–C. The posterior intercostal veins do not drain into the subclavian vein. The first posterior intercostal vein drains into the corresponding brachiocephalic vein. The second, third, and fourth posterior intercostal veins join to form the superior intercostal vein, which drains into the azygos vein on the right and the brachiocephalic vein on the left. The rest of the posterior intercostal veins drain into the azygos vein on the right and into the hemiazygos vein or accessory hemiazygos vein on the left.

27–E. The vagus nerve carries parasympathetic preganglionic fibers to the thoracic and abdominal viscerae. The left vagus nerve enters the thorax in front of the left subclavian artery and behind the left brachiocephalic vein. It passes behind the left bronchus, forms the pulmonary plexus, and continues to form an esophageal plexus. The vagus nerves lose their identity in the esophageal plexus. At the lower end of the esophagus, branches of the plexus reunite to form an anterior vagal trunk (anterior gastric nerve), which can be cut (vagotomy) to reduce gastric secretion.

28–D. The sternal angle is the junction of the manubrium and the body of the sternum. It is located at the level where the second rib articulates with the sternum, the trachea bifurcates into the right and left bronchi, the aortic arch both begins and ends, and the inferior border of the superior mediastinum is demarcated.

29–D. Before birth, the ductus arteriosus connects the bifurcation of the pulmonary trunk with the aortic arch and carries poorly oxygenated blood. It is functionally closed shortly after birth; however, anatomic closure requires several weeks or months. After birth, the ductus arteriosus becomes the ligamentum arteriosum, which connects the arch of the aorta to the left pulmonary artery.

30–E. Stenosis of the aorta can cause left ventricular hypertrophy. Right ventricular hypertrophy can occur as a result of pulmonary stenosis, pulmonary and tricuspid valve defects, or mitral valve stenosis.

31–C. The third rib articulates anteriorly with the body of the sternum and posteriorly with the transverse process and the body of the third thoracic vertebra and the body of the second thoracic vertebra.

32–D. The right primary bronchus is shorter than the left one and has a larger diameter. More foreign bodies enter it via the trachea because it is more vertical than the left primary bronchus. The right primary bronchus runs under the arch of the azygos vein and gives rise to the eparterial bronchus.

33–A. The right superior intercostal vein drains into the azygos vein; the left superior intercostal vein drains into the left brachiocephalic vein.

34–C. The left atrium lies behind the right atrium and forms the greater portion of the base or posterior surface of the heart. The right atrium also contributes to a lesser extent to the base.

35–C. The anterior and posterior pulmonary plexuses contain preganglionic parasympathetic and postganglionic sympathetic fibers.

36–A. The vagus nerve forms the esophageal plexus after it leaves the pulmonary plexus.

37–D. The left recurrent laryngeal nerve loops inferior to the arch of the aorta, left of the ligamentum arteriosum.

38–E. The greater splanchnic nerve arises from the fifth through ninth thoracic sympathetic ganglia in the posterior mediastinum, pierces the crus of the diaphragm, and ends in the celiac ganglion.

39–B. The phrenic nerve descends between the fibrous pericardium and the mediastinal pleura.

40–C. The gray rami connect the sympathetic trunk to every spinal nerve. The white rami are limited to the spinal cord segments between T1 and L2. The intercostal nerves are ventral primary rami of the thoracic nerves.

41–D. The marginal branch of the right coronary artery supplies the anterior wall of the right ventricle.

42–E. The right coronary artery gives rise to the sinus node artery, which supplies the sinoatrial node. The posterior interventricular artery, a branch of the right coronary artery, gives rise to a branch that supplies the atrioventricular node.

43–A. The left coronary artery arises from the left aortic sinus.

44–A. The left coronary artery branches to form the anterior interventricular artery.

45–B. The mitral valve is best heard through a stethoscope at the left fifth intercostal space at the midclavicular line.

46–C. Blood passes through the pulmonary valve as it leaves the right ventricle. The pulmonary valve can be auscultated over the second left intercostal space, just lateral to the sternum.

47–A. The right atrium receives its blood supply from the anterior cardiac veins.

48–D. The apex of the heart is formed by the tip of the left ventricle and is located in the left fifth intercostal space.

49–D. The left ventricle briefly becomes enlarged in response to coarctation of the aorta.

50–A. The sinoatrial node is in the myocardium of the posterior wall of the right atrium, near the opening of the superior vena cava.

51–C. The chordae tendineae are tendinous strands that extend from the papillary muscles to the cusps of the atrioventricular valve. The papillary muscles and chordae tendineae prevent the cusps from being everted into the atrium during ventricular contraction.

52–E. The anulus fibrosus is a fibrous ring surrounding the atrioventricular orifice, which is the attachment site of the cusps of the atrioventricular valve.

53–B. The septomarginal trabecula is a moderator band that extends from the interventricular septum to the base of the anterior papillary muscle of the right ventricle.

54–A. The crista terminalis is a muscular ridge that represents the junction between the sinus venarum and the remainder of the right atrium. It is indicated externally by the sulcus terminalis.

55–B. The septomarginal trabecula carries the right branch of the atrioventricular bundle from the septum to the opposite wall of the ventricle.

5

Abdomen

Anterior Abdominal Wall

I. Abdomen (Figure 5-1)

- is divided topographically by two transverse and two longitudinal planes into nine regions: **right and left hypochondriac; epigastric; right and left lumbar; umbilical; right and left inguinal (iliac);** and **hypogastric (pubic)**.
- is divided also by vertical and horizontal planes through the umbilicus into four quadrants: right and left upper quadrants and right and left lower quadrants.

II. Muscles of the Anterior Abdominal Wall

Muscle	Origin	Insertion	Nerve	Action
External oblique	External surface of lower eight ribs (5–12)	Anterior half of iliac crest; anterior–superior iliac spine; pubic tubercle; linea alba	Intercostal n. (T7–T11); subcostal n. (T12)	Compresses abdomen; flexes trunk; active in forced expiration
Internal oblique	Lateral two-thirds of inguinal ligament; iliac crest; thoracolumbar fascia	Lower four costal cartilages; linea alba; pubic crest; pectineal line	Intercostal n. (T7–T11); subcostal n. (T12); iliohypogastric and ilioinguinal nn. (L1)	Compresses abdomen; flexes trunk; active in forced expiration
Transverse	Lateral one-third of inguinal ligament; iliac crest; thoracolumbar fascia; lower six costal cartilages	Linea alba; pubic crest; pectineal line	Intercostal n. (T7–T12); subcostal n. (T12); iliohypogastric and ilioinguinal nn. (L1)	Compresses abdomen; depresses ribs
Rectus abdominis	Pubic crest and pubic symphysis	Xiphoid process and costal cartilages 5–7	Intercostal n. (T7–T11); subcostal n. (T12)	Depresses ribs; flexes trunk
Pyramidal	Pubic body	Linea alba	Subcostal n. (T12)	Tenses linea alba

147

Muscle	Origin	Insertion	Nerve	Action
Cremaster	Middle of inguinal ligament; lower margin of internal oblique muscle	Pubic tubercle and crest	Genitofemoral n.	Retracts testis

III. Fasciae and Ligaments of the Anterior Abdominal Wall

– are organized into superficial (**tela subcutanea**) and deep layers, the superficial having a thin fatty layer (**Camper's fascia**) and the deep having a membranous layer (**Scarpa's fascia**).

A. Superficial fascia

1. **Superficial layer of the superficial fascia (Camper's fascia)**
   – continues over the inguinal ligament to merge with the superficial fascia of the thigh.
   – continues over the pubis and perineum as the superficial layer of the superficial perineal fascia.

2. **Deep layer of the superficial fascia (Scarpa's fascia)**
   – is attached to the **fascia lata** just below the inguinal ligament.
   – continues over the pubis and perineum as the membranous layer (**Colles' fascia**) of the superficial perineal fascia.
   – continues over the penis as the **superficial fascia of the penis** and over the scrotum as the **tunica dartos,** which contains smooth muscle.
   – Rupture of the spongy urethra results in presence of extravasated urine between this fascia and the deep fascia of the abdomen.

B. Deep fascia

– covers the muscles and continues over the spermatic cord at the superficial inguinal ring as the **external spermatic fascia.**

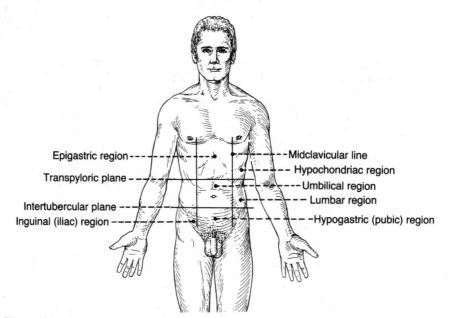

Epigastric region --- Midclavicular line

Transpyloric plane --- Hypochondriac region

--- Umbilical region

Intertubercular plane --- Lumbar region

Inguinal (iliac) region --- Hypogastric (pubic) region

Figure 5-1. Planes of subdivision of the abdomen.

– continues over the penis as the deep fascia of the penis (**Buck's fascia**) and over the pubis and perineum as the **deep perineal fascia.**

C. Linea alba

– is a **tendinous median raphe** between the two rectus abdominis muscles, extending from the xiphoid process to the pubic symphysis.
– is formed by the fusion of the aponeuroses of the external oblique, internal oblique, and transverse muscles of the abdomen.

D. Linea semilunaris

– is a **curved line** along the lateral border of the rectus abdominis.

E. Linea semicircularis (arcuate line)

– is a **crescent-shaped line** marking the termination of the posterior sheath of the rectus abdominis just below the level of the iliac crest.

F. Lacunar ligament

– represents the medial triangular expansion of the inguinal ligament to the pectineal line of the pubis.
– forms the medial border of the femoral ring and the floor of the inguinal canal.

G. Pectineal (Cooper's) ligament

– is a strong fibrous band that extends laterally from the lacunar ligament along the pectineal line of the pubis.

H. Inguinal ligament

– is the folded lower border of the aponeurosis of the external oblique muscle, extending between the anterior–superior iliac spine and the pubic tubercle.

I. Iliopectineal arcus or ligament

– is a **fascial partition** that separates the muscular (lateral) and vascular (medial) lacunae deep to the inguinal ligament.

1. The **muscular lacuna** transmits the iliopsoas muscle.

2. The **vascular lacuna** transmits the femoral sheath and its contents, including the femoral vessels, a femoral branch of the genitofemoral nerve, and the femoral canal.

J. Reflected inguinal ligament

– is formed by certain fibers of the inguinal ligament reflected from the pubic tubercle upward toward the **linea alba.**
– also has some reflection from the lacunar ligament.

K. Falx inguinalis (conjoint tendon)

– is formed by the aponeuroses of the internal oblique and transverse muscles of the abdomen and is inserted into the pubic tubercle.
– strengthens the posterior wall of the medial half of the **inguinal canal.**

L. Rectus sheath (Figure 5-2)

– is formed by fusion of the aponeuroses of the external oblique, internal oblique, and transverse muscles of the abdomen.
– encloses the rectus abdominis and sometimes the pyramidal muscle.
– also contains the superior and inferior epigastric vessels and the ventral primary rami of thoracic nerves 7 to 12.

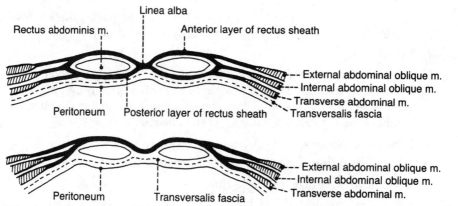

Figure 5-2. Arrangement of the rectus sheath above the umbilicus (*upper*) and below the arcuate line (*lower*).

1. **Anterior layer of the rectus sheath**

   a. **Above the arcuate line:** aponeuroses of the external and internal oblique muscles

   b. **Below the arcuate line:** aponeuroses of the external oblique, internal oblique, and transverse muscles

2. **Posterior layer of the rectus sheath**

   a. **Above the arcuate line:** aponeuroses of the internal oblique and transverse muscles

   b. **Below the arcuate line:** rectus abdominis is in contact with the transversalis fascia

IV. Inguinal Region

A. Inguinal (Hesselbach's) triangle

   – is bounded medially by the linea semilunaris (lateral edge of the rectus abdominis), laterally by the inferior epigastric vessels, and inferiorly by the inguinal ligament.

   – is an area of potential weakness and hence is a common site of a **direct inguinal hernia.**

B. Inguinal rings

1. **Superficial inguinal ring**

   – is a **triangular opening** in the aponeurosis of the external oblique muscle that lies just lateral to the pubic tubercle.

   – transmits the **spermatic cord** in the male and the **round ligament of the uterus** in the female.

2. **Deep inguinal ring**

   – lies in the transversalis fascia, just lateral to the inferior epigastric vessels.

   – is formed by embryonic extension of the processus vaginalis through the abdominal wall and subsequent passage of the testes through the transversalis fascia during their descent into the scrotum.

C. Inguinal canal

– begins at the deep inguinal ring and ends at the superficial ring.

– is much smaller in the female than in the male.

– transmits the **spermatic cord (or round ligament of the uterus)** and the **ilioinguinal nerve.**

1. **Anterior wall:** aponeuroses of the external oblique and internal oblique muscles

2. **Posterior wall:** aponeurosis of the transverse abdominal muscle and transversalis fascia

3. **Superior wall (roof):** arching fibers of the internal oblique and transverse muscles

4. **Inferior wall (floor):** inguinal and lacunar ligaments

V. Spermatic Cord, Scrotum, and Testis

A. Spermatic cord

– contains the **ductus deferens,** deferential vessels, testicular artery, pampiniform plexus of veins, lymphatics, and autonomic nerves of the testes.

– has several fasciae:

1. **External spermatic fascia,** derived from the aponeurosis of the external oblique muscle

2. **Cremasteric fascia** (cremaster muscle and fascia), originating in the internal oblique muscle

3. **Internal spermatic fascia,** derived from the transversalis fascia

B. Fetal structures

1. **Processus vaginalis testis**

– is a **peritoneal diverticulum** in the fetus that evaginates into the developing scrotum and forms the visceral and parietal layers of the **tunica vaginalis testis.**

– normally closes before birth or shortly thereafter and loses its connection with the peritoneal cavity.

– Its persistence may result in a **congenital indirect inguinal hernia.**

– Its occlusion may cause **fluid accumulation** (hydrocele processus vaginalis).

2. **Tunica vaginalis**

– is a **double serous membrane,** a peritoneal sac that covers the front and sides of the **testis and epididymis.**

– is derived from the abdominal peritoneum and forms the **innermost layer of the scrotum.**

3. **Gubernaculum testis**

– is the **fetal ligament** that connects the bottom of the fetal testis to the developing scrotum.

– appears to be important in **testicular descent** (pulls the testis down as it migrates).

– is homologous to the ovarian ligament and the round ligament of the uterus.

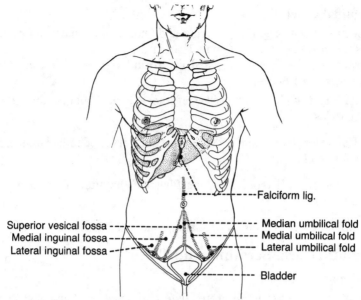

Falciform lig.

Superior vesical fossa
Medial inguinal fossa
Lateral inguinal fossa

Median umbilical fold
Medial umbilical fold
Lateral umbilical fold

Bladder

Figure 5-3. Peritoneal folds over the anterior abdominal wall.

VI. Inner Surface of the Anterior Abdominal Wall (Figure 5-3)

A. Supravesical fossa

– is a depression on the anterior abdominal wall between the medial and lateral umbilical folds of the peritoneum.

B. Medial inguinal fossa

– is a depression on the anterior abdominal wall between the medial and median umbilical folds of the peritoneum.
– lies lateral to the supravesical fossa.
– is the fossa where most **direct inguinal hernias** occur.

C. Lateral inguinal fossa

– is a depression on the anterior abdominal wall, lateral to the lateral umbilical fold of the peritoneum.

D. Umbilical folds or ligaments

1. Median umbilical fold

– extends from the apex of the bladder to the umbilicus and contains the **urachus** (the fibrous remains of the fetal allantois).

2. Medial umbilical fold

– extends from the side of the bladder to the umbilicus and contains the **obliterated umbilical artery** (a branch of the internal iliac artery).

3. Lateral umbilical fold

– extends from the medial side of the deep inguinal ring to the arcuate line and contains **inferior epigastric vessels.**

E. Transversalis fascia

– is the lining fascia of the entire abdominopelvic cavity between the parietal peritoneum and the inner surface of the abdominal muscles.

– continues with the diaphragmatic, psoas, iliac, pelvic, and quadratus lumborum fasciae.
– forms the **deep inguinal ring** and gives rise to the femoral sheath and the internal spermatic fascia.
– is directly in contact with the rectus abdominis below the arcuate line.

VII. Nerves of the Anterior Abdominal Wall

A. Subcostal nerve
– is the ventral ramus of the twelfth thoracic nerve and innervates the muscles of the anterior abdominal wall.
– Its **lateral cutaneous branch** innervates the skin of the side of the hip.

B. Iliohypogastric nerve
– arises from the **first lumbar nerve** and innervates the internal oblique and transverse muscles of the abdomen.
– divides into a **lateral cutaneous branch** to supply the skin of the lateral side of the buttocks and an **anterior cutaneous branch** to supply the skin above the pubis.

C. Ilioinguinal nerve
– arises from the **first lumbar nerve,** pierces the internal oblique muscle near the deep inguinal ring, and runs medially through the inguinal canal and then through the superficial inguinal ring.
– innervates the internal oblique and transverse muscles.
– gives rise to a **femoral branch,** which innervates the upper and medial parts of the thigh, and the **anterior scrotal nerve,** which innervates the skin of the root of the penis (or the skin of the mons pubis) and the anterior part of the scrotum (or the labium majus).

VIII. Lymphatic Drainage of the Anterior Abdominal Wall

A. Lymphatics in the region above the umbilicus drain into the axillary lymph nodes.

B. Lymphatics in the region below the umbilicus drain into the superficial inguinal nodes.

C. Superficial inguinal lymph nodes receive lymph from the lower abdominal wall, buttocks, penis, scrotum, labium majus, and the lower parts of the vagina and anal canal. Their efferent vessels enter primarily to the external iliac nodes and ultimately to the lumbar (aortic) nodes.

IX. Blood Vessels of the Anterior Abdominal Wall

A. Superior epigastric artery
– arises from the **internal thoracic artery,** enters the rectus sheath, and descends on the posterior surface of the rectus abdominis.
– anastomoses with the inferior epigastric artery within the rectus abdominis.

B. Inferior epigastric artery
– arises from the **external iliac artery** above the inguinal ligament, enters the rectus sheath, and ascends between the rectus abdominis and the posterior layer of the rectus sheath.

- anastomoses with the superior epigastric artery, providing collateral circulation between the subclavian and external iliac arteries.
- gives rise to the **cremasteric artery,** which accompanies the spermatic cord.

C. Deep circumflex iliac artery

- arises from the **external iliac artery** and runs laterally along the inguinal ligament and the iliac crest between the transverse and internal oblique muscles.
- Its ascending branch anastomoses with the **musculophrenic artery.**

D. Superficial epigastric arteries

- arise from the **femoral artery** and run superiorly toward the umbilicus over the inguinal ligament.
- anastomose with branches of the inferior epigastric artery.

E. Superficial circumflex iliac artery

- arises from the **femoral artery** and runs laterally upward, parallel to the inguinal ligament.
- anastomoses with the deep circumflex iliac and lateral femoral circumflex arteries.

F. Superficial (external) pudendal arteries

- arise from the **femoral artery,** pierce the cribriform fascia, and run medially to supply the skin above the pubis.

G. Thoracoepigastric veins

- are longitudinal venous connections between the lateral thoracic vein and the superficial epigastric vein.
- provide a collateral route for venous return if a caval or portal obstruction occurs.

X. Clinical Considerations

A. Inguinal hernia

- occurs superior to the inguinal ligament and medial to the pubic tubercle, whereas a **femoral hernia** occurs inferior to the ligament and lateral to the tubercle.

1. Indirect inguinal hernia

- passes through the deep inguinal ring, inguinal canal, and superficial inguinal ring and descends into the scrotum.
- lies lateral to the inferior epigastric vessels.
- is **congenital** and is associated with the persistence of the processus vaginalis.
- is found more commonly on the right side in men and is more common than a direct inguinal hernia.
- is covered by the peritoneum and the coverings of the spermatic cord.

2. Direct inguinal hernia

- occurs through the **posterior wall of the inguinal canal** (in the region of the inguinal triangle) but does not descend into the scrotum.
- lies medial to the inferior epigastric vessels and protrudes forward to (but rarely through) the superficial inguinal ring.

- is **acquired** (develops after birth) and is associated with weakness in the posterior wall of the inguinal canal lateral to the falx inguinalis.
- Its sac is formed by the peritoneum.

B. Umbilical hernia

- may occur due to failure of the midgut to return to the abdomen early in fetal life (**exomphalos or omphalocele**).
- may also occur as a **protrusion of the bowel or omentum** through the abdominal wall at the umbilicus, due to incomplete closure of the anterior abdominal wall after ligation of the umbilical cord at birth.

C. Epigastric hernia

- is a **protrusion of extraperitoneal fat or a small piece of greater omentum** through a defect in the linea alba above the umbilicus.

D. Cremasteric reflex

- is a **drawing up of the testis** by contraction of the cremaster muscle when the skin on the upper anteromedial side of the thigh is stroked.
- The efferent limb (of the reflex arc) is the **genital branch of the genitofemoral nerve**; the afferent limb is the **femoral branch of the genitofemoral nerve**.

Peritoneum and Peritoneal Cavity

I. Peritoneum

- is a **serous membrane** lined by mesothelial cells.
- consists of the **parietal peritoneum,** which lines the abdominal and pelvic walls and the inferior surface of the diaphragm, and the **visceral peritoneum,** which covers the viscerae.
- Its **parietal layer** is innervated by the phrenic, lower intercostal, subcostal, iliohypogastric, and ilioinguinal nerves; the **visceral layer** is innervated by visceral nerves, which travel along autonomic pathways, and is relatively insensitive to pain.

II. Peritoneal Reflections (Figures 5-4 and 5-5)

- support the viscera and provide pathways for associated neurovascular structures.

A. Omentum

1. Lesser omentum

- is a **double layer of peritoneum** extending from the porta hepatis of the liver to the lesser curvature of the stomach and the beginning of the duodenum.
- consists of the **hepatogastric and hepatoduodenal ligaments** and forms the anterior wall of the lesser sac of the peritoneal cavity.
- The left and right gastric vessels run between its two layers.
- Its right free margin contains the **proper hepatic artery, bile duct,** and **portal vein.**

2. Greater omentum

- hangs down like an apron from the greater curvature of the stomach, covering the transverse colon and other abdominal viscerae.

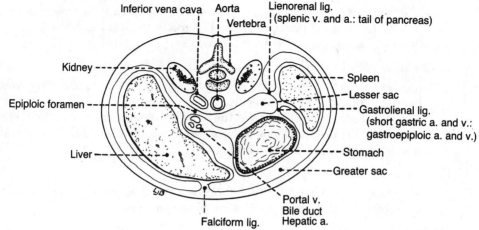

Figure 5-4. Horizontal section of the upper abdomen.

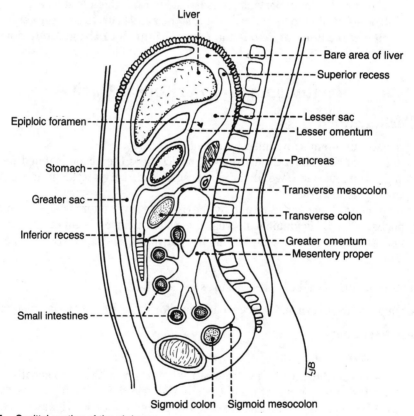

Figure 5-5. Sagittal section of the abdomen.

- is often referred to by surgeons as the "abdominal policeman" because it plugs the neck of a hernial sac, preventing the entrance of coils of the small intestine.
- adheres to areas of inflammation and wraps itself around the inflamed organs, thus preventing serious diffuse peritonitis.

B. Mesenteries

1. Mesentery of the small intestine (mesentery proper)

- is a **double fold of peritoneum** that suspends the jejunum and ileum from the posterior abdominal wall and transmits nerves and blood vessels to and from the small intestine.
- Its root extends from the duodenojejunal flexure to the right iliac fossa and is about 15 cm (6 inches) long.
- Its free border encloses the **small intestine,** which is about 6 m (20 feet) long.
- contains the superior mesenteric and intestinal (jejunal and ileal) vessels, nerves, and lymphatics.

2. Transverse mesocolon

- connects the posterior surface of the **transverse colon** to the posterior body wall.
- fuses with the greater omentum to form the **gastrocolic ligament.**
- contains the middle colic vessels, nerves, and lymphatics.

3. Sigmoid mesocolon

- is an inverted V-shaped peritoneal fold that connects the **sigmoid colon** to the pelvic wall and contains the sigmoid vessels.

4. Mesoappendix

- connects the **appendix** to the mesentery of the ileum and contains the appendicular vessels.

C. Peritoneal ligaments

1. Lienogastric (gastrosplenic) ligament

- extends from the left portion of the greater curvature of the stomach to the hilus of the spleen.
- contains the short gastric vessels and the left gastroepiploic vessels.

2. Lienorenal ligament

- runs from the hilus of the spleen to the left kidney.
- contains the splenic vessels and the tail of the pancreas.

3. Gastrophrenic ligament

- runs from the upper part of the greater curvature of the stomach to the diaphragm.

4. Gastrocolic ligament

- runs from the greater curvature of the stomach to the transverse colon.

5. Phrenicocolic ligament

- runs from the colic flexure to the diaphragm.

6. Falciform ligament

- is a **sickle-shaped peritoneal fold** connecting the **liver** to the diaphragm and the anterior abdominal wall.
- appears to demarcate the right lobe from the left lobe of the liver on the diaphragmatic surface.
- contains the **ligamentum teres hepatis** and the **paraumbilical vein,** which connects the left branch of the portal vein with the subcutaneous veins in the region of the umbilicus.

7. Ligamentum teres hepatis (round ligament of the liver)

- lies in the free margin of the falciform ligament and ascends from the umbilicus to the inferior (visceral) surface of the liver, lying in the fissure that forms the left boundary of the quadrate lobe of the liver.
- is formed after birth from the remnant of the **left umbilical vein,** which carries oxygenated blood from the placenta to the left branch of the portal vein in the fetus. (The right umbilical vein is obliterated during the embryonic period.)

8. Coronary ligament

- is a peritoneal reflection from the diaphragmatic surface of the liver onto the diaphragm and encloses a triangular area of the right lobe, the **bare area of the liver.**
- Its right and left extensions form the **right and left triangular ligaments,** respectively.

9. Ligamentum venosum

- is the fibrous remnant of the **ductus venosus.**
- lies in the fissure on the inferior surface of the liver, forming the left boundary of the **caudate lobe of the liver.**

D. Peritoneal folds

- are peritoneal reflections with free edges.

1. Umbilical folds

- are five folds of peritoneum below the umbilicus, including the median, medial, and lateral umbilical folds.

2. Rectouterine fold

- extends from the cervix of the uterus, along the side of the rectum, to the posterior pelvic wall, forming the rectouterine pouch (of Douglas).

3. Ileocecal fold

- extends from the terminal ileum to the cecum.

III. Peritoneal Cavity (see Figures 5-4 and 5-5)

- is a **potential space** between the parietal and visceral peritoneum.
- contains a film of fluid that lubricates the surface of the peritoneum and facilitates free movements of the viscerae.
- is a completely closed sac in the male but communicates with the exterior through the openings of the uterine tubes in the female.
- is divided into the lesser and greater sacs.

A. Lesser sac (omental bursa)

- is an irregular space that lies behind the liver, lesser omentum, stomach, and upper anterior part of the greater omentum.
- is a closed sac, except for its communication with the greater sac through the **epiploic foramen.**
- presents three recesses:

1. **Superior recess** lies behind the stomach, lesser omentum, and liver.

2. **Inferior recess** lies behind the stomach, extending into the layers of the greater omentum.

3. **Splenic recess** extends to the left at the hilus of the spleen.

B. Greater sac
 – extends across the entire breadth of the abdomen and from the diaphragm to the pelvic floor.
 – presents numerous recesses:

1. **Subphrenic (suprahepatic) recess**
    – is a peritoneal pocket between the diaphragm and the anterior and superior part of the liver.
    – is separated into right and left recesses by the **falciform ligament.**

2. **Subhepatic recess**
    – is a peritoneal pocket between the liver and the transverse colon.

3. **Hepatorenal recess**
    – is a deep peritoneal pocket between the liver anteriorly and the kidney and suprarenal gland posteriorly.

4. **Morison's pouch**
    – is formed by the right subhepatic recess and the hepatorenal recess.
    – communicates with the subphrenic recess, the lesser sac via the epiploic foramen, and the right paracolic gutter and thus the pelvic cavity.

5. **Paracolic recesses (gutters)**
    – lie lateral to the ascending colon (right paracolic gutter) and lateral to the descending colon (left paracolic gutter).

C. Epiploic (Winslow's) foramen
   – is a natural opening between the lesser and greater sacs.
   – is bounded superiorly by peritoneum on the caudate lobe of the liver, inferiorly by peritoneum on the first part of the duodenum, anteriorly by the free edge of the lesser omentum, and posteriorly by peritoneum covering the inferior vena cava.

Gastrointestinal Viscera

I. Stomach (Figure 5-6)

 – rests, in the supine position, on the **stomach bed,** which is formed by the pancreas, spleen, left kidney, left suprarenal gland, transverse colon and its mesocolon, and diaphragm.
 – is covered entirely by peritoneum and is located in the left hypochondriac and epigastric regions of the abdomen.
 – has **greater and lesser curvatures,** anterior and posterior walls, cardiac and pyloric openings, and cardiac and angular notches.
 – is divided into four regions: the **cardia, fundus, body,** and **pylorus.** The pylorus is divided into the **pyloric antrum** and **canal.** The **pyloric orifice** is surrounded by a thickened ring of gastric circular muscle, the **pyloric sphincter.**

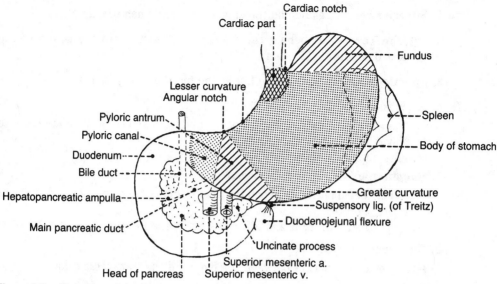

Figure 5-6. Stomach and duodenum.

- receives blood from the right and left gastric, right and left gastroepiploic, and short gastric arteries.
- When the stomach is contracted, longitudinal folds of mucous membrane, the **rugae,** appear.
- The **gastric canal,** a grooved channel along the lesser curvature formed by the rugae, directs fluids toward the pylorus.
- produces **acid** and **pepsin** in its fundus and body, and produces the hormone **gastrin** in its pyloric antrum.

II. Small Intestine (Figure 5-7)

- extends from the pyloric opening to the ileocecal junction.
- is the location of **complete digestion and absorption** of most of the products of digestion, as well as water, electrolytes, and minerals such as calcium and iron.
- consists of the **duodenum, jejunum,** and **ileum.**

A. Duodenum

- is a C-shaped tube surrounding the head of the pancreas, and is the shortest (25 cm [10 inches] long) but widest part of the small intestine.
- is retroperitoneal except for the beginning of the first part, which is connected to the liver by the **hepatoduodenal ligament.**
- receives blood from the celiac (foregut) and superior mesenteric (midgut) artery.
- is divided into four parts:

1. Superior (first) duodenum

- has a mobile or free section, termed the **duodenal cap** (because of its appearance on radiographs), into which the pylorus invaginates.

2. Descending (second) duodenum

- contains the **junction of the foregut and midgut,** where the **common bile and main pancreatic ducts** open.
- contains the **greater papilla,** on which terminal openings of the bile and main pancreatic ducts are located, and the **lesser papilla,** which lies 2 cm

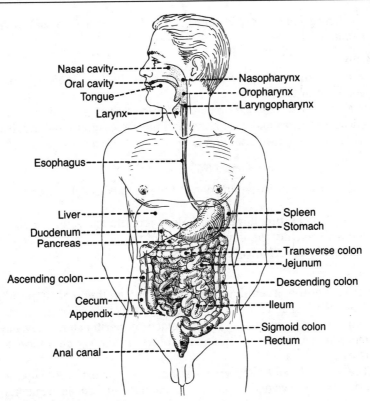

Figure 5-7. Diagram of the digestive system.

above the greater papilla and marks the site of entry of the **accessory pancreatic duct.**

3. **Transverse (third) portion**
   – is the longest part and crosses the inferior vena cava, aorta, and vertebral column to the left.
   – is crossed anteriorly by the superior mesenteric vessels.

4. **Ascending (fourth) portion**
   – ascends to the left of the aorta to the level of the second lumbar vertebra and terminates at the duodenojejunal junction, which is fixed in position by the **suspensory ligament (of Treitz),** a surgical landmark. This fibromuscular band is attached to the right crus of the diaphragm.

B. **Jejunum**
   – makes up the proximal two-fifths of the small intestine (the ileum makes up the distal three-fifths).
   – is emptier, larger in diameter, and thicker-walled than the ileum.
   – has the **plicae circulares** (circular folds), which are tall and closely packed.
   – contains no Peyer's patches (aggregations of lymphoid tissue).
   – has translucent areas called **windows** between the blood vessels of its mesentery.
   – has less prominent **arterial arcades** (anastomotic loops) in its mesentery than does the ileum.

– has longer **vasa recta** (straight arteries, or arteriae rectae) than those of the ileum.

C. Ileum

– is longer than the jejunum and occupies the **false pelvis** in the right lower quadrant of the abdomen.
– is characterized by the presence of **Peyer's patches** (lower portion), shorter plicae circulares and vasa recta, and more mesenteric fat and arterial arcades when compared with the jejunum.

III. Large Intestine (see Figure 5-7)

– extends from the ileocecal junction to the anus and is approximately 1.5 m (5 feet) long.
– consists of the **cecum, appendix, colon, rectum,** and **anal canal.**
– functions to convert the liquid contents of the ileum into semisolid feces by **absorbing fluid and electrolytes.**

A. Colon

– consists of the **ascending, transverse, descending,** and **sigmoid** segments.
– The ascending and descending segments are retroperitoneal, and the transverse and sigmoid segments are surrounded by peritoneum (they have their own mesenteries, the **transverse mesocolon** and the **sigmoid mesocolon,** respectively).
– The ascending and transverse segments are supplied by the superior mesenteric artery and the vagus nerve; the descending and sigmoid segments are supplied by the inferior mesenteric artery and the pelvic splanchnic nerves.
– is characterized by the following:

1. Teniae coli: three narrow bands of the outer longitudinal muscular coat

2. Sacculations or haustra: produced by the teniae, which are slightly shorter than the gut

3. Epiploic appendages: peritoneum-covered sacs of fat, attached in rows along the teniae

B. Cecum

– is the **blind pouch of the large intestine,** lies in the right iliac fossa, and is usually surrounded by peritoneum.

C. Appendix

– is a **narrow, hollow, muscular tube** with large aggregations of lymphoid tissue in its wall.
– Its base lies deep to **McBurney's point,** the junction of the lateral one-third of a line between the right anterior–superior iliac spine and the umbilicus, where pressure of the finger elicits tenderness in **acute appendicitis.**
– is suspended from the terminal ileum by a small mesentery, the **mesoappendix,** which contains the appendicular vessels.
– When inflamed, causes spasm and distension, resulting in **pain** that is referred to the epigastrium.

D. Rectum and anal canal

– extend from the sigmoid colon to the anus.
– are described as **pelvic organs.**

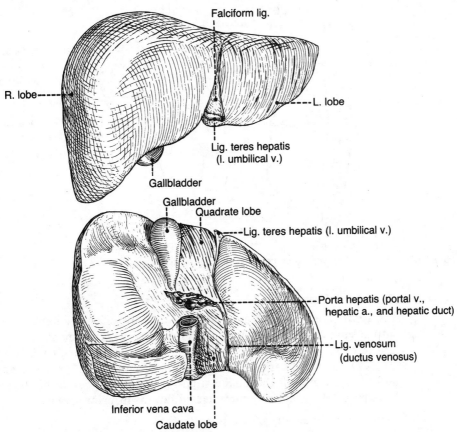

Figure 5-8. Anterior and visceral surfaces of the liver.

IV. Accessory Organs of the Digestive System

A. Liver (Figures 5-8 and 5-9)

- is the **largest visceral organ** and the **largest gland** in the body.
- plays an important role in **bile production and secretion, detoxification** (by filtering the blood to remove bacteria and foreign particles that have gained entrance from the intestine), **blood-clotting mechanisms,** and **storage** of glycogen, vitamin, iron, and copper. In the fetus, it is important in the manufacture of red blood cells.
- is surrounded by the peritoneum and is attached to the diaphragm by the **coronary and falciform ligaments** and the right and left **triangular ligaments.**
- has a **bare area** on the diaphragmatic surface, which is limited by layers of the coronary ligament but is devoid of peritoneum.
- receives blood from the hepatic artery and portal vein; its venous blood is drained by the hepatic veins in the inferior vena cava.
- appears to be divided into right and left lobes on the diaphragmatic surface by the coronary ligament. (These lobes serve as landmarks but do not correspond to the structural units or hepatic segments.)
- is divided, on the basis of hepatic drainage and blood supply, into the **right and left lobes** by the fossae for the gallbladder and the inferior vena cava.

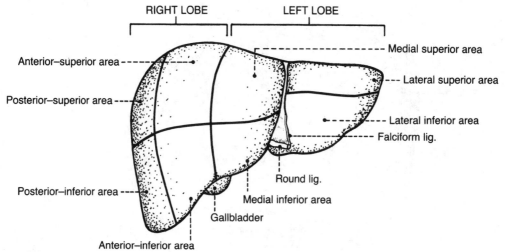

Figure 5-9. Divisions of the liver, based on hepatic drainage and blood supply.

1. Lobes of the liver

- The **right lobe** is divided into **anterior** and **posterior segments;** each in turn is subdivided into superior and inferior areas or segments.
- The **left lobe** is divided into **medial** and **lateral segments,** each of which is subdivided into superior and inferior areas (segments). Thus, the segments of the left lobe include the **medial superior (caudate lobe), medial inferior (quadrate lobe), lateral superior,** and **lateral inferior segments.**

2. Fissures and ligaments of the liver

- include an H-shaped group of fissures:

a. Fissure for the round ligament (**ligamentum teres hepatis**), located between the left lobe and the quadrate lobe

b. Fissure for the **ligamentum venosum,** located between the left lobe and the caudate lobe

c. Fossa for the **gallbladder,** located between the quadrate lobe and the major part of the right lobe

d. Fissure for the **inferior vena cava,** located between the caudate lobe and the major part of the right lobe

e. Porta hepatis (the crossbar of the H) for the hepatic ducts, the proper hepatic artery, and the branches of the portal vein

B. Gallbladder (Figure 5-10)

- is a **pear-shaped sac** lying on the inferior surface of the liver in a fossa between the right and quadrate lobes; has a capacity of about 30 ml.
- consists of the **fundus** (the rounded blind end); the **body** (the major part); and the **neck** (the narrow part), which gives rise to the **cystic duct** with spiral valves.
- receives bile, stores it, and concentrates it by absorbing water and salts.
- **contracts to expel bile** as a result of stimulation by the hormone **cholecystokinin,** which is produced by the duodenal mucosa when food arrives in the duodenum.

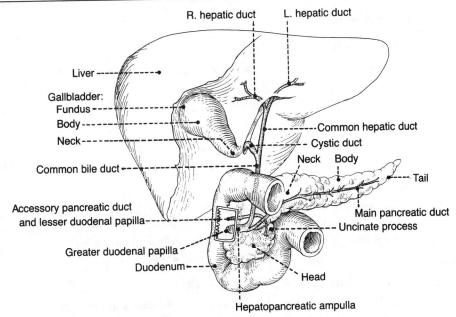

Figure 5-10. Extrahepatic bile passages and pancreatic ducts.

    – receives blood from the cystic artery, which arises from the right hepatic artery within the **cystohepatic triangle (of Calot),** which is formed by the visceral surface of the liver superiorly, the cystic duct inferiorly, and the common hepatic duct medially.

C. Pancreas (see Figure 5-10)

    – lies largely in the floor of the lesser sac in the epigastric and left hypochondriac regions, where it forms a major portion of the stomach bed.

    – is a retroperitoneal organ except for a small portion of its **tail,** which lies in the **lienorenal ligament.**

    – The **head** lies within the C-shaped concavity of the duodenum.

    – The **uncinate process** is a projection of the lower part of the head to the left behind the superior mesenteric vessels.

    – receives blood from branches of the splenic artery and from the superior and inferior pancreaticoduodenal arteries.

    – is both an **exocrine gland,** which produces digestive enzymes, and an **endocrine gland,** which secretes two hormones, insulin and glucagon.

    – The **main pancreatic duct (duct of Wirsung)** begins in the tail, runs to the right along the entire pancreas, and joins the bile duct to form the **hepatopancreatic ampulla (ampulla of Vater)** before entering the second part of the duodenum at the greater papilla.

    – The **accessory pancreatic duct (Santorini's duct)** begins in the lower portion of the head and empties at the lesser duodenal papilla about 2 cm above the greater papilla.

    – If tumors are present in the head, bile flow will be obstructed, resulting in **jaundice.**

D. Duct system for bile passage

1. Right and left hepatic ducts

- are formed by union of the **intrahepatic ductules** from each lobe of the liver.
- **drain bile** from the corresponding halves of the liver.

2. Common hepatic duct

- is formed by union of the right and left hepatic ducts.
- is accompanied by the proper hepatic artery and the portal vein.

3. Cystic duct

- has spiral folds (valves) to keep it constantly open, and thus bile can pass upward into the **gallbladder** when the common bile duct is closed.
- runs alongside the hepatic duct before joining the common hepatic duct.
- is a common site of impaction of **gallstones.**

4. Common bile duct (ductus choledochus)

- is formed by union of the common hepatic duct and the cystic duct.
- is located lateral to the proper hepatic artery and anterior to the portal vein in the right free margin of the lesser omentum.
- descends behind the first part of the duodenum and runs through the head of the pancreas.
- joins the main pancreatic duct to form the **hepatopancreatic duct (hepato-pancreatic ampulla),** which enters the second part of the duodenum at the greater papilla.
- The **sphincter of Boyden** is a circular muscle layer around its lower end.

5. Hepatopancreatic duct or ampulla (ampulla of Vater)

- is formed by union of the common bile duct and the main pancreatic duct and enters the second part of the duodenum at the greater papilla. This represents the junction of the embryonic foregut and midgut.
- The **sphincter of Oddi** is a circular muscle layer around it in the greater duodenal papilla.

V. Spleen

- is a **large lymphatic organ** lying against the diaphragm and ribs 9 to 11 in the left hypochondriac region.
- is developed in the dorsal mesogastrium and supported by the **lienogastric and lienorenal ligaments.**
- is composed of **white pulp,** which consists of lymphatic nodules and diffuse lymphatic tissue, and **red pulp,** which consists of venous sinusoids connected by splenic cords.
- is **hematopoietic** in early life and later functions in worn-out **red blood cell destruction.**
- **filters blood** (removes particulate matter and cellular residue from the blood), **stores red corpuscles,** and **produces lymphocytes and antibodies.**
- is supplied by the splenic artery and is drained by the splenic vein.
- may be removed surgically with minimal effect on body function.

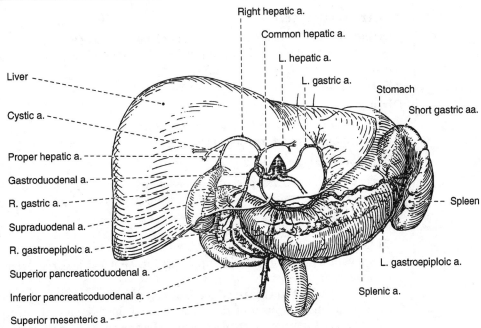

Figure 5-11. Branches of the celiac trunk.

VI. Celiac and Mesenteric Arteries

A. Celiac trunk (Figure 5-11)

– arises from the front of the abdominal aorta immediately below the aortic hiatus of the diaphragm, between the right and left crura.

– divides into the left gastric, splenic, and common hepatic arteries.

1. Left gastric artery

– is the **smallest branch** of the celiac trunk.

– runs upward and to the left toward the cardia, giving rise to **esophageal and hepatic branches,** and then turns to the right and runs along the lesser curvature within the lesser omentum to anastomose with the right gastric artery.

2. Splenic artery

– is the **largest branch** of the celiac trunk.

– runs a highly tortuous course along the superior border of the pancreas and enters the lienorenal ligament.

– gives rise to:

a. A number of pancreatic branches, including the **dorsal pancreatic artery**

b. A few **short gastric arteries,** which pass through the lienogastric ligament to reach the fundus of the stomach

c. The **left gastroepiploic artery,** which reaches the greater omentum through the lienogastric ligament and runs along the greater curvature of the stomach to distribute to the stomach and greater omentum

3. Common hepatic artery

– runs to the right along the upper border of the pancreas and divides into the proper hepatic artery, the gastroduodenal artery, and probably the right gastric artery.

a. Proper hepatic artery

– ascends in the free edge of the lesser omentum and divides, near the porta hepatis, into the **left and right hepatic arteries;** the right hepatic artery gives rise to the **cystic artery,** which supplies the gallbladder.

– gives rise, near its beginning, to the right gastric artery.

b. Right gastric artery

– runs to the pylorus and then along the lesser curvature of the stomach and anastomoses with the left gastric artery.

c. Gastroduodenal artery

– descends behind the first part of the duodenum, giving off the supraduodenal artery to its superior aspect and a few retroduodenal arteries to its inferior aspect.

– divides into two major branches:

(1) The **right gastroepiploic artery** runs to the left along the greater curvature of the stomach, supplying the stomach and the greater omentum.

(2) The **superior pancreaticoduodenal artery** passes between the duodenum and the head of the pancreas and further divides into the anterior–superior pancreaticoduodenal artery and the posterior–superior pancreaticoduodenal artery.

B. Superior mesenteric artery (Figure 5-12)

– arises from the aorta behind the neck of the pancreas.

– descends across the uncinate process of the pancreas and the third part of the duodenum and then enters the root of the mesentery behind the transverse colon to run to the right iliac fossa.

– gives rise to the following branches:

1. Inferior pancreaticoduodenal artery

– passes to the right and divides into the anterior–inferior pancreaticoduodenal artery and the posterior–inferior pancreaticoduodenal artery, which anastomose with the corresponding branches of the superior pancreaticoduodenal artery.

2. Middle colic artery

– enters the transverse mesocolon and divides into the **right branch,** which anastomoses with the right colic artery, and the **left branch,** which anastomoses with the ascending branch of the left colic artery. The branches of the mesenteric arteries form an anastomotic channel, the **marginal artery,** along the large intestine.

3. Ileocolic artery

– descends behind the peritoneum toward the right iliac fossa and ends by dividing into the **ascending colic artery,** which anastomoses with the right colic artery, **anterior and posterior cecal arteries, the appendicular artery,** and **ileal branches.**

4. Right colic artery

– arises from the superior mesenteric artery or the ileocolic artery.

– runs to the right behind the peritoneum and divides into **ascending and descending branches,** distributing to the ascending colon.

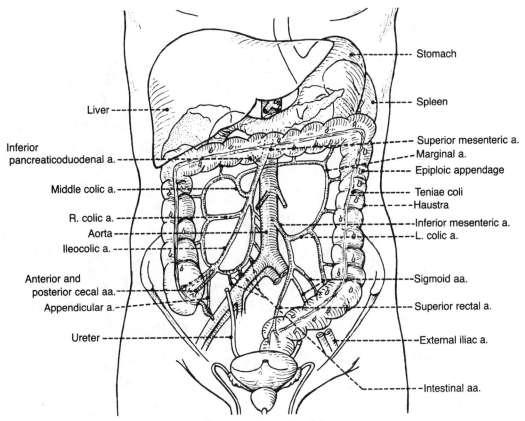

Figure 5-12. Branches of the superior and inferior mesenteric arteries.

5. Intestinal arteries

– are 12 to 15 in number and supply the jejunum and ileum.

– branch and anastomose to form a series of arcades in the mesentery.

C. Inferior mesenteric artery (see Figure 5-12)

– passes to the left behind the peritoneum and distributes to the descending and sigmoid colons and the upper portion of the rectum.

– gives rise to:

1. Left colic artery

– runs to the left behind the peritoneum toward the descending colon and divides into **ascending and descending branches.**

2. Sigmoid arteries

– are two to three in number, run toward the sigmoid colon in its mesentery, and divide into **ascending and descending branches.**

3. Superior rectal artery

– is the **termination of the inferior mesenteric artery,** descends into the pelvis, divides into two branches that follow the sides of the rectum, and anastomoses with the middle and inferior rectal arteries. (The middle and inferior rectal arteries arise from the internal iliac and internal pudendal arteries, respectively.)

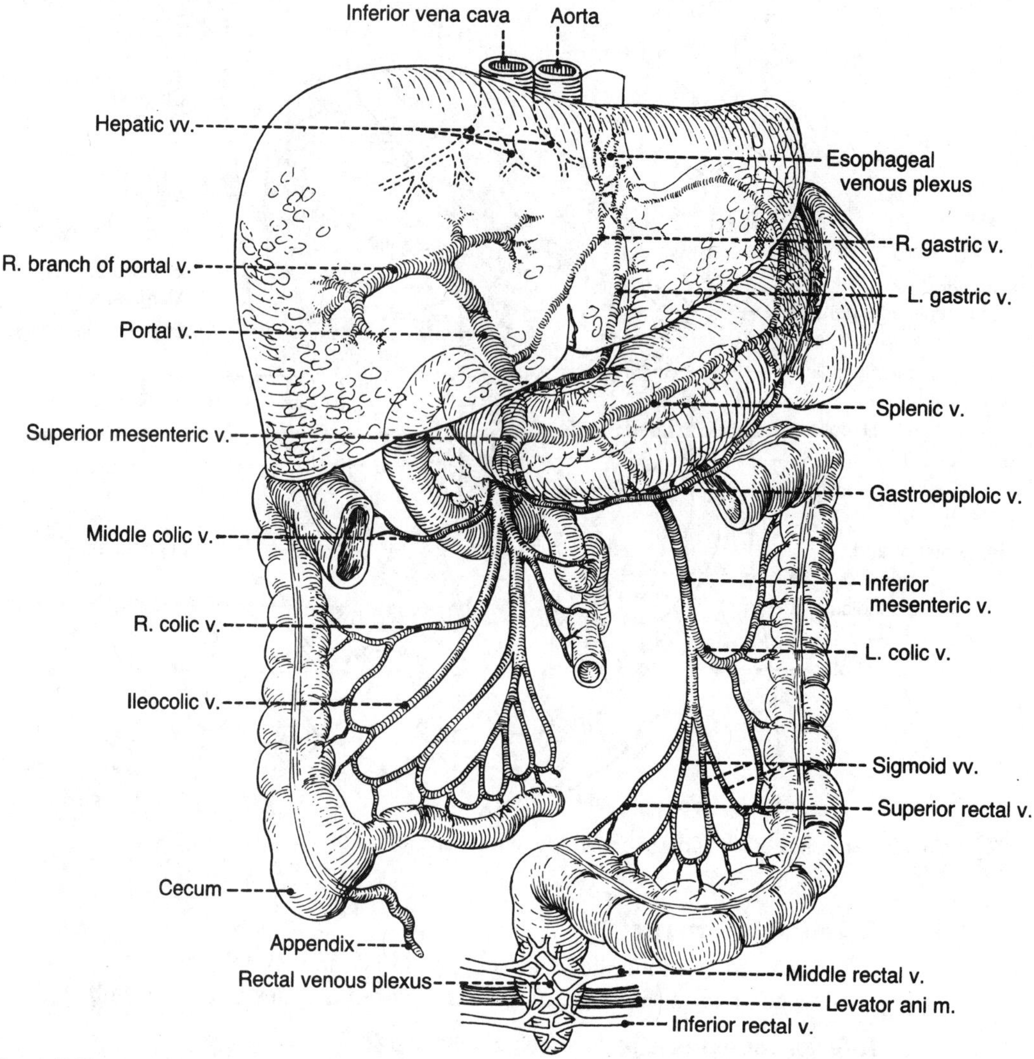

Figure 5-13. Portal venous system.

VII. Portal Venous System (Figure 5-13)

A. Portal vein

- drains the abdominal part of the gut, spleen, pancreas, and gallbladder and is 8 cm (3.2 inches) long.
- is formed by the union of the **splenic vein** and the **superior mesenteric vein** posterior to the neck of the pancreas. The **inferior mesenteric vein** joins either the splenic or the superior mesenteric vein or the junction of these two veins.
- receives the **left gastric (or coronary) vein.**
- carries deoxygenated blood containing nutrients.
- carries twice as much blood as the hepatic artery and maintains a higher blood pressure than in the inferior vena cava.
- ascends behind the bile duct and hepatic artery within the free margin of the lesser omentum.

1. **Superior mesenteric vein**
   – accompanies the superior mesenteric artery on its right side in the root of the mesentery.
   – crosses the third part of the duodenum and the uncinate process of the pancreas and terminates posterior to the neck of the pancreas by joining the splenic vein, thereby forming the portal vein.
   – Its tributaries are some of the veins that accompany the branches of the superior mesenteric artery.

2. **Splenic vein**
   – is formed by the union of tributaries from the spleen and receives the short gastric, left gastroepiploic, and pancreatic veins.

3. **Inferior mesenteric vein**
   – is formed by the union of the superior rectal and sigmoid veins and receives the left colic vein.

4. **Left gastric (coronary) vein**
   – drains normally into the portal vein.
   – Its **esophageal tributaries** anastomose with the esophageal veins of the azygos system at the lower part of the esophagus and thereby enter the systemic venous system.

5. **Paraumbilical veins**
   – are found in the **falciform ligament** and are virtually closed; however, they dilate in **portal hypertension.**
   – connect the left branch of the portal vein with the small subcutaneous veins in the region of the umbilicus, which are radicles of the superior epigastric, inferior epigastric, thoracoepigastric, and superficial epigastric veins.

B. **Important portal–caval (systemic) anastomoses**
   – occur between:

   1. The left gastric vein and the esophageal vein of the azygos system

   2. The superior rectal vein and the middle and inferior rectal veins

   3. The paraumbilical veins and radicles of the epigastric (superficial and inferior) veins

   4. The retroperitoneal veins draining the colon and twigs of the renal, suprarenal, and gonadal veins

VIII. Clinical Considerations

A. **Gastric ulcer**
   – erodes the mucosa and penetrates the gastric wall to various depths.
   – may perforate into the lesser sac and erode the pancreas and the splenic artery, causing **fatal hemorrhage.**

B. **Duodenal ulcer**
   – penetrates the wall of the superior duodenum, erodes the gastroduodenal artery, and is commonly located in the duodenal cap.

C. **Meckel's diverticulum**
   – is an evagination of the **terminal part of the ileum** located on the antimesenteric side of the ileum; it occurs in about 2% of the population.

– may contain **gastric and pancreatic tissues** in its wall.

– represents persistent portions of the **embryonic yolk stalk** (vitelline, or the omphalomesenteric duct) that are present in some adults.

– is clinically important because **bleeding** may occur from an ulcer in its wall.

D. Liver cirrhosis

– is a condition in which liver cells are progressively destroyed and replaced by fibrous tissue that surrounds the intrahepatic blood vessels and biliary radicles, impeding the circulation of blood through the liver.

– causes **portal hypertension,** resulting in esophageal varices, hemorrhoids, and caput medusae.

E. Portal hypertension

– results from **thrombosis of the portal vein** or **liver cirrhosis.**

– causes a dilatation of veins in the lower part of the esophagus, forming **esophageal varices.** Their rupture results in vomiting of blood (**hematemesis).**

– results in **caput medusae** (dilated veins radiating from the umbilicus), which occurs because the paraumbilical veins enclosed in the free margin of the falciform ligament anastomose with branches of the epigastric (superficial and inferior) veins around the umbilicus.

– may result in **hemorrhoids** because of enlargement of veins around the anal canal.

– can be reduced by diverting blood from the portal to the caval system; this is done by anastomosing the splenic vein to the renal vein or by creating a communication between the portal vein and the inferior vena cava.

F. Gallstones

– are formed by solidification of bile constituents.

– are composed chiefly of **cholesterol crystals,** usually mixed with bile pigments and calcium.

– are most common in fat, fertile (multiparous) females over forty years old (**4-F individual).**

– may become lodged in the **fundus of the gallbladder,** which may ulcerate through the wall of the gallbladder into the transverse colon or into the duodenum. In the former case, they are passed naturally to the rectum, but in the latter case they may be held up at the **ileocecal junction,** producing an **intestinal obstruction.**

– may become lodged in the **common bile duct,** obstructing bile flow to the duodenum and leading to **jaundice.**

– may become lodged in the **hepatopancreatic ampulla,** blocking both the biliary and the pancreatic duct systems. In this case, bile may enter the pancreatic duct system, causing aseptic or noninfectious **pancreatitis.**

G. Megacolon (Hirschsprung's disease)

– is caused by the **absence of enteric ganglia in the lower part of the colon,** which leads to dilatation of the colon proximal to the inactive segment.

– is of **congenital** origin and is usually diagnosed during infancy and childhood; symptoms are constipation, abdominal distension, and vomiting.

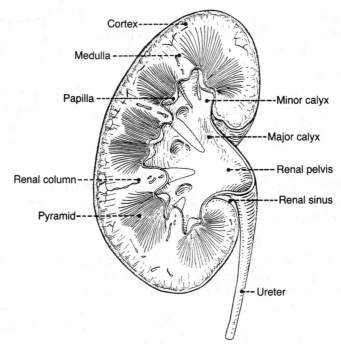

Figure 5-14. Frontal section of the kidney.

Retroperitoneal Viscera, Diaphragm, and Posterior Abdominal Wall

I. Kidney, Ureter, and Suprarenal Gland

A. Kidney (Figure 5-14)

- is retroperitoneal and extends from L1–L4 in the erect position. The right kidney lies a little lower than the left, owing to the large size of the right lobe of the liver, and usually is related to rib 12 posteriorly. The left kidney is related to ribs 11 and 12 posteriorly.
- is invested by a firm fibrous **capsule.**
- is surrounded by a mass of fat and fibrous fascia (**the renal fascia**), which divides the fat into two regions. The **perirenal fat** lies in the **perinephric space** between the capsule of the kidney and the renal fascia, and the **pararenal fat** lies external to the renal fascia.
- has an indentation, the **hilus,** on its medial border, through which the ureter, renal vessels, and nerves enter or leave the organ.
- consists of the **medulla** and the **cortex,** containing 1 to 2 million **nephrons,** which are the anatomical and functional units of the kidney. Each nephron consists of a **renal corpuscle, a proximal convoluted tubule, Henle's loop, and a distal convoluted tubule.**
- Its arterial segments—**superior, anterosuperior, anteroinferior, inferior, and posterior segments**—are of surgical importance.

1. Cortex

- forms the outer part of the kidney and also projects into the medullary region between the renal pyramids as **renal columns.**

– contains renal corpuscles and proximal and distal convoluted tubules. (The **renal corpuscle** consists of a tuft of capillaries (the glomerulus), surrounded by a **glomerular capsule,** which is the invaginated blind end of the nephron.)

2. **Medulla**

– forms the inner part of the kidney and consists of 8 to 12 **renal pyramids,** which contain straight tubules (**Henle's loops**) and **collecting tubules.** An apex of the renal pyramid, the **renal papilla,** fits into the cup-shaped **minor calyx** on which the collecting tubules open.

3. **Minor calyces**

– receive urine from the collecting tubules and empty into two or three **major calyces,** which in turn empty into an upper dilated portion of the ureter, the **renal pelvis.**

B. **Ureter**

– is a **muscular tube** that extends from the kidney to the **urinary bladder.**
– is retroperitoneal, descends on the psoas muscle, is crossed by the gonadal vessels, and crosses the bifurcation of the common iliac artery.
– may be obstructed where it joins the renal pelvis (**ureteropelvic junction**), where it crosses the pelvic brim over the distal end of the common iliac artery, or where it enters the wall of the urinary bladder.
– receives blood from the aorta and from the renal, gonadal, common and internal iliac, umbilical, superior and inferior vesical, and middle rectal arteries.

C. **Suprarenal (adrenal) gland**

– is a retroperitoneal organ lying on the superomedial aspect of the kidney, and is surrounded by a capsule and renal fascia.
– is pyramidal on the right and semilunar on the left.
– receives arteries from three sources: the superior suprarenal artery from the **inferior phrenic artery;** the middle suprarenal from the abdominal **aorta;** and the inferior suprarenal artery from the **renal artery.**
– Its **medulla** is derived from embryonic neural crest cells, receives preganglionic sympathetic nerve fibers directly, and secretes epinephrine and norepinephrine.
– Its **cortex** is essential to life and produces steroid hormones, including mineralocorticoids (aldosterone), glucocorticoids (such as cortisone), and sex hormones.
– It is drained via the suprarenal vein, which empties into the **inferior vena cava** on the right and the **renal vein** on the left.

II. Posterior Abdominal Blood Vessels and Lymphatics

A. **Aorta** (Figure 5-15)

– passes through the **aortic hiatus** in the diaphragm at the level of T12, descends anterior to the vertebral bodies, and bifurcates into the **right and left common iliac arteries** anterior to L4.
– gives rise to:

1. **Inferior phrenic arteries**

– arise from the aorta immediately below the aortic hiatus and give rise to the **superior suprarenal arteries.**
– diverge across the crura of the diaphragm, the left artery passing posterior to the esophagus and the right passing posterior to the inferior vena cava.

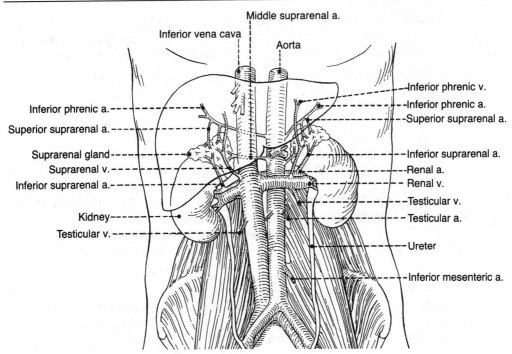

Inferior vena cava

Middle suprarenal a.

Aorta

Inferior phrenic a.

Superior suprarenal a.

Suprarenal gland

Suprarenal v.

Inferior suprarenal a.

Kidney

Testicular v.

Inferior phrenic v.

Inferior phrenic a.

Superior suprarenal a.

Inferior suprarenal a.

Renal a.

Renal v.

Testicular v.

Testicular a.

Ureter

Inferior mesenteric a.

Figure 5-15. Abdominal aorta and its branches.

2. Middle suprarenal arteries

– arise from the aorta and run laterally on the crura of the diaphragm just superior to the renal arteries.

3. Renal arteries

– arise from the aorta inferior to the origin of the superior mesenteric artery. The right artery is longer and a little lower than the left and passes posterior to the inferior vena cava; the left artery passes posterior to the left renal vein.

– give rise to the **inferior suprarenal and ureteric arteries.**

– divide into the superior, anterosuperior, anteroinferior, inferior, and posterior segmental branches.

4. Testicular or ovarian arteries

– descend retroperitoneally and run laterally on the psoas major muscle and across the ureter.

a. The **testicular artery** accompanies the ductus deferens into the scrotum, where it supplies the spermatic cord, epididymis, and testis.

b. The **ovarian artery** enters the suspensory ligament of the ovary, supplies the ovary, and anastomoses with the ovarian branch of the uterine artery.

5. Lumbar arteries

– consist of four or five pairs that arise from the back of the aorta.

– run posterior to the sympathetic trunk, the inferior vena cava (on the right side), the psoas major muscle, the lumbar plexus, and the quadratus lumborum.

– divide into smaller anterior branches (to supply adjacent muscles) and larger posterior branches, which accompany the dorsal primary rami of the corresponding spinal nerves and divide into spinal and muscular branches.

6. Celiac artery and superior and inferior mesenteric arteries

B. Inferior vena cava

– is formed on the right side of L5 by the union of the two **common iliac veins,** below the bifurcation of the aorta.
– is longer than the abdominal aorta and ascends at the right side of the aorta.
– passes through the **opening for the inferior vena cava** in the central tendon of the diaphragm at the level of T8 and enters the right atrium of the heart.
– receives the right gonadal, suprarenal, and inferior phrenic veins. On the left side, these veins usually drain into the left renal vein.
– also receives the three (left, middle, and right) **hepatic veins.** The middle and left hepatic veins frequently unite for about 1 cm before entering the vena cava.

C. Cisterna chyli

– is the lower dilated end of the **thoracic duct** and lies posterior to and just to the right of the aorta, usually between two crura of the diaphragm.
– is formed by the **intestinal and lumbar lymph trunks.**

III. Nerves of the Posterior Abdominal Wall

A. Lumbar plexus (Figure 5-16)

– is formed by the union of the ventral rami of the first three lumbar nerves and a part of the fourth lumbar nerve.
– lies anterior to the transverse processes of the lumbar vertebrae within the substance of the psoas muscle.

1. Subcostal nerve (T12)

– runs behind the lateral lumbocostal arch and in front of the quadratus lumborum; penetrates the transverse abdominal muscle to run between it and the internal oblique muscle.
– innervates the **external oblique, internal oblique, transverse, rectus abdominis, and pyramidalis muscles.**

2. Iliohypogastric nerve (L1)

– emerges from the lateral border of the psoas muscle and runs in front of the quadratus lumborum; pierces the transverse abdominal muscle near the iliac crest to run between this muscle and the internal oblique muscle.
– pierces the internal oblique muscle and then continues medially deep to the external oblique muscle.
– innervates the internal oblique and transverse muscles of the abdomen and divides into an **anterior cutaneous branch,** which innervates the skin above the pubis, and a **lateral cutaneous branch,** which innervates the skin of the gluteal region.

3. Ilioinguinal nerve (L1)

– runs in front of the quadratus lumborum; pierces the transverse and then the internal oblique muscle to run between the internal and external oblique aponeuroses.

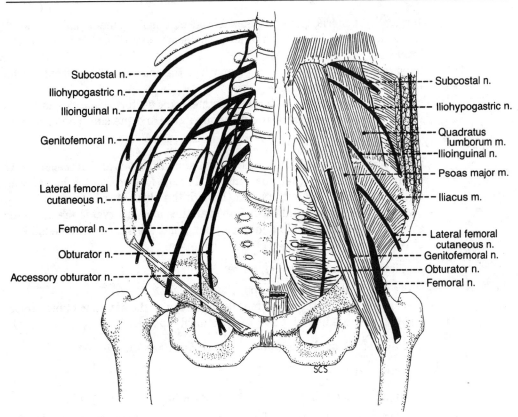

Figure 5-16. Lumbar plexus.

- accompanies the **spermatic cord (or the round ligament of the uterus),** continues through the inguinal canal, and emerges through the superficial inguinal ring.
- innervates the internal oblique and transverse muscles and gives off **femoral cutaneous branches** to the upper medial part of the thigh and **anterior scrotal or labial branches.**

4. **Genitofemoral nerve (L1–L2)**
   - emerges on the front of the psoas muscle and descends on its anterior surface.
   - divides into a **genital branch,** which enters the inguinal canal through the deep inguinal ring to supply the cremaster muscle and the scrotum (or labium majus), and a **femoral branch,** which supplies the skin of the femoral triangle.

5. **Lateral femoral cutaneous nerve (L2–L3)**
   - emerges from the lateral side of the psoas muscle and runs in front of the iliacus and behind the inguinal ligament.
   - innervates the skin of the anterior and lateral thigh.

6. **Femoral nerve (L2–L4)**
   - emerges from the lateral border of the psoas major and descends in the groove between the psoas and iliacus.

- enters the femoral triangle deep to the inguinal ligament and lateral to the femoral vessels outside the femoral sheath, and divides into numerous branches.
- innervates the skin of the thigh and leg, the muscles of the front of the thigh, and the hip and knee joints.
- innervates the quadriceps femoris, pectineal, and sartorius muscles and gives rise to the **anterior femoral cutaneous nerve** and the **saphenous nerve.**

7. **Obturator nerve (L2–L4)**

   - arises from the second, third, and fourth lumbar nerves and descends along the medial border of the psoas muscle. It runs forward on the lateral wall of the pelvis and enters the thigh through the **obturator foramen.**
   - divides into **anterior and posterior branches** and innervates the adductor group of muscles, the hip and knee joints, and the skin of the medial side of the thigh.

8. **Accessory obturator nerve (L3–L4)**

   - is present in about 9% of the population.
   - descends medial to the psoas muscle, passes over the superior pubic ramus, and supplies the hip joint and the pectineal muscle.

B. **Autonomic nerves in the abdomen** (Figure 5-17; see Figure 4-13)

   1. **Autonomic ganglia**

      a. **Sympathetic chain (paravertebral) ganglia**

         - are composed primarily of ascending and descending preganglionic sympathetic (GVE) fibers and visceral afferent fibers.
         - also contain cell bodies of the postganglionic sympathetic fibers.

      b. **Collateral (prevertebral) ganglia**

         - include the celiac, superior mesenteric, aorticorenal, and inferior mesenteric ganglia, usually located near the origin of the respective arteries.
         - are formed by cell bodies of the postganglionic sympathetic fibers.
         - receive preganglionic sympathetic fibers by way of the **greater, lesser, and least splanchnic nerves.**

   2. **Splanchnic nerves**

      a. **Thoracic splanchnic nerves**

         - contain preganglionic sympathetic (GVE) fibers with cell bodies located in the lateral horn (intermediolateral cell column) of the spinal cord, and also contain general visceral afferent (GVA) fibers with cell bodies located in the dorsal root ganglia.
         - The greater splanchnic nerve enters the celiac ganglion, the lesser splanchnic nerve enters the aorticorenal ganglion, and the least splanchnic nerve joins the renal plexus.

      b. **Lumbar splanchnic nerves**

         - arise from the lumbar sympathetic trunks and join the celiac, mesenteric, aortic, and superior hypogastric plexuses.
         - contain preganglionic sympathetic and GVA fibers.

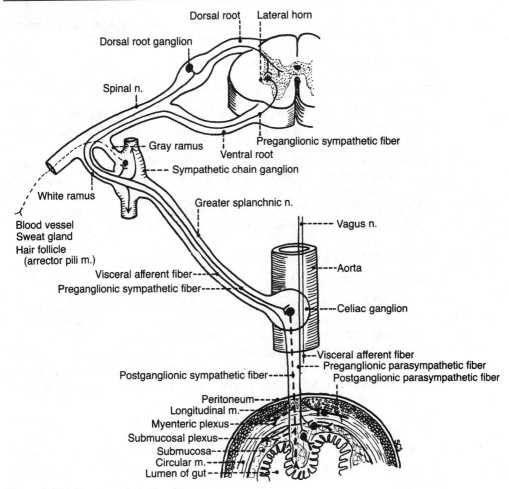

Figure 5-17. Nerve supply to the viscera.

3. Autonomic plexuses

a. Celiac plexus

- is formed by splanchnic nerves and branches from the vagus nerves.
- also contains the **celiac ganglia,** which receive the greater splanchnic nerves.
- lies on the front of the crura of the diaphragm and on the abdominal aorta at the origins of the celiac trunk and the superior mesenteric and renal arteries.
- extends its branches along the branches of the celiac trunk and forms the **subsidiary plexuses,** which are named according to the arteries along which they pass, such as gastric, splenic, hepatic, suprarenal, and renal plexuses.

b. Aortic plexus

- extends from the celiac plexus along the front of the aorta.
- extends its branches along the arteries and forms plexuses that are named accordingly—superior mesenteric, testicular (or ovarian), and inferior mesenteric.

– continues along the aorta and forms the **superior hypogastric plexus** just below the bifurcation of the aorta.

c. Superior and inferior hypogastric plexuses

4. Enteric division

– consists of the **myenteric (Auerbach's) plexus,** which is located chiefly between the longitudinal and circular muscle layers, and the **submucosal (Meissner's) plexus,** which is located in the submucosa.
– Both consist of preganglionic and postganglionic parasympathetic fibers, postganglionic sympathetic fibers, GVA fibers, and cell bodies of postganglionic parasympathetic fibers.
– Their sympathetic nerves inhibit gastrointestinal motility and secretion; parasympathetic nerves stimulate gastrointestinal motility and secretion.

IV. The Diaphragm and Its Openings

A. Diaphragm (Figure 5-18)

– arises from the xiphoid process (sternal part), lower six costal cartilages (costal part), medial and lateral lumbocostal arches (lumbar part), vertebrae L1–L3 for the right crus, and vertebrae L1–L2 for the left crus.
– inserts into the **central tendon** and is the principal muscle of inspiration.
– receives somatic motor fibers solely from the phrenic nerve; its central part receives sensory fibers from the phrenic nerve, whereas the peripheral part receives sensory fibers from the intercostal nerves.
– receives blood from the musculophrenic, pericardiophrenic, superior phrenic, and inferior phrenic arteries.
– descends when it contracts, causing an increase in thoracic volume by increasing the vertical length of the thorax and thus resulting in decreased thoracic pressure.
– ascends when it relaxes, causing a decrease in thoracic volume with increased thoracic pressure.

1. Right crus

– is larger and longer than the left crus.
– originates from vertebrae L1–L3 (the left crus originates from L1–L2).
– splits to enclose the esophagus.

2. Medial arcuate ligament (medial lumbocostal arch)

– extends from the body of L1 to the transverse process of L1 and passes over the psoas muscle and the sympathetic trunk.

3. Lateral arcuate ligament (lateral lumbocostal arch)

– extends from the transverse process of L1 to rib 12 and passes over the quadratus lumborum.

B. Apertures through the diaphragm

1. Vena caval hiatus (vena caval foramen)

– lies in the central tendon of the diaphragm at the level of T8 and transmits the inferior vena cava and the right phrenic nerve.

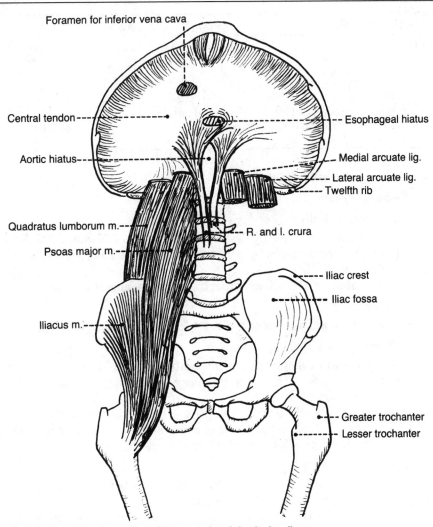

Figure 5-18. Diaphragm and muscles of the posterior abdominal wall.

2. Esophageal hiatus
– lies in the muscular part of the diaphragm (right crus) at the level of T10 and transmits the esophagus and anterior and posterior trunks of the vagus nerves.

3. Aortic hiatus
– lies behind or between two crura at the level of T12 and transmits the aorta, thoracic duct, greater splanchnic nerve, and azygos vein.

V. Muscles of the Posterior Abdominal Wall

Muscle	Origin	Insertion	Nerve	Action
Quadratus lumborum	Transverse processes of L3–L5; iliolumbar ligament; iliac crest	Lower border of last rib; transverse processes of L1–L3	Subcostal n.; L1–L3	Depresses rib 12; flexes trunk laterally

Muscle	Origin	Insertion	Nerve	Action
Psoas major	Transverse processes, intervertebral disks and bodies of T12–L5	Lesser trochanter	L2–L3	Flexes thigh and trunk
Psoas minor	Bodies and intervertebral disks of T12–L1	Pectineal line; iliopectineal eminence	L1	Aids in flexing of trunk

VI. Clinical Considerations

A. Renal disorders

1. **Horseshoe kidney**
   - is an anomalous kidney developed as a result of **fusion of the lower ends of the two kidneys** by a band of tissue extending across the vertebral column.
   - may obstruct the urinary tract by its **impingement on the ureters.**

2. **Renal ectopia**
   - is **abnormal position of a kidney,** frequently in the pelvis (**pelvic kidney**).

3. **Nephroptosis**
   - is **downward displacement or floating of the kidney** caused by loss of supporting fat. (The fat surrounding the kidney is a major factor in its stabilization.)
   - occurs frequently among truck drivers, horseback riders, and motorcyclists.
   - may cause a **kink in the ureter** or **compression of the ureter by an aberrant inferior polar artery,** resulting in hydronephrosis.

4. **Hydronephrosis**
   - is **distention of the renal pelvis and calyces** with accumulating urine as a result of obstruction of the ureter.
   - is caused by **renal calculi,** which may lodge within and obstruct a ureter or occlude a renal calyx.
   - may occur in a pregnant woman because the **developing fetus exerts pressure on the ureter** as it crosses the pelvic brim, obstructing urine flow.

5. **Pheochromocytoma**
   - is a **tumor of chromaffin tissue of the suprarenal medulla** and is characterized by excessive secretion of epinephrine and norepinephrine, resulting in hypertension that may be paroxysmal (episodic) and associated attacks of palpitations, headache, nausea and vomiting, dyspnea, profuse sweating, and tremor.

B. Renal transplantation
   - is performed through a transabdominal (traditionally retroperitoneal) approach to the kidney, by connecting the donor renal vessels to the recipient's external iliac vessels and suturing the donor ureter into the urinary bladder.

Review Test

Directions: Each of the numbered items or incomplete statements in this section is followed by answers or by completions of the statement. Select the ONE lettered answer or completion that is BEST in each case.

1. Which of the following fetal vessels becomes the round ligament of the liver after birth?

(A) Ductus venosus
(B) Paraumbilical vein
(C) Umbilical artery
(D) Ductus arteriosus
(E) Umbilical vein

2. A 36-year-old woman presents at the outpatient clinic with yellow pigmentation of skin and sclerae. Which of the following conditions would most likely cause obstructive jaundice?

(A) Aneurysm of the splenic artery
(B) Perforated ulcer of the stomach
(C) Damage to the pancreas during splenectomy
(D) Cancer of the head of the pancreas
(E) Cancer of the body of the pancreas

3. The transversalis fascia contributes to which of the following structures on the anterior abdominal wall?

(A) Superficial inguinal ring
(B) Deep inguinal ring
(C) Inguinal ligament
(D) Sac of a direct inguinal hernia
(E) Anterior wall of the inguinal canal

4. A lesion of which of the following nerves could cause damage to the preganglionic parasympathetic fibers to the liver?

(A) Phrenic nerves
(B) Lumbar splanchnic nerves
(C) Pelvic splanchnic nerves
(D) Vagus nerves
(E) Greater splanchnic nerves

5. Cell bodies of efferent and afferent nerve fibers in visceral branches of the sympathetic trunk are found in which of the following respective locations?

(A) Collateral ganglia; ventral horn of the spinal cord
(B) Intermediolateral cell column of the spinal cord; dorsal root ganglia
(C) Dorsal root ganglia; ventral horn of the spinal cord
(D) Sympathetic chain ganglia; terminal ganglia
(E) Intermediolateral cell column of the spinal cord; sympathetic chain ganglia

6. If an abdominal infection spread retroperitoneally, which of the following structures would most likely be affected?

(A) Stomach
(B) Transverse colon
(C) Jejunum
(D) Descending colon
(E) Spleen

7. The nerve fibers that innervate the suprarenal medulla to secrete noradrenaline are

(A) preganglionic sympathetic fibers
(B) postganglionic sympathetic fibers
(C) somatic motor fibers
(D) postganglionic parasympathetic fibers
(E) preganglionic parasympathetic fibers

8. Which of the following structures is part of, or is formed by, the internal oblique muscle?

(A) Lacunar ligament
(B) Inguinal ligament
(C) Cremaster muscle
(D) External spermatic fascia
(E) Internal spermatic fascia

9. Which of the following structures is formed by remnants of the embryonic urachus?

(A) Medial umbilical fold
(B) Round ligament of the uterus
(C) Inguinal ligament
(D) Median umbilical fold
(E) Lateral umbilical fold

10. Which of the following pairs of veins terminates in the same vein?

(A) Left and right ovarian veins
(B) Left and right testicular veins
(C) Left and right inferior phrenic veins
(D) Left and right suprarenal veins
(E) Left and right third lumbar veins

11. Which of the following structures crosses anterior to the inferior vena cava?

(A) Right sympathetic trunk
(B) Third right lumbar vein
(C) Third part of the duodenum
(D) Right renal artery
(E) Cisterna chyli

12. A patient presents with excruciating pain in his epigastrium caused by peritoneal irritation from gastric secretions and a perforated gastric ulcer. Which of the following nerves contains the fibers that convey this sharp, stabbing pain?

(A) Vagus nerve
(B) Greater splanchnic nerve
(C) Lower intercostal nerve
(D) White rami communicantes
(E) Gray rami communicantes

13. Which of the following statements concerning the inferior epigastric artery is true?

(A) It lies medial to a direct inguinal hernia
(B) It lies lateral and posterior to an indirect inguinal hernia
(C) It is a branch of the internal iliac artery
(D) It is a route of collateral circulation in coarctation of the aorta
(E) It anastomoses with the musculophrenic artery

14. Which of the following statements concerning the inguinal ligament is true?

(A) It is formed by the inferior free edge of the internal oblique muscle
(B) It extends between the anterior–inferior iliac spine and the ischial tuberosity
(C) It forms the roof of the inguinal canal
(D) It forms the floor of the inguinal canal
(E) It forms the lateral boundary of the inguinal triangle

15. A 42-year-old man presents at the emergency room with portal hypertension. Which of the following surgical procedures would be the most practical method of shunting portal blood around the liver?

(A) Connecting the superior mesenteric vein to the inferior mesenteric vein
(B) Connecting the portal vein to the superior vena cava
(C) Connecting the portal vein to the left renal vein
(D) Connecting the splenic vein to the left renal vein
(E) Connecting the superior rectal vein to the superior mesenteric vein

16. Rapid occlusion of which of the following arteries would result in ischemia of the suprarenal glands?

(A) Aorta; splenic and inferior phrenic arteries
(B) Renal, splenic, and inferior mesenteric arteries
(C) Aorta; inferior phrenic and renal arteries
(D) Superior mesenteric, inferior mesenteric, and renal arteries
(E) Aorta; hepatic and renal arteries

17. A radiograph of a 32-year-old woman reveals a perforation in the posterior wall of the stomach in which gastric contents have spilled into the lesser sac. An abdominal surgeon opened the lienogastric ligament to reach the lesser sac and accidentally cut an artery. Which of the following vessels would most likely be injured?

(A) Splenic artery
(B) Gastroduodenal artery
(C) Left gastric artery
(D) Right gastric artery
(E) Left gastroepiploic artery

18. Which of the following statements concerning the portal vein is true?

(A) It is formed posterior to the neck of the pancreas by the union of the splenic and renal veins
(B) It ascends anterior to the bile duct and the proper hepatic artery
(C) It passes anterior to the epiploic foramen in the free edge of the lesser omentum
(D) It has no tributaries superior to its beginning
(E) It carries as much blood as the hepatic artery

19. A tumor in the uncinate process of the pancreas may compress the

(A) common bile duct
(B) superior mesenteric vein
(C) portal vein
(D) main pancreatic duct
(E) hepatopancreatic ampulla (ampulla of Vater)

20. Which of the following structures would most likely be injured during an appendectomy performed at McBurney's point?

(A) Deep circumflex femoral artery
(B) Inferior epigastric artery
(C) Iliohypogastric nerve
(D) Genitofemoral nerve
(E) Spermatic cord

Directions: Each of the numbered items or incomplete statements in this section is negatively phrased, as indicated by a capitalized word such as NOT, LEAST, or EXCEPT. Select the ONE lettered answer or completion that is BEST in each case.

21. During pancreatectomy, the surgeon should ligate all of the following arteries EXCEPT

(A) splenic artery
(B) gastroduodenal artery
(C) superior mesenteric artery
(D) left gastric artery
(E) dorsal pancreatic artery

22. Loss of pain sensation from the abdominal viscera would be caused by a lesion of all of the following structures EXCEPT

(A) splanchnic nerves
(B) roots of the brachial plexus
(C) sympathetic nerves to the viscerae
(D) vagus nerve fibers
(E) white rami communicantes

23. All of the following statements concerning the portal venous system are true EXCEPT

(A) blood pressure in the portal vein is higher than that in the inferior vena cava
(B) the lower end of the esophagus is the area with the highest risk of venous varices as a result of portal hypertension
(C) numerous valves permit distension of the portal vein
(D) portal hypertension can cause caput medusae and hemorrhoids
(E) the portal vein carries at least twice as much blood as the hepatic artery

24. All of the following comparisons between the ileum and the jejunum are correct EXCEPT

(A) the ileum has fewer plicae circulares
(B) the ileum has fewer mesenteric arterial arcades
(C) less digestion and absorption of nutrients occur in the ileum
(D) the ileum has shorter vasa recta
(E) the mesentery of the ileum contains more fat

25. All of the following veins belong to the portal system EXCEPT

(A) the right colic vein
(B) the superior rectal vein
(C) the splenic vein
(D) the right suprarenal vein
(E) the left gastroepiploic vein

26. The aponeurosis of the external oblique muscle contributes to all of the following EXCEPT

(A) the falx inguinalis
(B) the anterior layer of the rectus sheath
(C) the anterior wall of the inguinal canal
(D) the inguinal ligament
(E) the external spermatic fascia

27. All of the following arteries follow the mesentery to reach the organs they supply EXCEPT

(A) left gastric artery
(B) superior mesenteric artery
(C) middle colic artery
(D) sigmoid arteries
(E) dorsal pancreatic artery

28. All of the following structures are related to embryonic or fetal blood vessels EXCEPT

(A) lateral umbilical fold
(B) medial umbilical fold
(C) ligamentum venosum
(D) ligamentum teres hepatis
(E) ligamentum arteriosum

29. An obstruction of the superior mesenteric vein would most likely enlarge all of the following veins EXCEPT

(A) the middle colic vein
(B) the right colic vein
(C) the inferior pancreaticoduodenal vein
(D) the ileocolic vein
(E) the left colic vein

30. A tumor located at the porta hepatis would block all of the following structures that enter or leave the liver EXCEPT

(A) the right hepatic artery
(B) hepatic nerves
(C) the common hepatic artery
(D) the left hepatic duct
(E) branches of the portal vein

31. All of the following statements concerning the suprarenal glands are true EXCEPT

(A) each suprarenal gland is associated with more arteries than veins
(B) thoracic splanchnic nerves carry postganglionic sympathetic fibers to the medulla of each gland
(C) each gland is surrounded by an extension of the renal fascia
(D) each gland lies on the superomedial aspect of the kidney
(E) the right suprarenal vein empties into the inferior vena cava

32. All of the following statements concerning the liver are true EXCEPT

(A) it receives blood from the hepatic artery and portal vein
(B) the quadrate lobe is the medial inferior segment of the left lobe
(C) the caudate lobe drains into the left hepatic duct
(D) it receives autonomic nerve fibers from the celiac plexus
(E) its function is to store and concentrate bile

33. All of the following statements concerning the common bile duct and the main pancreatic duct are true EXCEPT

(A) the common bile duct has the sphincter of Boyden, a circular muscle layer around its lower end
(B) the common bile duct traverses the head of the pancreas
(C) the main pancreatic duct carries two hormones, insulin and glucagon
(D) the main pancreatic duct begins in the tail and runs to the right along the entire pancreas
(E) the common bile duct lies anterior to the portal vein in the right free edge of the lesser omentum

34. To cut off the blood supply to the ureters, a surgeon has ligated all of the following arteries EXCEPT

(A) the renal artery
(B) the gonadal artery
(C) the middle rectal artery
(D) the inferior phrenic artery
(E) the common iliac artery

35. A portal–caval anastomosis exists between all of the following pairs of veins EXCEPT

(A) hepatic veins and inferior vena cava
(B) superior rectal vein and middle rectal vein
(C) left gastric vein and esophageal vein of the azygos system
(D) paraumbilical vein and superficial epigastric vein
(E) retrocolic veins and twigs of the renal vein

36. All of the following statements concerning the descending colon are true EXCEPT

(A) it has teniae coli and epiploic appendices
(B) it is a retroperitoneal organ
(C) it receives parasympathetic preganglionic fibers from the vagus nerve
(D) the inferior mesenteric artery provides most of its blood supply
(E) it converts the liquid contents of the ileum into semisolid feces by absorbing water

37. All of the following statements concerning the lesser omentum are true EXCEPT

(A) its right free edge forms a boundary of the epiploic foramen
(B) its right free edge contains the common bile duct, proper hepatic artery, and portal vein
(C) it consists of the hepatoduodenal and hepatogastric ligaments
(D) it is attached to the second part of the duodenum
(E) it forms the anterior wall of the lesser sac

38. All of the following statements concerning the inguinal canal are true EXCEPT

(A) it extends from the anterior–superior iliac spine to the pubic tubercle
(B) it begins at the deep inguinal ring in the transversalis fascia
(C) its anterior wall is primarily formed by the aponeurosis of the transverse muscle and the transversalis fascia
(D) it transmits the round ligament of the uterus or spermatic cord
(E) it ends at the superficial inguinal ring in the aponeurosis of the external oblique muscle

39. All of the following statements concerning a direct inguinal hernia are true EXCEPT

(A) it enters the inguinal canal through the posterior wall of the canal
(B) it lies lateral to the inferior epigastric artery
(C) it has a peritoneal covering
(D) it develops after birth
(E) it does not usually descend into the scrotum

40. All of the following statements concerning the ilioinguinal nerve are correct EXCEPT

(A) it runs in front of the quadratus lumborum
(B) it accompanies the spermatic cord during its course
(C) its genital branch supplies the cremaster muscle
(D) it emerges through the superficial inguinal ring
(E) it gives rise to an anterior scrotal branch

Directions: Each set of matching questions in this section consists of a list of four to twenty-six lettered options (some of which may be in figures) followed by several numbered items. For each numbered item, select the ONE lettered option that is most closely associated with it. To avoid spending too much time on matching sets with large numbers of options, it is generally advisable to begin each set by reading the list of options. Then, for each item in the set, try to generate the correct answer and locate it in the option list, rather than evaluating each option individually. Each lettered option may be selected once, more than once, or not at all.

Questions 41–45

Match each statement below with the vein it describes.

(A) Hepatic vein
(B) Portal vein
(C) Superior mesenteric vein
(D) Coronary vein
(E) Inferior mesenteric vein

41. Lies in front of the uncinate process of the pancreas

42. Drains blood directly from the lesser curvature of the stomach

43. Usually empties into the splenic vein

44. Lies immediately anterior to the epiploic foramen

45. Drains into the inferior vena cava

Questions 46–50

Match each statement below with the appropriate abdominal structure or component.

(A) Linea alba
(B) Linea semilunaris
(C) Linea semicircularis
(D) Transversalis fascia
(E) Falx inguinalis

46. Defines the lateral margin of the rectus abdominis

47. Contacts with the rectus abdominis muscle below the arcuate line

48. The tendinous medial line extending from the xiphoid process to the pubic symphysis

49. Composed of the aponeuroses of the internal oblique and transverse muscles of the abdomen

50. The crescent-shaped line marking the termination of the posterior layer of the rectus sheath

Questions 51–55

Match each statement below with the most appropriate ligament.

(A) Lienorenal ligament
(B) Lienogastric ligament
(C) Gastrophrenic ligament
(D) Falciform ligament
(E) Hepatoduodenal ligament

51. Contains a small portion of the tail of the pancreas

52. Contains the bile duct

53. Contains a paraumbilical vein

54. Contains short gastric arteries

55. Contains splenic vessels

Questions 56–60

Match each statement below with the most appropriate artery.

(A) Right gastric artery
(B) Left gastroepiploic artery
(C) Splenic artery
(D) Gastroduodenal artery
(E) Cystic artery

56. Located within the lienogastric ligament

57. Gives rise to the superior pancreaticoduodenal artery

58. A direct branch of the celiac artery

59. Runs along the lesser curvature of the stomach

60. Runs along the superior border of the pancreas

Questions 61–65

Match each structure below with the appropriate lettered component of this computed tomogram of the upper abdomen.

61. Inferior vena cava

62. Aorta

63. Crus of the diaphragm

64. Gallbladder

65. Liver

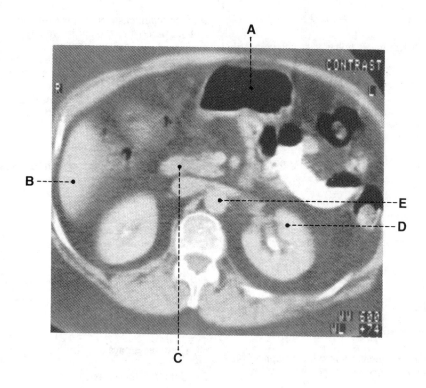

Answers and Explanations

1–E. The left umbilical vein becomes the round ligament of the liver after birth.

2–D. Because the bile duct traverses the head of the pancreas, cancer in the head of the pancreas would obstruct the bile duct, resulting in jaundice.

3–B. The deep inguinal ring lies in the transversalis fascia, just lateral to the inferior epigastric vessels. The superficial inguinal ring is in the aponeurosis of the external oblique muscle. The inguinal ligament and the anterior wall of the inguinal canal are formed by the aponeurosis of the external oblique muscle. The sac of a direct inguinal hernia is formed by the peritoneum.

4–D. The vagus nerves carry preganglionic parasympathetic fibers to the liver.

5–B. The cell bodies of the visceral efferent fibers are located in the intermediolateral cell column (or lateral horn) of the spinal cord; the cell bodies of visceral afferent fibers are located in the dorsal root ganglia.

6–D. The descending colon is a retroperitoneal organ.

7–A. The suprarenal medulla is the only organ that receives preganglionic sympathetic fibers.

8–C. The cremaster muscle and fascia arise from the internal oblique muscle.

9–D. The median umbilical fold contains the urachus, which is the fibrous remnant of the embryonic allantois.

10–E. The right and left third lumbar veins drain into the inferior vena cava.

11–C. The third part of the duodenum (transverse portion) crosses anterior to the inferior vena cava.

12–C. The parietal peritoneum receives pain fibers though the phrenic, lower intercostal, subcostal, iliohypogastric, and ilioinguinal nerves; however, the visceral peritoneum is innervated by visceral nerves and is relatively insensitive to pain. Excruciating pain caused by irritation of the parietal peritoneum by gastric contents is carried by somatic nerves, the lower intercostal nerves. The greater splanchnic nerves and white rami communicantes contain pain fibers which are visceral efferent fibers. The gray rami communicantes contain no sensory fibers.

13–D. In coarctation of the aorta, the epigastric artery provides a route of collateral circulation by anastomosing with the superior epigastric branch of the internal thoracic artery. The inferior epigastric artery is a branch of the external iliac artery. It lies lateral to a direct inguinal hernia and medial to an indirect inguinal hernia.

14–D. The inguinal ligament forms the floor of the inguinal canal and the inferior boundary of the inguinal triangle. It extends from the anterior–superior iliac spine to the pubic tubercle.

15–D. Portal hypertension can be reduced by diverting blood from the portal to the caval system; this is done by connecting the splenic vein to the left renal vein or by creating a communication between the portal vein and the inferior vena cava.

16–C. The suprarenal gland receives arteries from three sources. The superior suprarenal artery arises from the inferior phrenic artery, the middle suprarenal artery arises from the abdominal aorta, and the inferior suprarenal artery arises from the renal artery.

17–E. The left gastroepiploic artery runs through the lienogastric ligament to reach the greater omentum.

18–C. The portal vein ascends posterior to the bile duct and the proper hepatic artery and passes anterior to the epiploic foramen in the free edge of the lesser omentum. It is formed posterior to the neck of the pancreas by the union of the splenic and superior mesenteric veins. Tributaries of the portal vein include the left and right gastric veins and the posterior–superior pancreaticoduodenal vein. The portal vein carries twice as much blood as the hepatic artery.

19–B. The superior mesenteric vein crosses the uncinate process of the pancreas anteriorly near the duodenojejunal junction and joins the splenic vein posterior to the neck of the pancreas to form the portal vein. The common bile duct traverses the head of the pancreas. The main pancreatic duct runs through the body, neck, and head of the pancreas and joins the bile duct to form the hepatopancreatic ampulla, which enters the second part of the duodenum at the greater papilla.

20–C. The iliohypogastric nerve runs medially and inferiorly between the internal oblique and transverse abdominal muscles at McBurney's point, the point midway between the anterior–superior iliac spine and the umbilicus.

21–D. The left gastric artery does not supply the pancreas. The splenic artery gives rise to a number of pancreatic branches, including the dorsal pancreatic artery. The superior pancreaticoduodenal artery arises from the gastroduodenal artery, and the inferior pancreaticoduodenal artery arises from the superior mesenteric artery.

22–D. The vagus nerve contains sensory fibers associated with reflexes; it does not contain pain fibers.

23–C. The portal vein and its tributaries have no valves, or, if present, they are insignificant.

24–B. The ileum has more mesenteric arterial arcades than the jejunum. The plicae circulares (circular folds) in the upper part of the ileum are less prominent than those in the jejunum, but the lower part of the ileum has no plicae circulares.

25–D. The suprarenal vein belongs to the systemic (or caval) venous system and drains into the inferior vena cava on the right and the renal vein on the left. The right colic vein empties into the superior mesenteric vein, which joins with the splenic vein to form the portal vein. The left gastroepiploic vein empties into the splenic vein, and the superior rectal vein empties into the inferior mesenteric vein, which enters the splenic vein, the superior mesenteric vein, or the junction of the two.

26–A. The falx inguinalis is formed by the aponeuroses of the internal oblique and transverse muscles of the abdomen.

27–E. The pancreas is a retroperitoneal organ, except for a small portion of its tail; thus, the dorsal pancreatic artery arising from the splenic artery runs behind the peritoneum.

28–A. The lateral umbilical fold (ligament) contains the inferior epigastric artery and vein, which are adult blood vessels.

29–E. The lower left colic vein is a tributary of the inferior mesenteric vein.

30–C. The common hepatic artery gives rise to the gastroduodenal artery and the proper hepatic artery, which divides into the right and left hepatic arteries.

31–B. The suprarenal medulla is the only organ that receives preganglionic sympathetic fibers. The chromaffin cells in the suprarenal medulla may be considered modified postganglionic sympathetic neurons.

32–E. The liver has important roles in bile production and secretion. The gallbladder receives bile, concentrates it by absorbing water and salts, and stores it until delivered to the duodenum.

33–C. The endocrine part of the pancreas secretes the hormones insulin and glucagon, which are transported through the bloodstream.

34–D. The ureter receives blood from the aorta and from the renal, gonadal, common iliac, internal iliac, umbilical, superior and inferior vesical, and middle rectal arteries.

35–A. The hepatic veins and inferior vena cava are both systemic (caval) veins.

36–C. The descending colon receives parasympathetic preganglionic fibers from the pelvic splanchnic nerves, which arise from the sacral segments of the spinal cord (S2–S4).

37–D. The hepatoduodenal ligament is attached to the beginning of the first part of the duodenum.

38–C. The anterior wall of the inguinal canal is formed by the aponeuroses of the external oblique and internal oblique muscles. The posterior wall is formed by the aponeurosis of the transverse abdominal muscle and the transversalis fascia.

39–B. A direct inguinal hernia lies medial to the inferior epigastric vessels, whereas an indirect hernia lies lateral to the inferior epigastric vessels.

40–C. The genitofemoral nerve descends on the anterior surface of the psoas muscle and gives rise to a genital branch, which enters the inguinal canal through the deep inguinal ring to supply the cremaster muscle, and a femoral branch, which supplies the skin of the femoral triangle.

41–C. The superior mesenteric vessels lie in front of the uncinate process of the pancreas.

42–D. The coronary (left gastric) vein drains blood from the lesser curvature of the stomach into the portal vein.

43–E. The inferior mesenteric vein usually empties into the splenic vein; however, it may empty into the superior mesenteric vein or the junction of the superior mesenteric and splenic veins.

44–B. The portal vein lies immediately anterior to the epiploic foramen in the free margin of the lesser omentum.

45–A. The hepatic veins drain into the inferior vena cava.

46–B. The linea semilunaris is a curved line along the lateral margin of the rectus abdominis.

47–D. The transversalis fascia contacts with the rectus abdominis muscle below the arcuate line, marking the lower limit of the posterior layer of the rectus sheath.

48–A. The linea alba is the tendinous medial line extending from the xiphoid process to the pubic symphysis.

49–E. The aponeuroses of the internal oblique and transverse muscles form the falx inguinalis.

50–C. The linea semicircularis is the crescent-shaped line marking the termination of the posterior layer of the rectus sheath.

51–A. The lienorenal ligament contains the splenic vessels and a small portion of the tail of the pancreas.

52–E. The hepatoduodenal ligament is part of the lesser omentum and contains the bile duct, the hepatic artery, and the portal vein.

53–D. The falciform ligament contains a paraumbilical vein that connects the left branch of the portal vein with subcutaneous veins around the umbilicus.

54–B. The lienogastric (gastrosplenic) ligament contains several short gastric and left gastroepiploic vessels.

55–A. The lienorenal ligament contains the splenic vessels.

56–B. The lienogastric ligament contains the left gastroepiploic and short gastric vessels.

57–D. The gastroduodenal artery is divided into the superior pancreaticoduodenal and right gastroepiploic arteries.

58–C. The splenic, left gastric, and common hepatic arteries are the direct branches of the celiac trunk.

59–A. The right gastric artery arises either from the proper hepatic artery or from the common hepatic artery and runs between layers of the lesser omentum along the lesser curvature of the stomach.

60–C. The splenic artery follows a tortuous course along the superior border of the pancreas.

61–C.

62–E.

63–D.

64–A.

65–B.

6

Perineum and Pelvis

Perineal Region

I. Perineum

- is a **diamond-shaped space** that has the same boundaries as the inferior aperture of the pelvis.
- is bounded by the **pubic symphysis** anteriorly, the **ischiopubic rami** anterolaterally, the **ischial tuberosities** laterally, the **sacrotuberous ligaments** posterolaterally, and the **tip of the coccyx** posteriorly.
- Its floor is skin and fascia, and its roof is formed by the **pelvic diaphragm** with its fascial covering.
- is divided into an anterior **urogenital triangle** and a posterior **anal triangle** by a line connecting the two **ischial tuberosities**.

II. Urogenital Triangle (Figures 6-1 and 6-2)

A. Superficial perineal space (pouch)

- lies between the **inferior fascia of the urogenital diaphragm (perineal membrane)** and the membranous layer of the superficial perineal fascia (**Colles' fascia**).
- contains the superficial transverse perineal muscle, the ischiocavernosus muscles and crus of the penis or clitoris, the bulbospongiosus muscles and the bulb of the penis or the vestibular bulbs, the central tendon of the perineum, the greater vestibular glands (in the female), branches of the internal pudendal vessels, and the perineal nerve and its branches.

1. Colles' fascia

- is the **deep membranous layer** of the superficial perineal fascia.
- is continuous with **Scarpa's fascia of the anterior abdominal wall.**
- in the male, is continuous with the **tunica dartos** and passes over the penis as the **superficial fascia of the penis.**
- forms the lower boundary of the superficial perineal pouch.

2. Perineal membrane

- is the **inferior fascia of the urogenital diaphragm** and lies between the external genitalia and the urogenital diaphragm.
- is thickened anteriorly to form the **transverse ligament of the perineum.**
- is perforated by the **urethra** and attached to the **ischiopubic rami.**

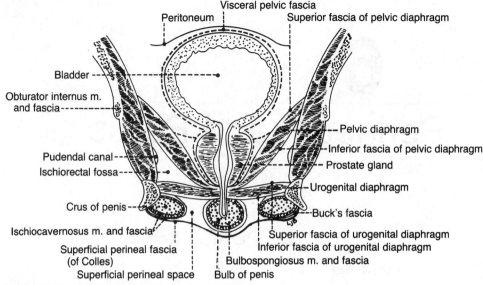

Figure 6-1. Frontal section of the male perineum and pelvis.

– forms the inferior boundary of the deep perineal pouch and the superior boundary of the superficial pouch.

3. **Muscles of the superficial perineal space** (Figures 6-3 and 6-4)

   a. **Ischiocavernosus muscles**

      – arise from the inner surface of the ischial tuberosities and the ischiopubic rami.
      – insert into the **corpus cavernosum** (the crus of the penis or clitoris).
      – are innervated by the perineal branch of the pudendal nerve.

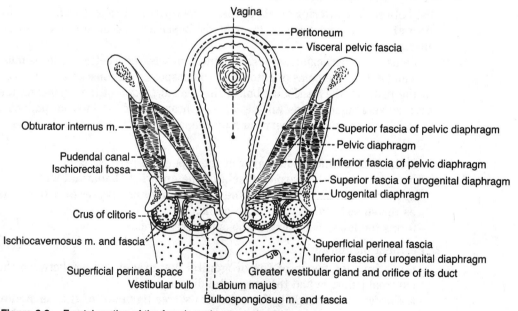

Figure 6-2. Frontal section of the female perineum and pelvis.

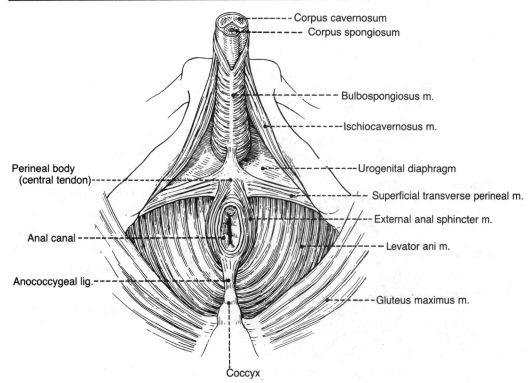

Figure 6-3. Muscles of the male perineum.

> – **maintain erection** of the penis by compressing the crus and the deep dorsal vein of the penis, thereby retarding venous return.

   b. Bulbospongiosus muscles
   - – arise from the perineal body and, in the male, the fibrous raphe of the bulb of the penis.
   - – insert into the **corpus spongiosum.**
   - – are innervated by the perineal branch of the pudendal nerve.
   - – in the male, **compress the bulb,** impeding venous return from the penis and thereby **maintaining erection.** Contraction (along with contraction of the ischiocavernosus) **constricts the corpus spongiosum,** thereby expelling the last drops of urine or the final semen in ejaculation.
   - – in the female, compress the erectile tissue of the vestibular bulbs and **constrict the vaginal orifice.**

   c. Superficial transverse perineal muscle
   - – arises from the ischial rami and tuberosities.
   - – inserts into the **central tendon (perineal body).**
   - – is innervated by the perineal branch of the **pudendal nerve.**
   - – **stabilizes the central tendon.**

4. **Perineal body (central tendon of the perineum)**
   - – is a **fibromuscular mass** located in the center of the perineum between the anal canal and the vagina (or the bulb of the penis).
   - – serves as a site of attachment for the superficial and deep transverse perineal, bulbospongiosus, levator ani, and external anal sphincter muscles.

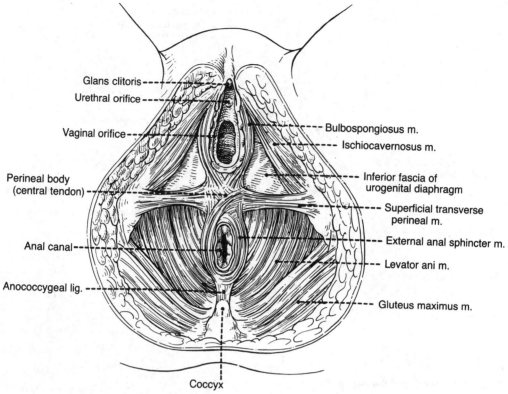

Figure 6-4. Muscles of the female perineum.

5. Greater vestibular (Bartholin's) glands

– lie in the superficial perineal pouch in the female, under cover of or behind the vestibular bulbs.
– are homologous to the **bulbourethral glands in the male.**
– are compressed during coitus and secrete mucus that **lubricates the vagina.**
– Ducts open into the **vestibule** between the **labium minora** and the **hymen.**

B. Deep perineal space (pouch)

– lies between the **superior and inferior fasciae of the urogenital diaphragm.**
– contains the deep transverse perineal muscle and sphincter urethrae, the membranous part of the urethra, the bulbourethral glands (in the male), and branches of the internal pudendal vessels and pudendal nerve.

1. Muscles of the deep perineal space

a. Deep transverse perineal muscle

– arises from the inner surface of the **ischial rami.**
– inserts into the medial tendinous raphe and the perineal body; in the female, it also inserts into the **wall of the vagina.**
– is innervated by the perineal branches of the pudendal nerve.
– stabilizes the perineal body and **supports the prostate gland or the vagina.**

b. Sphincter urethrae

- arises from the inferior pubic ramus.
- inserts into the median raphe and perineal body.
- is innervated by the perineal branch of the pudendal nerve.
- **encircles the membranous urethra** in the male and constricts it.
- In the female, its inferior part is attached to the anterolateral wall of the vagina, forming a **urethrovaginal sphincter** that compresses both the urethra and vagina.

2. Urogenital diaphragm

- consists of the deep transverse perineal muscle and the sphincter urethrae, and is invested by superior and inferior fasciae.
- stretches between the two pubic rami and ischial rami.
- Its inferior fascia gives attachment to the **bulb of the penis.**
- is pierced by the urethra and the vagina in the female and by the membranous urethra in the male.
- does not reach the pubic symphysis anteriorly.

3. Bulbourethral (Cowper's) glands

- lie among the fibers of the sphincter urethrae in the deep perineal pouch in the male, on the posterolateral sides of the membranous urethra.
- Ducts pass through the inferior fascia of the urogenital diaphragm to open into the bulbous portion of the **spongy (penile) urethra.**

III. Anal Triangle

A. Ischiorectal (ischioanal) fossa (see Figures 6-1 and 6-2)

- contains **ischioanal fat;** the **inferior rectal nerves and vessels,** which are branches of the internal pudendal vessels and the pudendal nerve; and a perineal branch of the **posterior femoral cutaneous nerve.**
- has the following **boundaries:**

1. Anterior: the posterior borders of the superficial and deep transverse perineal muscles

2. Posterior: the gluteus maximus muscle and the sacrotuberous ligament

3. Superomedial: the sphincter ani externus and levator ani muscles

4. Lateral: the obturator fascia covering the obturator internus muscle

5. Floor: the skin over the anal triangle

B. Muscles of the anal triangle (Figure 6-5)

1. Obturator internus

- arises from the inner surface of the **obturator membrane.**
- Its tendon passes around the lesser sciatic notch to insert into the medial surface of the greater trochanter of the femur.
- is innervated by the nerve to the obturator.
- **laterally rotates the thigh.**

2. Sphincter ani externus

- arises from the tip of the coccyx and the anococcygeal ligament.
- inserts into the central tendon of the perineum.

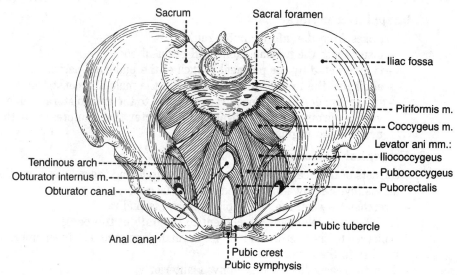

Figure 6-5. Muscles of the perineum and pelvis.

- is innervated by the inferior rectal nerve.
- **closes the anus.**

3. **Levator ani muscle**
   - arises from the body of the pubis, the arcus tendineus of the levator ani (a thickened part of the obturator fascia), and the ischial spine.
   - inserts into the coccyx and the anococcygeal raphe or ligament.
   - is innervated by the branches of the anterior rami of sacral nerves S3 and S4 and the perineal branch of the pudendal nerve.
   - **supports and raises the pelvic floor.**
   - consists of the **puborectalis, pubococcygeus, and iliococcygeus.**
   - Its most anterior fibers, which are also the most medial, are called the levator prostatae or pubovaginalis.

4. **Coccygeus**
   - arises from the ischial spine and the sacrospinous ligament.
   - inserts into the coccyx and the lower part of the sacrum.
   - is innervated by branches of the third and fourth sacral nerves.
   - **supports and raises the pelvic floor.**

IV. External Genitalia and Associated Structures

A. Fasciae and ligaments

1. **Fundiform ligament of the penis**
   - arises from the linea alba and the membranous layer of the superficial fascia of the abdomen.
   - splits into left and right parts, **encircles the body of the penis,** and blends with the superficial penile fascia.
   - enters the septum of the scrotum.

2. **Suspensory ligament of the penis (or the clitoris)**
   - arises from the pubic symphysis and the arcuate pubic ligament and inserts into the deep fascia of the penis or to the body of the clitoris.
   - lies deep to the fundiform ligaments.

3. Deep fascia of the penis (Buck's fascia)

- is a continuation of the deep perineal fascia.
- is continuous with the fascia covering the external oblique muscle and the rectus sheath.

4. Tunica albuginea

- is a **fibrous connective tissue layer** that envelops both the corpora cavernosa and the corpus spongiosum.
- is very dense around the **corpora cavernosa,** thereby greatly impeding venous return and resulting in the extreme turgidity of these structures when the erectile tissue becomes engorged with blood.
- is more elastic surrounding the **corpus spongiosum,** which, therefore, does not become excessively turgid on erection and permits passage of the ejaculate.

5. Tunica vaginalis

- is a **double serous membrane,** a peritoneal sac on the end of the processus vaginalis that covers the front and sides of the testis and epididymis.
- is a closed sac, derived from the abdominal peritoneum, forming the **innermost layer of the scrotum.**
- consists of a parietal layer adjacent to the internal spermatic fascia and a visceral layer adherent to the testis and epididymis.

B. Male external genitalia

1. Scrotum

- is a cutaneous pouch consisting of skin and the underlying dartos.
- Its skin is relatively thin with little or no fat, which is important in maintaining a temperature lower than the rest of the body.
- contains the **testis** and its covering and the **epididymis.**
- The **tunica dartos** is continuous with the superficial penile fascia and the superficial perineal fascia, is formed by fusion of the superficial and deep layers of superficial fascia, consists largely of smooth muscle fibers, contains no fat, and functions in temperature regulation.
- is contracted and wrinkled when cold (or sexually stimulated), bringing the testis into close contact with the body to conserve heat; is relaxed when warm and hence is flaccid and distended to dissipate heat.
- receives blood from the external pudendal arteries and the posterior scrotal branches of the internal pudendal arteries. It also receives branches of the testicular and cremaster arteries.
- is innervated by the anterior scrotal branch of the **ilioinguinal nerve,** the genital branch of the **genitofemoral nerve,** the posterior scrotal branch of the perineal branch of the **pudendal nerve,** and the perineal branch of the **posterior femoral cutaneous nerve.**

2. Penis (Figure 6-6)

- consists of three masses of **vascular erectile tissue,** the paired corpora cavernosa and the midline corpus spongiosum, which are bounded by tunica albuginea.
- The **root** includes two crura and the bulb of the penis; the **body** contains the single corpus spongiosum and the paired corpora cavernosa.

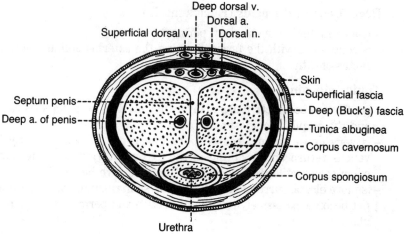

Figure 6-6. Cross-section of the penis.

- The **glans penis** is formed by the terminal part of the corpus spongiosum and is covered by a free fold of skin, the **prepuce**. The **frenulum** of the prepuce is a median ventral fold passing from the deep surface of the prepuce.
- has a prominent margin of the glans penis, the **corona;** a median slit near the tip of the glans, the **external urethral orifice;** and a terminal dilated part of the urethra in the glans, the **fossa navicularis**.

C. Female external genitalia

1. Labia majora

- are two **longitudinal folds of skin** that run downward and backward from the **mons pubis.**
- are joined anteriorly by the **anterior labial commissure.**
- Their outer surfaces are covered with pigmented skin containing sebaceous and sweat glands and, after puberty, are covered with hair.
- are homologous to the **scrotum** of the male.
- contain the terminations of the round ligaments of the uterus.

2. Labia minora

- unlike the labia majora, are hairless and contain no fat.
- are divided into two parts, upper (lateral) and lower (medial).

a. The **lateral parts,** above the clitoris, fuse to form the **prepuce of the clitoris.**

b. The **medial parts,** below the clitoris, fuse to form the **frenulum of the clitoris.**

3. Vestibule of the vagina (urogenital sinus)

- is the space or cleft between the labia minora.
- The openings for the urethra, the vagina, and the ducts of the greater vestibular glands are in its floor.

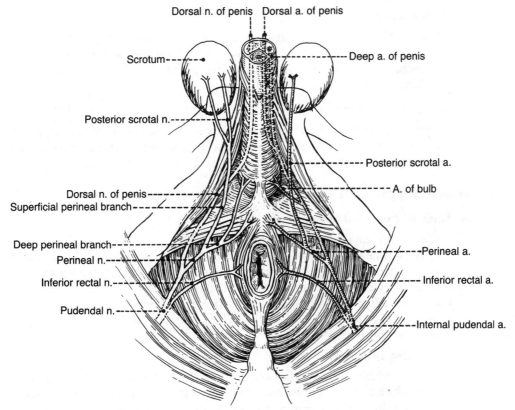

Dorsal n. of penis   Dorsal a. of penis

Scrotum --

Deep a. of penis

Posterior scrotal n. --

Posterior scrotal a.

Dorsal n. of penis --
Superficial perineal branch --

A. of bulb

Deep perineal branch --
Perineal n. --

Perineal a.

Inferior rectal n. --

Inferior rectal a.

Pudendal n. --

Internal pudendal a.

Figure 6-7. Internal pudendal artery and pudendal nerve and branches.

4. Clitoris

– is homologous to the **penis** in the male; consists of **erectile tissue** and is enlarged as a result of engorgement with blood.
– consists of two crura, two corpora cavernosa, and a glans but has no corpus spongiosum. The **glans clitoris** is derived from the corpora cavernosa and is covered by a sensitive epithelium.

V. Nerve Supply of the Perineal Region (Figure 6-7)

A. Pudendal nerve (S2–S4)

– passes through the greater sciatic foramen between the piriformis and coccygeus muscles.
– crosses the ischial spine and enters the perineum with the internal pudendal artery through the lesser sciatic foramen.
– enters the **pudendal canal,** gives rise to the inferior rectal nerve and the perineal nerve, and terminates as the **dorsal nerve of the penis (or clitoris).**

B. Inferior rectal nerve

– arises within the pudendal canal, divides into several branches, crosses the ischiorectal fossa, and innervates the sphincter ani externus and the skin around the anus.

C. Perineal nerve

– arises within the pudendal canal and divides into a **deep branch,** which supplies all of the perineal muscles, and a **superficial (posterior scrotal or labial) branch,** which in turn divides into two branches to supply the scrotum or labia majora.

D. Dorsal nerve of the penis (or clitoris)

– pierces the perineal membrane, runs between the two layers of the suspensory ligament of the penis or clitoris, and runs deep to the deep fascia on the dorsum of the penis or clitoris to innervate the skin, prepuce, and glans.

VI. Blood Supply of the Perineal Region (see Figure 6-7)

A. Internal pudendal artery

– arises from the internal iliac artery.
– leaves the pelvis by way of the greater sciatic foramen below the piriformis and coccygeus and immediately enters the perineum through the lesser sciatic foramen by hooking around the ischial spine.
– is accompanied by the pudendal nerve during its course.
– passes along the lateral wall of the ischiorectal fossa in the pudendal canal.
– gives rise to the following:

1. Inferior rectal artery

– arises within the pudendal canal, pierces the wall of the pudendal canal, and breaks into several branches, which cross the ischiorectal fossa to **muscles and skin around the anal canal.**

2. Perineal arteries

– supply the superficial perineal muscles and give rise to **transverse perineal branches** and **posterior scrotal branches.**

3. Artery of the bulb

– arises within the deep perineal space, pierces the perineal membrane, and supplies the bulb of the penis and the bulbourethral glands in the male and the vestibular bulbs and the greater vestibular gland in the female.

4. Urethral artery

– pierces the perineal membrane, enters the corpus spongiosum of the penis or clitoris, and continues to the **glans penis or clitoris.**

5. Deep arteries of the penis or clitoris

– are terminal branches of the internal pudendal artery.
– pierce the perineal membrane and run through the center of the **corpus cavernosum of the penis or clitoris.**

6. Dorsal arteries of the penis or clitoris

– run on each side of the deep dorsal vein and deep to the deep fascia (Buck's fascia) and superficial to the tunica albuginea to supply the glans and the prepuce.
– pierce the perineal membrane and pass through the suspensory ligament of the penis or clitoris.

B. External pudendal artery

 – arises from the femoral artery, emerges through the saphenous ring, and passes medially over the spermatic cord or the round ligament of the uterus to supply the **skin above the pubis, penis, and scrotum or labium majus.**

C. Veins of the penis

1. Deep dorsal vein of the penis

 – is an unpaired vein that begins in the sulcus behind the glans and lies in the dorsal midline deep to the deep fascia and superficial to the tunica albuginea.
 – leaves the perineum through the gap between the **arcuate pubic ligament** and the **transverse perineal ligament.**
 – passes through the suspensory ligament of the penis below the arcuate pubic ligament and drains into the prostatic and pelvic venous plexuses.

2. Superficial dorsal vein of the penis

 – runs toward the pubic symphysis between the superficial and deep fasciae on the dorsum of the penis and divides into the right and left branches, which end in the **external (superficial) pudendal veins.** The external pudendal vein drains into the greater saphenous vein.

VII. Clinical Considerations

A. Extravasated urine

 – may result from **rupture of the spongy urethra** below the urogenital diaphragm; urine may pass into the superficial perineal space.
 – cannot spread laterally into the thigh because the inferior fascia of the urogenital diaphragm (the perineal membrane) and the superficial fascia of the perineum are firmly attached to the ischiopubic rami and are connected with the deep fascia of the thigh (fascia lata).
 – cannot spread posteriorly into the anal region because the perineal membrane and Colles' fascia are continuous with each other around the superficial transverse perineal muscles.
 – spreads inferiorly into the scrotum, anteriorly around the penis, and superiorly into the abdominal wall.

B. Hydrocele

 – is an **accumulation of fluid in the cavity of the tunica vaginalis of the testis** or along the spermatic cord.

C. Varicocele

 – occurs when **enlargement (varicosity) of the veins of the spermatic cord** forms an appearance of a "bag of worms," accompanied by a constant pulling and dragging, and frequently causing oligospermia.
 – is more common on the left side, probably due to a malignant tumor of the left kidney, which blocks the exit of the testicular vein.

D. Vasectomy

 – is **surgical excision of a portion of the vas deferens** (ductus deferens) through the scrotum.
 – stops the passage of spermatozoa, but does not reduce the amount of ejaculate greatly nor does it diminish sexual desire.

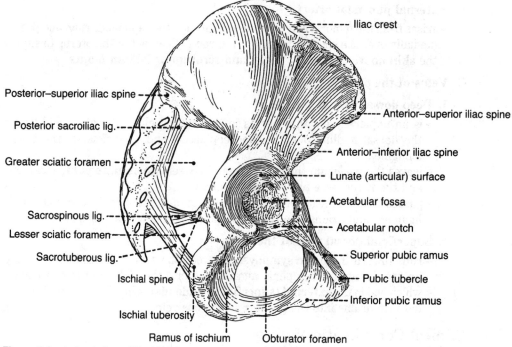

Posterior–superior iliac spine

Posterior sacroiliac lig.

Greater sciatic foramen

Sacrospinous lig.

Lesser sciatic foramen

Sacrotuberous lig.

Ischial spine

Ischial tuberosity

Ramus of ischium

Iliac crest

Anterior–superior iliac spine

Anterior–inferior iliac spine

Lunate (articular) surface

Acetabular fossa

Acetabular notch

Superior pubic ramus

Pubic tubercle

Inferior pubic ramus

Obturator foramen

Figure 6-8. Lateral view of the hip bone.

E. Mediolateral episiotomy

– is a **surgical incision through the posterolateral vaginal wall,** just lateral to the perineal body, to enlarge the birth canal and thus prevent uncontrolled tearing during parturition. In a **median episiotomy,** the incision is carried posteriorly in the midline through the posterior vaginal wall and the central tendon (perineal body).

Pelvis

I. Bony Pelvis (Figures 6-8, 6-9, and 6-10)

A. Pelvis

– is the **basin-shaped ring of bone** formed by the two **hip bones,** the **sacrum,** and the **coccyx.** (The hip, or coxal bone, consists of the ilium, ischium, and pubis.)

– is divided by the **pelvic brim** or iliopectineal line into the **pelvis major (false pelvis)** above and the **pelvis minor (true pelvis)** below.

– Its outlet is closed by the **coccygeus** and **levator ani** muscles, which form the **floor of the pelvis.**

– The normal anatomic position is tilted, thus:

1. The anterior–superior iliac spine and the pubic tubercles are in the same vertical plane.

2. The coccyx is in the same horizontal plane as the upper margin of the pubic symphysis.

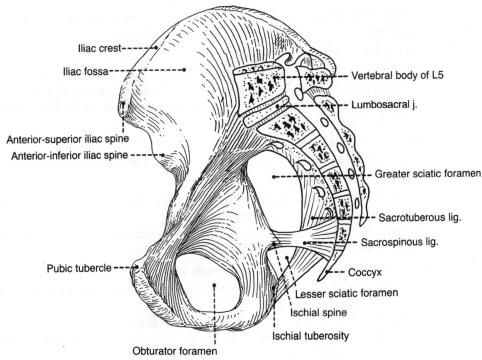

Figure 6-9. Medial view of the hip bone.

3. The axis of the pelvic cavity running through the central point of the inlet and the outlet almost parallels the curvature of the sacrum.

B. **Upper pelvic aperture (pelvic inlet or pelvic brim)**
  – is the **superior rim of the pelvic cavity;** is bounded posteriorly by the promontory of the sacrum and the anterior border of the ala of the sacrum (**sacral part**), laterally by the arcuate or iliopectineal line of the ilium (**iliac part**), and anteriorly by the pectineal line, the pubic crest, and the superior margin of the pubic symphysis (**pubic part**).

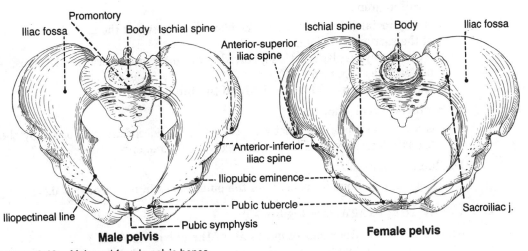

Figure 6-10. Male and female pelvic bones.

– is measured using transverse, oblique, and anteroposterior (conjugate) diameters.

C. Lower pelvic aperture (pelvic outlet)

– is a **diamond-shaped aperture** bounded posteriorly by the sacrum and coccyx, laterally by the ischial tuberosities and sacrotuberous ligaments, and anteriorly by the pubic symphysis, arcuate ligament, and rami of the pubis and ischium.
– is closed by the pelvic and urogenital diaphragms.

D. Pelvis major (false pelvis)

– is the expanded portion of the bony pelvis above the pelvic brim.

E. Pelvis minor (true pelvis)

– is the **cavity of the pelvis** below the pelvic brim (or superior aperture) and above the pelvic outlet (or inferior aperture).
– Its outlet is closed by the coccygeus and levator ani muscles and the perineal fascia, which form the floor of the pelvis.

F. Differences between the female and male pelvis

– The bones of the female pelvis are usually smaller, lighter, and thinner than those of the male.
– The inlet is transversely oval in the female and heart-shaped in the male.
– The outlet is larger in the female than in the male because of the everted ischial tuberosities in the female.
– The cavity is wider and shallower in the female than in the male.
– The subpubic angle or pubic arch is larger and the greater sciatic notch is wider in the female than in the male.
– The female sacrum is shorter and wider than the male sacrum.
– The obturator foramen is oval or triangular in the female and round in the male.

II. Joints of the Pelvis (see Figures 6-9 and 6-10)

A. Lumbosacral joint

– is the joint between vertebra L5 and the base of the sacrum, joined by an intervertebral disk and supported by the iliolumbar ligaments.

B. Sacroiliac joint

– is a **synovial joint** of an irregular plane type between the articular surfaces of the sacrum and ilium.
– is covered by cartilage and is supported by the anterior, posterior, and interosseous sacroiliac ligaments.
– **transmits the weight of the body to the hip bone.**

C. Sacrococcygeal joint

– is a **cartilaginous joint** between the sacrum and coccyx, reinforced by the anterior, posterior, and lateral sacrococcygeal ligaments.

D. Pubic symphysis

– is a **cartilaginous joint** between the pubic bones in the median plane.

III. Pelvic Diaphragm (see Figure 6-5)

– forms the **pelvic floor** and **supports all of the pelvic viscerae.**
– is formed by the **levator ani and coccygeus** muscles and their fascial coverings.

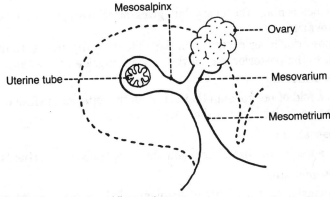

Figure 6-11. Sagittal section of the broad ligament.

- lies posterior and deep to the urogenital diaphragm as well as medial and deep to the ischiorectal fossa.
- on contraction, **raises the entire pelvic floor.**
- flexes the anorectal canal during **defecation** and helps the voluntary control of **micturition.**
- helps to direct the fetal head toward the birth canal at **parturition.**

IV. Ligaments of the Female Pelvis

A. Broad ligament of the uterus (Figures 6-11 and 6-12)

- consists of **two layers of peritoneum** and extends from the lateral margin of the uterus to the lateral pelvic wall.
- contains the uterine tube, uterine vessels, round ligament of the uterus, ovarian ligament, ureter, uterovaginal nerve plexus, and lymphatic vessels.

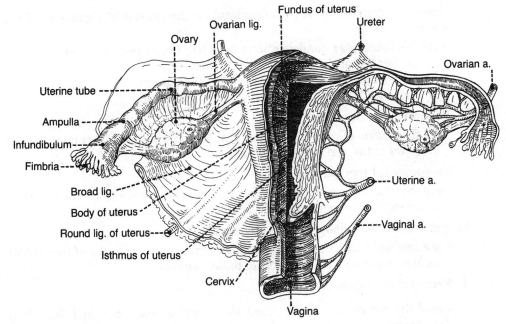

Figure 6-12. Female reproductive organs.

– does not contain the ovary but gives attachment to the ovary through the **mesovarium.**

– Its posterior layer curves from the isthmus of the uterus, as the **rectouterine fold,** to the posterior wall of the pelvis alongside the rectum.

1. **Mesovarium**

– is a fold of peritoneum that connects the anterior surface of the **ovary** with the posterior layer of the broad ligament.

2. **Mesosalpinx**

– is a fold of the broad ligament that suspends the **uterine tube.**

3. **Mesometrium**

– is a major part of the broad ligament below the mesosalpinx and mesovarium.

B. Round ligament of the uterus

– is attached to the uterus in front of and below the attachment of the uterine tube and represents the remains of the lower part of the **gubernaculum.**

– runs within the layers of the broad ligament, contains smooth muscle fibers, and holds the fundus of the uterus forward, keeping the uterus anteverted and anteflexed.

– enters the inguinal canal at the deep inguinal ring, emerges from the superficial inguinal ring, and becomes lost in the subcutaneous tissue of the labium majus.

C. Ovarian ligament

– is a **fibromuscular cord** that extends from the uterine end of the ovary to the side of the uterus below the uterine tube through the broad ligament.

– runs within the layers of the broad ligament.

D. Suspensory ligament of the ovary

– is a **band of peritoneum** that extends upward from the ovary to the pelvic wall and transmits the ovarian vessels, nerves, and lymphatics.

E. Lateral or transverse cervical (cardinal or Mackenrodt's) ligaments of the uterus

– are **fibromuscular condensations of pelvic fascia** from the cervix and the lateral fornices of the vagina to the pelvic walls.

– extend laterally below the base of the broad ligament.

– contain smooth muscle fibers and **support the uterus.**

F. Pubocervical ligaments

– are firm bands of connective tissue that extend from the posterior surface of the pubis to the cervix of the uterus.

G. Sacrocervical ligaments

– are firm fibromuscular bands of pelvic fascia that extend from the lower end of the sacrum to the cervix and the upper end of the vagina.

H. Pubovesical (or puboprostatic) ligaments

– are condensations of the pelvic fascia that extend from the neck of the bladder (or the prostate gland in the male) to the pelvic bone.

I. Rectouterine ligaments

– **hold the cervix back and upward** and sometimes elevate a shelf-like fold of peritoneum (**rectouterine fold**), which passes from the isthmus of the uterus to the posterior wall of the pelvis lateral to the rectum. It corresponds to the **sacrogenital fold** in the male.

V. Ureter and Urinary Bladder

A. Ureter

– is a **muscular tube** that **transmits urine** by peristaltic waves.

– has **three constrictions** along its course: at its origin where the pelvis of the ureter joins the ureter, where it crosses the pelvic brim, and at its junction with the bladder.

– crosses the **pelvic brim** in front of the bifurcation of the common iliac artery and descends retroperitoneally on the lateral pelvic wall.

– lies 1 to 2 cm lateral to the cervix of the uterus in the female; it passes posterior and medial to the ductus deferens and lies in front of the seminal vesicle in the male.

– in the female, is accompanied in its course by the uterine artery, which runs above and anterior to it. Because of this location, it is sometimes injured by a clamp during surgical procedures, and may be ligated and sectioned by mistake during a hysterectomy.

B. Urinary bladder

– is situated below the peritoneum and is slightly lower in the female than in the male.

– The anterior end is the **apex,** and its posteroinferior triangular portion is the **fundus or base.**

– The **neck of the bladder,** the area where the fundus and the inferolateral surfaces come together, leads into the **urethra.**

– The **trigone** is bounded by the two orifices of the ureters and the internal urethral orifice.

– The musculature (bundles of smooth muscle fibers) as a whole is known as the **detrusor muscle of the bladder.**

– extends upward above the pelvic brim as it fills; may reach as high as the umbilicus if fully distended.

– receives blood from the superior and inferior vesical arteries (and from the vaginal artery in the female).

– Its venous blood is drained by the **prostatic (or vesical) plexus** of veins, which empties into the internal iliac vein.

– is innervated by nerve fibers from the vesical and prostatic plexuses, which are extensions from the inferior hypogastric plexuses. The parasympathetic nerve originating from the cord segment S2–S4 causes the musculature (detrusor) of the bladder wall to contract, relaxes the internal sphincter, and promotes emptying.

C. Micturition (urination)

– is initiated by stimulating **stretch receptors in the detrusor muscle** in the bladder wall by the increasing volume (about 300 ml for adult) of urine.

– Relaxation of the levator ani allows the neck of the bladder to descend, eliciting **a reflex contraction of the detrusor muscle.**

– Afferent impulses from the bladder wall transmit to the spinal cord (S2–S4) via the pelvic splanchnic nerves.

– Efferent impulses pass via the preganglionic parasympathetic fibers in the pelvic splanchnic nerves and the inferior hypogastric plexus to the bladder

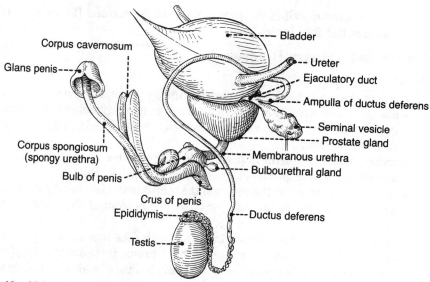

Figure 6-13. Male reproductive organs.

wall, where they synapse with postganglionic neurons, causing contraction of the detrusor muscle.

– Sympathetic fibers cause relaxation of the **sphincter vesicae (internal urethral sphincter).**

– Somatic motor fibers in the pudendal nerve cause relaxation of the **external urethral sphincter.**

– can be assisted by contraction of the abdominal muscles, which increases the intra-abdominal and pelvic pressures.

– In the male, the **bulbospongiosus muscle** expels the last few drops of urine from the urethra.

VI. Male Genital Organs (Figures 6-13 and 6-14)

A. Testis

– develops retroperitoneally and descends into the scrotum retroperitoneally.

– is covered by the **tunica albuginea,** which lies beneath the visceral layer of the **tunica vaginalis.**

– **produces spermatozoa** and **secretes sex hormones.**

– is supplied by the testicular artery from the abdominal aorta and is drained by veins of the pampiniform plexus.

– Its lymph vessels ascend with the testicular vessels and drain into the lumbar (aortic) nodes; lymphatic vessels in the scrotum drain into the superficial inguinal nodes.

B. Epididymis

– consists of a head, body, and tail. The tail contains a **convoluted duct** about 6 m (20 feet) long.

– Its functions include **maturation and storage of spermatozoa,** as well as **propulsion of the spermatozoa** into the ductus deferens.

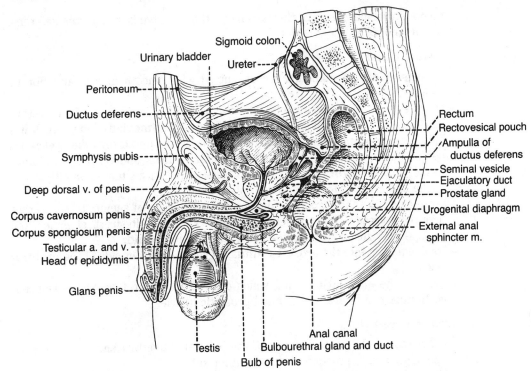

Urinary bladder
Peritoneum---
Ductus deferens---
Symphysis pubis---
Deep dorsal v. of penis---
Corpus cavernosum penis---
Corpus spongiosum penis---
Testicular a. and v.-
Head of epididymis---
Glans penis---

Sigmoid colon
Ureter---

Rectum
Rectovesical pouch
Ampulla of ductus deferens
Seminal vesicle
Ejaculatory duct
Prostate gland
Urogenital diaphragm
External anal sphincter m.

Testis
Bulb of penis
Anal canal
Bulbourethral gland and duct

Figure 6-14.   Sagittal section of the male pelvis.

C. Ductus deferens

- is a thick-walled tube with a relatively small lumen.
- enters the pelvis at the deep inguinal ring at the lateral side of the inferior epigastric artery.
- crosses the medial side of the umbilical artery, ureter, and obturator nerve and vessels during its course.
- Its dilated part is termed the **ampulla.**
- receives innervation mainly from sympathetic nerves of the hypogastric plexus.

D. Ejaculatory ducts

- are formed by the union of the ductus deferens with the ducts of the seminal vesicles. Peristaltic contractions of the muscular layer of the ductus deferens and the ejaculatory ducts propel spermatozoa with seminal fluid into the urethra.
- open into the prostatic urethra on the **seminal colliculus** just lateral to the blind **prostatic utricle** (see VI G).

E. Seminal vesicles

- are enclosed by dense endopelvic fascia and are **lobulated glandular structures** that are diverticula of the ductus deferens.
- lie inferior and lateral to the ampullae of the ductus deferens against the fundus (base) of the bladder.
- produce the alkaline constituent of the **seminal fluid.**
- Their lower ends become narrow and form ducts that join the ampullae of the ductus deferens to form the **ejaculatory ducts.**

– do not store spermatozoa; this is done by the epididymis, the ductus deferens, and its ampulla.

F. Prostate gland

– consists chiefly of glandular tissue, mixed with smooth muscle and fibrous tissue.
– has five lobes: the **anterior lobe** (or isthmus), which lies in front of the urethra and is devoid of glandular substance; the **middle (median) lobe,** which lies between the urethra and the ejaculatory ducts; the **posterior lobe,** which lies behind the urethra and below the ejaculatory ducts and contains glandular tissue; and the **right and left lateral lobes,** which are situated on either side of the urethra and form the main mass of the gland.
– secretes a fluid that produces the characteristic odor of semen; this, together with the secretion from the seminal vesicles and the bulbourethral glands, and the spermatozoa, constitute the **semen or seminal fluid.**
– Its ducts open into the **prostatic sinus,** a groove on either side of the **urethral crest.**
– receives the **ejaculatory duct,** which opens into the urethra on the **seminal colliculus** just lateral to the blind prostatic utricle.

G. Urethral crest

– is located on the posterior wall of the **prostatic urethra** and has numerous openings for the prostatic ducts on either side.
– The **seminal colliculus (verumontanum)** is an ovoid-shaped enlargement of the urethral crest upon which the two ejaculatory ducts and the prostatic utricle open.
– At the summit of the colliculus is the **prostatic utricle,** which is an invagination (a blind pouch) about 5 mm deep; it is analogous to the uterus and vagina in the female.

H. Prostatic sinus

– is a groove between the urethral crest and the wall of the prostatic urethra and receives the ducts of the prostate gland.

I. Erection

– depends on **parasympathetic nerve stimulation,** which dilates the arteries supplying the erectile tissue, and thus causes engorgement of the corpora cavernosa and corpus spongiosum, compressing the veins and thus impeding venous return.
– is also maintained by **contraction of the bulbospongiosus and ischiocavernosus muscles,** which compresses the erectile tissues of the bulb and the crus.

J. Ejaculation

– Friction to the glans penis and other sexual stimuli result in **excitation of sympathetic fibers,** leading to contraction of the smooth muscle of the epididymal ducts, the ductus deferens, the seminal vesicles, and the prostate in turn.
– The smooth muscle contracts, pushing spermatozoa and the secretions of both the seminal vesicles and prostate into the prostatic urethra, where they join secretions from the bulbourethral and penile urethral glands. All of these secretions are ejected together from the penile urethra as a result of the rhythmic contractions of the bulbospongiosus, which compresses the urethra.

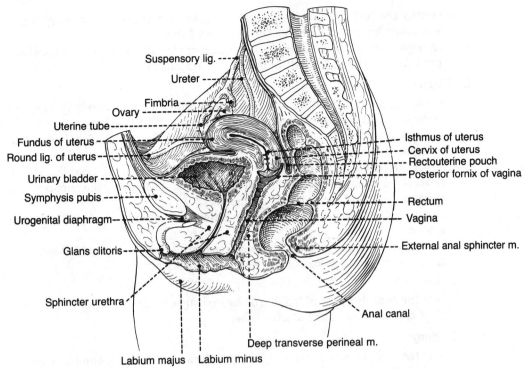

Figure 6-15. Sagittal section of the female pelvis.

– The sphincter of the bladder contracts and prevents the entry of urine into the prostatic urethra and the reflux of the semen into the bladder.

VII. Female Genital Organs (Figure 6-15; see Figure 6-12)

A. Ovaries

– lie on the posterior aspect of the **broad ligament** on the side wall of the pelvic minor and are bounded by the external and internal iliac vessels.
– are not covered by the peritoneum, and thus the ovum or oocyte is expelled into the peritoneal cavity and then into the uterine tube.
– are not enclosed in the broad ligament, but their anterior surface is attached to the posterior surface of the broad ligament by the **mesovarium.**
– The surface is covered by **germinal (columnar) epithelium,** which is modified from the developmental peritoneal covering of the ovary.
– are supplied primarily by the ovarian arteries, which are contained in the suspensory ligament and anastomose with branches of the uterine artery.
– are drained by the ovarian veins; the right ovarian vein joins the inferior vena cava, and the left ovarian vein joins the left renal vein.

B. Uterine tubes

– extend from the uterus to the uterine end of the ovaries and **connect the uterine cavity to the peritoneal cavity.**
– are each subdivided into four parts: the **uterine part,** the **isthmus,** the **ampulla** (the longest and widest part), and the **infundibulum** (the funnel-shaped termination formed of **fimbriae**).

– **convey the fertilized or unfertilized oocytes to the uterus** by ciliary action and muscular contraction, which takes 3 to 4 days.

– transport spermatozoa in the opposite direction; **fertilization** takes place within the tube, usually in the **infundibulum or ampulla.**

C. Uterus

– is the organ of gestation, in which the fertilized oocyte normally becomes embedded and the developing organism grows until its birth.

– is normally **anteverted** with respect to the vagina and **anteflexed** at the junction of its cervix and body.

– is supported by the pelvic diaphragm; the urogenital diaphragm; the round, broad, lateral, or transverse cervical (cardinal) ligaments; and the pubocervical, sacrocervical, and rectouterine ligaments.

– is supplied primarily by the uterine artery and secondarily by the ovarian artery.

– Its anterior surface rests on the posterosuperior surface of the bladder.

– is divided into four parts for the purpose of description:

1. Fundus

– is the **rounded part** of the uterus located superior and anterior to the plane of the entrance of the uterine tube.

2. Body

– is the main part of the uterus located inferior to the fundus and superior to the isthmus. The uterine cavity is triangular in the **coronal section** and is continuous with the lumina of the uterine tube and with the internal os.

3. Isthmus

– is the **constricted part** of the uterus located between the body and cervix of the uterus. It corresponds to the internal os.

4. Cervix

– is the inferior narrow part of the uterus that projects into the vagina and divides into the following regions:

a. Internal os, the junction of the cervical canal with the uterine body

b. Cervical canal, the cavity of the cervix between the internal and external ostia

c. External os, the opening of the cervical canal into the vagina

D. Vagina

– extends between the vestibule and the cervix of the uterus.

– is located at the lower end of the birth canal.

– serves as the **excretory channel** for the products of menstruation; also serves to receive the penis during coitus.

– The **fornix** is the recess between the cervix and the wall of the vagina.

– Its opening into the vestibule is partially closed by a membranous crescentic fold, the **hymen.**

– is supported by the levator ani; the transverse cervical, pubocervical, and sacrocervical ligaments (upper part); the urogenital diaphragm (middle part); and the perineal body (lower part).

– receives blood from the vaginal branches of the uterine artery and of the internal iliac artery.

– has lymphatic drainage in two directions: The lymphatics from the upper three-fourths drain into the internal iliac nodes; those from the lower one-fourth, below the hymen, drain downward to the perineum and thus into the superficial inguinal nodes.

VIII. Rectum and Anal Canal

A. Rectum

– is the part of the **large intestine** that extends from the sigmoid colon to the anal canal and follows the curvature of the sacrum and coccyx.

– Its lower dilated part, the **ampulla,** lies immediately above the pelvic diaphragm and **stores the feces.**

– has a peritoneal covering on its anterior, right, and left sides for the proximal third; only on its front for the middle third; and no covering for the distal third.

– Its mucous membrane and the circular muscle layer form three folds, the **transverse folds of the rectum (Houston's valves).**

– receives blood from the superior, middle, and inferior rectal arteries and the middle sacral artery. (The superior rectal artery pierces the muscular wall and courses in the submucosal layer and anastomoses with branches of the inferior rectal artery. The middle rectal artery supplies the posterior part of the rectum.)

– Its venous blood returns to the portal venous system via the superior rectal vein and to the caval (systemic) system via the middle and inferior rectal veins. (The middle rectal vein drains primarily the muscular layer of the lower part of the rectum and upper part of the anal canal.)

– receives parasympathetic nerve fibers by way of the pelvic splanchnic nerve.

B. Anal canal

– lies below the pelvic diaphragm and ends at the **anus.**

– is divided into an upper two-thirds (**visceral portion**), which belongs to the intestine, and a lower one-third (**somatic portion**), which belongs to the perineum with respect to mucosa, blood supply, and nerve supply.

– The **anal columns** are 5 to 10 longitudinal folds of mucosa in its upper half (each column contains a small artery and a small vein).

– The **anal valves,** crescent-shaped mucosal folds, connect the lower ends of the anal columns.

– The **anal sinuses** are a series of pouch-like recesses at the lower end of the anal column in which the anal glands open.

– The **internal anal sphincter** is a thickening of the circular smooth muscle in the lower part of the rectum that is separated from the **external anal sphincter** (skeletal muscle has three parts: subcutaneous, superficial, and deep) by the intermuscular groove called **Hilton's white line.**

– The **pectinate line,** a serrated line following the anal valves and crossing the bases of the anal columns, is a point of demarcation.

1. The epithelium is **columnar** or **cuboidal** above the pectinate line and **stratified squamous** below it.

2. Venous drainage above the pectinate line goes into the **portal venous system** mainly via the superior rectal vein; below the pectinate line, it goes into the **caval system** via the middle and inferior rectal veins.

3. The lymphatic vessels drain into the **internal iliac nodes** above the line and into the **superficial inguinal nodes** below it.

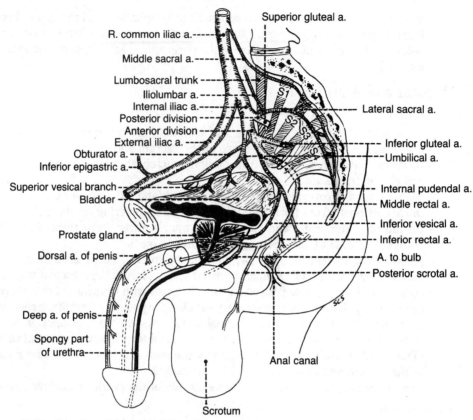

Figure 6-16. Branches of the internal iliac artery.

4. The sensory innervation above the line is through fibers from the pelvic plexus and thus is of the **visceral** type; the sensory innervation below it is by **somatic** nerve fibers of the pudendal nerve (which are very sensitive).

5. **Internal hemorrhoids** occur above the pectinate line, and **external hemorrhoids** occur below it.

C. Defecation

– is initiated by **distension of the rectum,** which has filled from the sigmoid colon.

– The intra-abdominal pressure is increased by contraction of the diaphragm, the abdominal muscles, and the levator ani.

– The **puborectalis** relaxes, so decreasing the angle between the ampulla of the rectum and the upper portion of the anal canal.

– The smooth muscle of the wall of the rectum contracts, the internal anal sphincter relaxes, and the external anal sphincter relaxes to pass the feces.

– After evacuation, the contraction of the puborectalis and the anal sphincters closes the anal canal.

IX. Blood Vessels of the Pelvis (Figure 6-16)

A. Internal iliac artery

– arises from the bifurcation of the common iliac artery, in front of the sacroiliac joint, and is crossed in front by the ureter at the pelvic brim.

– is commonly divided into a **posterior division,** which gives rise to the iliolumbar, lateral sacral, and superior gluteal arteries; and an **anterior division,** which gives rise to the inferior gluteal, internal pudendal, umbilical, obturator, inferior vesical, middle rectal, and uterine arteries.

1. **Iliolumbar artery**
   – runs superolaterally to the iliac fossa, passing deep to the psoas major.
   – divides into an **iliac branch** supplying the iliacus muscle and the ilium and a **lumbar branch** supplying the psoas major and quadratus lumborum muscles.

2. **Lateral sacral artery**
   – passes medially in front of the sacral plexus, giving rise to **spinal branches,** which enter the anterior sacral foramina to supply the spinal meninges and the roots of the sacral nerves, and the muscles and skin overlying the sacrum.

3. **Superior gluteal artery**
   – usually runs between the lumbosacral trunk and the first sacral nerve.
   – leaves the pelvis through the **greater sciatic foramen** above the piriformis muscle to supply muscles in the buttocks.

4. **Inferior gluteal artery**
   – runs between the first and second or between the second and third sacral nerves.
   – leaves the pelvis through the **greater sciatic foramen,** inferior to the piriformis.

5. **Internal pudendal artery**
   – leaves the pelvis through the greater sciatic foramen, passing between the piriformis and coccygeus muscles, and enters the perineum through the **lesser sciatic foramen.**

6. **Umbilical artery**
   – runs forward along the lateral pelvic wall and along the side of the bladder.
   – Its proximal part gives rise to the **superior vesical artery** to the superior part of the bladder and, in the male, to the **artery of the ductus deferens,** which supplies the ductus deferens, the seminal vesicles, the lower part of the ureter, and the bladder.
   – Its distal part is obliterated and continues forward as the **medial umbilical ligament.**

7. **Obturator artery**
   – usually arises from the internal iliac artery, but in about 20% to 30% of the population it arises from the inferior epigastric artery. It then passes close to or across the femoral canal to reach the obturator foramen and hence is susceptible to damage during hernia operations.
   – runs through the upper part of the obturator foramen, divides into **anterior and posterior branches,** and supplies the muscles of the thigh.
   – Its **posterior branch** gives rise to an acetabular branch, which enters the joint through the acetabular notch and reaches the head of the femur by way of the ligamentum capitis femoris.

8. **Inferior vesical artery**
- occurs in the male and corresponds to the **vaginal artery** in the female.
- supplies the fundus of the bladder, prostate gland, seminal vesicles, ductus deferens, and lower part of the ureter.

9. **Vaginal artery**
- arises from the uterine or internal iliac artery.
- gives rise to numerous branches to the anterior and posterior wall of the vagina and makes longitudinal anastomoses in the median plane to form the **anterior and posterior azygos arteries of the vagina.**

10. **Middle rectal artery**
- runs medially to supply mainly the muscular layer of the lower part of the rectum and the upper part of the anal canal.
- also supplies the prostate gland and seminal vesicles (or vagina) and the ureter.

11. **Uterine artery**
- is homologous to the **artery of the ductus deferens** in the male.
- arises from the internal iliac artery or in common with the vaginal or middle rectal artery.
- runs medially in the base of the broad ligament to reach the junction of the cervix and the body of the uterus and runs in front of and above the ureter near the lateral fornix of the vagina.
- divides into a large **superior branch,** supplying the body and fundus of the uterus, and a smaller **vaginal branch,** supplying the cervix and vagina.
- takes a tortuous course along the lateral margin of the uterus and ends by anastomosing with the ovarian artery.

B. **Median sacral artery**
- is an unpaired artery, arising from the posterior aspect of the abdominal aorta just before its bifurcation.
- descends in front of the sacrum, supplying the posterior portion of the rectum, and ends in the **coccygeal body,** which is a small cellular and vascular mass located in front of the tip of the coccyx.

C. **Superior rectal artery**
- is the direct continuation of the inferior mesenteric artery.

D. **Ovarian artery**
- arises from the abdominal aorta, crosses the proximal end of the external iliac artery to enter the pelvic minor, and reaches the ovary through the suspensory ligament of the ovary.

X. Nerve Supply to the Pelvis

A. **Sacral plexus**
- is formed by the fourth and fifth lumbar ventral rami (the lumbosacral trunk) and the first four sacral ventral rami.
- lies largely on the internal surface of the piriformis muscle in the pelvis.

1. **Superior gluteal nerve (L4–S1)**
- leaves the pelvis through the greater sciatic foramen, above the piriformis.
- innervates the gluteus medius, gluteus minimis, and tensor fascia lata muscles.

2. **Inferior gluteal nerve (L5–S2)**
   – leaves the pelvis through the greater sciatic foramen, below the piriformis.
   – innervates the gluteus maximus muscle.

3. **Sciatic nerve (L4–S3)**
   – is the **largest nerve in the body** and is composed of **peroneal** and **tibial** parts.
   – leaves the pelvis through the greater sciatic foramen below the piriformis.
   – enters the thigh in the hollow between the ischial tuberosity and the greater trochanter of the femur.

4. **Nerve to the obturator internus muscle (L5–S2)**
   – leaves the pelvis through the greater sciatic foramen below the piriformis.
   – enters the perineum through the lesser sciatic foramen.
   – innervates the obturator internus and superior gemellus muscles.

5. **Nerve to the quadratus femoris muscle (L5–S1)**
   – leaves the pelvis through the greater sciatic foramen, below the piriformis.
   – descends deep to the gemelli and obturator internus muscles and ends in the deep surface of the quadratus femoris, supplying the quadratus femoris and the inferior gemellus muscles.

6. **Posterior femoral cutaneous nerve (S1–S3)**
   – leaves the pelvis through the greater sciatic foramen below the piriformis.
   – lies alongside the sciatic nerve and descends on the back of the knee.
   – gives rise to several **inferior cluneal nerves** and **perineal branches**.

7. **Pudendal nerve (S2–S4)**
   – leaves the pelvis through the greater sciatic foramen below the piriformis.
   – enters the perineum through the lesser sciatic foramen and the pudendal canal in the lateral wall of the ischiorectal fossa.

8. **Branches distributed to the pelvis**
   – include the nerve to the piriformis muscle (S1–S2), the nerves to the levator ani and coccygeus muscles (S3–S4), the nerve to the sphincter ani externus muscle, and the pelvic splanchnic nerves (S2–S4).

B. **Autonomic nerves**

   1. **Superior hypogastric plexus**
      – is the continuation of the aortic plexus below the aortic bifurcation and receives the lower two lumbar splanchnic nerves.
      – lies behind the peritoneum, descends in front of the fifth lumbar vertebra, and ends by bifurcation into the **right and left hypogastric nerves** in front of the sacrum.
      – contains preganglionic and postganglionic sympathetic fibers, visceral afferent fibers, and few, if any, parasympathetic fibers, which may run a recurrent course through the inferior hypogastric plexus.

   2. **Hypogastric nerve**
      – is the lateral extension of the superior hypogastric plexus and lies in the extraperitoneal connective tissue lateral to the rectum.
      – provides branches to the sigmoid colon and the descending colon.
      – is joined by the pelvic splanchnic nerves to form the inferior hypogastric or pelvic plexus.

3. Inferior hypogastric (pelvic) plexus

- is formed by the union of **hypogastric, pelvic splanchnic, and sacral splanchnic nerves** and lies against the posterolateral pelvic wall, lateral to the rectum, vagina, and base of the bladder.
- contains **pelvic ganglia,** in which both sympathetic and parasympathetic preganglionic fibers synapse. Hence, it consists of preganglionic and post-ganglionic sympathetic fibers, preganglionic and postganglionic parasympathetic fibers, and visceral afferent fibers.
- gives rise to subsidiary plexuses, including the middle rectal plexus, utero-vaginal plexus, vesical plexus, differential plexus, and prostatic plexus.

4. Sacral splanchnic nerves

- consist primarily of preganglionic sympathetic fibers that come off the chain and synapse in the inferior hypogastric (pelvic) plexus.

5. Pelvic splanchnic nerves (nervi erigentes)

- are the only splanchnic nerves that carry parasympathetic fibers. (All other splanchnic nerves are sympathetic.)
- arise from the sacral segment of the spinal cord (S2–S4).
- contribute to the formation of the pelvic (or inferior hypogastric) plexus, and supply the descending colon, sigmoid colon, and other viscerae in the pelvis and perineum.

XI. Clinical Considerations

A. Uterine prolapse

- is the **protrusion of the cervix** of the uterus into the vagina close to the vestibule.
- causes a **bearing-down sensation** in the womb and an increased frequency of and burning sensation on urination.
- occurs as a result of advancing age and is characterized by increased relaxation and loss of tonus of the muscular and fascial structures that constitute the support of the pelvic viscera.
- may be surgically corrected; however, during prolapse surgery, the ureter may be mistaken for the uterine artery and erroneously ligated. (The uterine artery crosses cranially and anterior to the ureter.)

B. Hemorrhoids

- are dilated internal and external venous plexuses around the rectum and anal canal.

1. Internal hemorrhoids occur above the pectinate line and are covered by mucous membrane; their pain fibers are carried by autonomic nerves.

2. External hemorrhoids are situated below the pectinate line, are covered by skin, and are more painful than internal hemorrhoids because their pain fibers are carried by the **inferior rectal nerves.**

C. Hysterectomy

- is **surgical removal of the uterus,** performed either through the abdominal wall or through the vagina.

D. Prostatectomy

– is **surgical removal of a part of the hypertrophied prostate gland**
– Hypertrophy occurs most in the middle or median lobe, obstructing the internal urethral orifice and thus leading to nocturia (excessive urination at night), dysuria (difficulty or pain in urination), and urgency (sudden desire to urinate).

E. Vaginal examination

– is an examination of pelvic structures through the vagina.

1. Inspection with a speculum

– Purpose is to observe the vaginal walls, the posterior fornix as the site of culdocentesis (aspiration of fluid from the rectouterine excavation by puncture of the vaginal wall), the uterine cervix, and the cervical os.

2. Digital examination

– Purpose is to palpate the urethra and bladder through the anterior fornix; the perineal body, rectum, coccyx, and sacrum through the posterior fornix; and the ovaries, uterine tubes, ureters, and ischial spines through the lateral fornices.

3. Bimanual examination

– is performed by placing the fingers of one hand in the vagina and exerting pressure on the lower abdomen with the other hand.
– enables physicians to determine the size and position of the uterus, to palpate the ovaries and uterine tubes, and to detect pelvic inflammation or neoplasms.

F. Rectal examination

– is used to determine the size and consistency of the **prostate gland.**
– is also used to palpate the bladder, seminal vesicle, and ampulla of the ductus deferens anteriorly; the coccyx and sacrum posteriorly; and the ischiorectal fossa (abscess) laterally.

Review Test

Directions: Each of the numbered items or incomplete statements in this section is followed by answers or by completions of the statement. Select the ONE lettered answer or completion that is BEST in each case.

1. Carcinoma of the uterus can spread directly to the labia majora in lymphatics that follow the

(A) pubic arcuate ligament
(B) suspensory ligament of the ovary
(C) cardinal ligament
(D) suspensory ligament of the clitoris
(E) round ligament of the uterus

2. Tenderness and swelling of the left testicle may be produced by thrombosis in the

(A) internal pudendal vein
(B) left renal vein
(C) internal iliac vein
(D) inferior epigastric vein
(E) external pudendal vein

3. If a stab wound injures structures that leave the pelvis above the piriformis muscle, which of the following structures is most likely to be damaged?

(A) Sciatic nerve
(B) Internal pudendal artery
(C) Superior gluteal nerve
(D) Inferior gluteal artery
(E) Posterior femoral cutaneous nerve

4. Which of the following ligaments normally is found in the inguinal canal?

(A) Suspensory ligament of the ovary
(B) Ovarian ligament
(C) Broad ligament
(D) Round ligament of the uterus
(E) Pubovesical ligament

5. Pelvic splanchnic nerves carry preganglionic efferent fibers that synapse in

(A) terminal ganglia on or near the viscerae
(B) sympathetic chain ganglia
(C) collateral ganglia
(D) dorsal root ganglia
(E) the ganglion impar

6. As the uterine artery passes from the anterior division of the internal iliac artery to the uterus, it crosses a structure that is sometimes mistakenly ligated during pelvic surgery. This structure is the

(A) ovarian artery
(B) ovarian ligament
(C) mesovarium
(D) ureter
(E) round ligament of the uterus

7. Which of the following statements concerning the pelvic vasculature is true?

(A) The median sacral artery is a branch of the internal iliac artery
(B) Branches of the uterine artery usually anastomose with branches of the ovarian arteries
(C) The prostatic plexus of veins connects with veins on the external part of the body
(D) The inferior vesical arteries are branches of the obturator artery
(E) The internal pudendal artery is a branch of the posterior division of the internal iliac artery

8. A lesion of the sacral splanchnic nerves would primarily damage

(A) postganglionic parasympathetic fibers
(B) postganglionic sympathetic fibers
(C) preganglionic sympathetic fibers
(D) preganglionic parasympathetic fibers
(E) postganglionic sympathetic and parasympathetic fibers

9. Which of the following statements concerning the ischiorectal fossa is true?

(A) It is bounded anteriorly by the transverse ligament of the urogenital diaphragm
(B) It is partially bounded posteriorly by the gluteus maximus
(C) The pudendal canal runs along its medial wall
(D) The levator ani separates it from the urogenital triangle
(E) It contains a perineal branch of the fifth lumbar nerve

10. Which of the following structures constitutes the superior boundary of the superficial perineal space?

(A) Pelvic diaphragm
(B) Colles' fascia
(C) Superficial layer of the superficial fascia
(D) Deep layer of the superficial fascia
(E) Perineal membrane

11. Which one of the following groups of structures is found in the urogenital diaphragm?

(A) Deep transverse perineal muscles; bulbourethral glands; membranous urethra
(B) Deep transverse perineal and bulbospongiosus muscles; part of the spongy urethra
(C) Arteries to the bulb; ischiocavernosus muscles; bulbourethral glands
(D) Superficial transverse perineal and sphincter urethrae muscles; prostatic urethra
(E) Sphincter urethrae; bulbourethral and great vestibular glands

12. Which of the following statements concerning the external anal sphincter is true?

(A) It is primarily innervated and controlled by autonomic nerves
(B) It contains mostly smooth muscles
(C) It has deep, superficial, and subcutaneous components
(D) It has lateral fibers that interdigitate with fibers of the obturator internus muscle
(E) It extends superiorly as far as the lower end of the sigmoid colon

13. Which of the following statements concerning ducts from the prostate gland is true?

(A) They open into the membranous part of the urethra
(B) They open onto the seminal colliculus
(C) They open onto the cavernous urethra
(D) They open on either side of the urethral crest
(E) They open into the prostatic utricle

14. Which of the following statements concerning the duct of the seminal vesicle is true?

(A) It joins the duct of a bulbourethral gland to form the ejaculatory duct
(B) It opens into the membranous urethra
(C) It widens just before it opens into the urinary bladder to form the ejaculatory duct
(D) It widens to form the ampulla of the ductus deferens
(E) It unites with the ductus deferens to form a single ejaculatory duct

15. Which of the following statements concerning the round ligament of the uterus is true?

(A) It does not contain smooth muscle fibers
(B) It inserts into the pecten pubis
(C) It traverses the inguinal canal
(D) It contains the uterine artery
(E) It runs medial to the inferior epigastric vessels

16. Destruction of the urogenital diaphragm would most likely cause paralysis of which of the following muscles?

(A) Sphincter urethrae
(B) Coccygeus
(C) Superficial transverse perineal muscle
(D) Levator ani
(E) Obturator internus

17. A benign tumor located near a gap between the arcuate pubic ligament and the transverse perineal ligament might compress which of the following structures?

(A) Dorsal nerve of the penis
(B) Deep dorsal vein of the penis
(C) Superficial dorsal vein
(D) Dorsal artery of the penis
(E) Deep artery of the penis

18. If an obstetrician performed a median episiotomy that damaged the perineal body, function of which of the following muscles might be impaired?

(A) Ischiocavernosus and sphincter urethrae
(B) Deep transverse perineal and obturator internus
(C) Bulbospongiosus and superficial transverse perineal
(D) Superficial sphincter of anus and sphincter urethrae
(E) Bulbospongiosus and ischiocavernosus

19. A 22-year-old man has a gonorrheal infection that has infiltrated the space located between the inferior fascia of the urogenital diaphragm and the superficial perineal fascia. Which of the following structures might be inflamed with it?

(A) Prostate gland
(B) Bulbourethral gland
(C) Membranous part of the male urethra
(D) Superficial transverse perineal muscle
(E) Sphincter urethrae

20. Which of the following ducts opens into the prostatic urethra on the seminal colliculus?

(A) Duct of the seminal vesicles
(B) Duct of the prostate gland
(C) Ejaculatory duct
(D) Duct of the bulbourethral gland
(E) Duct of the epididymis

21. The normal position of the uterus is

(A) anteflexed and anteverted
(B) retroflexed and anteverted
(C) anteflexed and retroverted
(D) retroverted and retroflexed
(E) anteverted and retroverted

22. Which of the following statements concerning perineal structures is true?

(A) The dorsal artery of the penis lies superficial to Buck's fascia
(B) Buck's fascia intervenes between the bulb of the penis and Colles' fascia
(C) The superficial perineal pouch lies between the pelvic fascia and the perineal membrane
(D) The deep artery of the penis is found in the corpus spongiosum
(E) The deep perineal pouch contains the greater vestibular gland

23. A 62-year-old man is incapable of penile erection after rectal surgery with prostatectomy. The patient is most likely to have suffered lesion of which of the following nerves?

(A) Dorsal nerve of the penis
(B) Perineal nerve
(C) Hypogastric nerve
(D) Sacral splanchnic nerve
(E) Pelvic splanchnic nerve

24. Which of the following structures is drained by the lumbar (aortic) lymph nodes?

(A) Perineum
(B) Lower part of the vagina
(C) External genitalia
(D) Ovary
(E) Lower part of the anterior abdominal wall

Questions 25–29

Answer questions 25–29 using the diagram below.

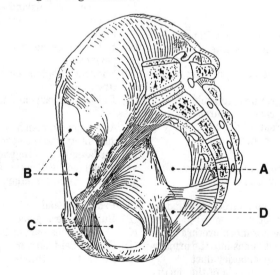

25. Which of the following structures passes through C in the diagram?

(A) Iliolumbar artery
(B) Umbilical artery
(C) Ilioinguinal nerve
(D) Obturator nerve
(E) Nerve to the obturator internus muscle

26. Which of the following structures passes through D in the diagram?

(A) Pudendal nerve
(B) Posterior cutaneous nerve of the thigh
(C) Tendon of the obturator externus muscle
(D) Inferior gluteal artery
(E) Piriformis muscle

Directions: Each of the numbered items or incomplete statements in this section is negatively phrased, as indicated by a capitalized word such as NOT, LEAST, or EXCEPT. Select the ONE lettered answer or completion that is BEST in each case.

27. All of the following structures pass through A in the diagram EXCEPT

(A) the piriformis muscle
(B) the superior gluteal nerve
(C) the sciatic nerve
(D) the pudendal nerve
(E) the tendon of the obturator internus muscle

28. All of the following structures pass through B in the diagram EXCEPT

(A) the femoral vein
(B) the iliopsoas muscle
(C) the external pudendal artery
(D) the femoral canal
(E) the lateral femoral cutaneous nerve

29. All of the following structures pass through both A and D in the diagram EXCEPT

(A) the internal pudendal artery
(B) the pudendal nerve
(C) the nerve to the obturator internus muscle
(D) the tendon of the obturator internus muscle
(E) the internal pudendal vein

30. A 16-year-old boy presents at the emergency room with rupture of the penile urethra. Extravasated urine from this injury can spread into all of the following structures EXCEPT

(A) the scrotum
(B) the penis
(C) the superficial perineal space
(D) the lower abdominal wall
(E) the thigh

31. All of the following structures may be palpated during vaginal examination EXCEPT

(A) apex of the urinary bladder
(B) ischial spines
(C) ureter
(D) ovary
(E) uterine cervix

32. All of the following structures lie in the broad ligament for all or part of their course EXCEPT

(A) ovarian ligament
(B) uterine artery
(C) round ligament of the uterus
(D) uterine tube
(E) suspensory ligament of the ovary

33. All of the following structures can be palpated on rectal examination in the male EXCEPT

(A) prostate gland
(B) bulb of the penis
(C) ureters
(D) seminal vesicles
(E) ampulla of the ductus deferens

34. All of the following statements concerning the levator ani are true EXCEPT

(A) the puborectal sling (puborectalis) relaxes during defecation
(B) the iliococcygeus usually is the most developed muscle in the levator ani
(C) the pubococcygeus may be torn or damaged during parturition, allowing the descent of the pelvic viscera
(D) it forms the lateral wall of the ischiorectal fossa
(E) it forms the major part of the pelvic diaphragm

35. All of the following structures form part of the boundary of the pelvic inlet EXCEPT

(A) the promontory of the sacrum
(B) the anterior border of the ala of the sacrum
(C) the pectineal line
(D) the iliac crest
(E) the pubic crest

36. To completely anesthetize the skin of the urogenital triangle, sensation must be blocked in all of the following nerves EXCEPT

(A) the ilioinguinal nerve
(B) the iliohypogastric nerve
(C) the posterior cutaneous nerve of the thigh
(D) the pudendal nerve
(E) the genitofemoral nerve

37. All of the following structures form part of the boundary of the perineum EXCEPT

(A) the pubic arcuate ligament
(B) the tip of the coccyx
(C) ischial tuberosities
(D) the sacrospinous ligament
(E) the sacrotuberous ligament

38. All of the following statements concerning the scrotum are true EXCEPT

(A) it is innervated anteriorly by the ilioinguinal nerve
(B) it receives its blood supply from the testicular artery
(C) it is partitioned into two sacs by a septum of superficial fascia
(D) lymphatic drainage from the scrotum is primarily into superficial inguinal lymph nodes
(E) it has a dartos layer of fascia and muscle that is continuous with Colles' layer of superficial fascia in the perineum

39. All of the following structures help support the uterus EXCEPT

(A) the sacrogenital folds
(B) the round ligament of the uterus
(C) lateral cervical ligaments
(D) the pelvic diaphragm
(E) rectouterine folds

40. When performing a mediolateral episiotomy during breech delivery, an obstetrician would make an incision through all of the following structures EXCEPT

(A) the vaginal wall
(B) the superficial transverse perineal muscle
(C) the bulbospongiosus
(D) the levator ani
(E) the perineal membrane

41. All of the following arteries send branches to the labia majora EXCEPT

(A) the internal iliac artery
(B) the obturator artery
(C) the uterine artery
(D) the internal pudendal artery
(E) the external pudendal artery

42. All of the following nerves from the lumbosacral plexus leave the abdominal or pelvic cavity EXCEPT

(A) the ilioinguinal nerve
(B) the genitofemoral nerve
(C) the lumbosacral trunk
(D) the femoral nerve
(E) the lateral femoral cutaneous nerve

43. All of the following structures lie in the superficial perineal space EXCEPT

(A) ischiocavernosus muscles
(B) superficial transverse perineal muscles
(C) greater vestibular glands
(D) bulbourethral glands
(E) vestibular bulbs

44. Superficial inguinal nodes receive lymph from all of the following structures EXCEPT

(A) the lower part of the anal canal
(B) the labium majus
(C) the clitoris
(D) the testis
(E) the scrotum

45. All of the following statements concerning the prostate gland are correct EXCEPT

(A) the prostatic utricle opens at the apex of the seminal colliculus
(B) the middle lobe of the prostate gland is posterior to the urethra
(C) the uvula of the male bladder often is accentuated by hypertrophy of the middle lobe of the prostate
(D) the lateral lobes form the main mass of the prostate gland
(E) the prostatic utricle is the terminal end of the duct of the prostate gland

46. All of the following structures cross the pelvic brim EXCEPT

(A) the ovarian artery
(B) the ureter
(C) the round ligament of the uterus
(D) the uterine artery
(E) the lumbosacral trunk

47. All of the following occur during ejaculation EXCEPT

(A) the urethral sphincter at the neck of the bladder closes
(B) the prostate gland, seminal vesicles, and bulbourethral glands contract
(C) smooth muscles in the efferent ducts and the ductus deferens contract
(D) semen is propelled into the urethra
(E) urine leaves the bladder

48. All of the following statements concerning the bulbourethral (Cowper's) glands are true EXCEPT

(A) they lie in the deep perineal space
(B) they are embedded in the sphincter urethrae
(C) they produce semen and sperm
(D) their ducts open into the bulbous portion of the penile urethra
(E) they lie on the posterolateral side of the membranous urethra

49. All of the following statements concerning the sphincter urethrae are true EXCEPT

(A) it is striated muscle
(B) it is innervated by the perineal nerve
(C) its innervation originates in spinal cord segments S2–S4
(D) it is enclosed in the pelvic fascia
(E) it forms a part of the urogenital diaphragm

50. All of the following statements concerning the pudendal nerve are true EXCEPT

(A) it passes through the lesser sciatic foramen
(B) it innervates the scrotum
(C) it gives off branches that innervate the labia majora
(D) it can be blocked by injecting an anesthetic near the inferior margin of the ischial spine
(E) it arises from the lumbar plexus

Directions: Each set of matching questions in this section consists of a list of four to twenty-six lettered options (some of which may be in figures) followed by several numbered items. For each numbered item, select the ONE lettered option that is most closely associated with it. To avoid spending too much time on matching sets with large numbers of options, it is generally advisable to begin each set by reading the list of options. Then, for each item in the set, try to generate the correct answer and locate it in the option list, rather than evaluating each option individually. Each lettered option may be selected once, more than once, or not at all.

Questions 51–55

Match each description below with the most appropriate muscle.

(A) Bulbospongiosus
(B) Ischiocavernosus
(C) Sphincter urethrae
(D) Levator ani
(E) Obturator internus

51. Helps form a major uterine support

52. Covers, or is close to, the major vestibular glands

53. Lies on the surface of the crus

54. Forms the lateral wall of the ischiorectal fossa

55. Embedded in this muscle (in males) is an accessory reproductive gland

Questions 56–60

Match each description below with the most appropriate structure.

(A) Prostate gland
(B) Seminal vesicle
(C) Great vestibular gland
(D) Bulbourethral gland
(E) Anal gland

56. Lies on the posterolateral aspect of the bladder

57. Lies lateral to the membranous urethra

58. Lies in the superficial perineal space

59. Has numerous ducts that open into the urethra

60. Has ducts that empty into the bulb of the penis

Questions 61–65

Match each of the following statements with the most appropriate organ or structure.

(A) Ovary
(B) Uterus
(C) Uterine tube
(D) Vagina
(E) Clitoris

61. Ovum fertilization occurs here

62. Is supported by the cardinal ligaments

63. Opens into the peritoneal cavity

64. Is bounded by the external and internal iliac vessels

65. Is attached to the pubic symphysis by the suspensory ligament

Answers and Explanations

1–E. The round ligament of the uterus runs laterally from the uterus through the deep inguinal ring, inguinal canal, and superficial inguinal ring and becomes lost in the subcutaneous tissues of the labium majus. Thus, carcinoma of the uterus can spread directly to the labium majus by traveling in lymphatics that follow the ligament.

2–B. A tender swollen left testis may be produced by thrombosis in the left renal vein because the left testicular vein drains into the left renal vein.

3–C. The superior gluteal nerve leaves the pelvis through the greater sciatic foramen, above the piriformis. The sciatic nerve, internal pudendal vessels, inferior gluteal vessels and nerve, and posterior femoral cutaneous nerve leave the pelvis below the piriformis.

4–D. The round ligament of the uterus is found in the inguinal canal along its course.

5–A. The pelvic splanchnic nerves carry preganglionic parasympathetic (GVE) fibers that synapse in the ganglia of the inferior hypogastric plexus and in minute ganglia in the muscular walls of the pelvic organs.

6–D. The ureter runs under the uterine artery near the cervix; thus, the ureter is sometimes mistakenly ligated during pelvic surgery.

7–B. Branches of the uterine artery usually anastomose with branches of the ovarian arteries. The median sacral artery is a branch of the abdominal aorta. The prostatic plexus of veins is connected with the external pudendal vein and the superficial and deep dorsal veins of the penis. The inferior vesical artery is a branch of the internal iliac artery. The internal pudendal and inferior gluteal arteries are branches of the anterior division of the internal iliac artery.

8–C. The sacral splanchnic nerves consist primarily of preganglionic sympathetic neurons.

9–B. The ischiorectal fossa is bounded anteriorly by the superficial transverse perineal and deep transverse perineal muscles, and posteriorly by the gluteus maximus and the sacrotuberous ligament. It contains the inferior rectal nerve and vessels and the pudendal canal, which runs along the inner surface of its lateral wall.

10–E. The superior (deep) boundary of the superficial perineal space is the perineal membrane (inferior fascia of the urogenital diaphragm). Colles' fascia is the deep membranous layer of the superficial perineal fascia.

11–A. The urogenital diaphragm in the deep perineal space consists of the deep transverse perineal muscles and the sphincter urethrae; it is pierced in the male by the membranous urethra, which is accompanied by the bulbourethral (Cowper's) glands.

12–C. The external anal sphincter has deep, superficial, and subcutaneous components.

13–D. Ducts from the prostate gland open into the prostatic sinus, which is a groove on either side of the urethral crest. The prostate gland receives the ejaculatory duct, which opens into the prostatic urethra on the seminal colliculus just lateral to the prostatic utricle.

14–E. A duct from a seminal vesicle joins the ductus deferens to form an ejaculatory duct.

15–C. The round ligament of the uterus enters the inguinal canal through the deep inguinal ring, emerges through the superficial inguinal ring, and becomes lost in the subcutaneous tissue of the labium majus. It contains smooth muscle fibers and runs lateral to the inferior epigastric vessels.

16–A. The urogenital diaphragm consists of the sphincter urethrae and deep transverse perineal muscles.

17–B. The deep dorsal vein of the penis enters the pelvis through a gap between the arcuate pubic ligament and the transverse perineal ligament.

18–C. The perineal body (central tendon of the perineum) is a fibromuscular node at the center of the perineum. It provides attachment for the bulbospongiosus, the superficial and deep transverse perineal, and the sphincter ani externus muscles.

19–D. The superficial transverse perineal muscle is located in the superficial perineal space between the inferior fascia of the urogenital diaphragm and the membranous layer of the superficial perineal fascia (Colles' fascia). The bulbourethral (Cowper's) glands and the membranous urethra are found in the deep perineal pouch.

20–C. The ejaculatory duct opens into the prostatic urethra on the seminal colliculus. The ducts of the seminal vesicles and the ductus deferens form the ejaculatory duct. The ducts of the prostate gland open into the prostatic sinus, which is a groove on each side of the urethral crest. The duct of the bulbourethral gland opens into the lumen of the bulbous portion of the penile urethra.

21–A. The normal position of the uterus is anteflexed and anteverted.

22–B. The dorsal artery of the penis lies deep to Buck's fascia. The superficial perineal pouch lies between the membranous layer of superficial perineal fascia (Colles' fascia) and the inferior fascia of the urogenital diaphragm (perineal membrane). The deep artery of the penis runs through the corpus cavernosum.

23–E. The pelvic splanchnic nerve contains preganglionic parasympathetic fibers, whereas the sacral splanchnic nerve contains preganglionic sympathetic fibers. Parasympathetic fibers are responsible for erection, whereas sympathetic fibers are involved with ejaculation. The right and left hypogastric nerves contain primarily sympathetic fibers and visceral sensory fibers. The dorsal nerve of the penis and the perineal nerve provide sensory nerve fibers.

24–D. The lymphatic vessels from the ovary ascend with the ovarian vessels in the suspensory ligament and terminate in the lumbar (aortic) nodes. Lymphatic vessels from the perineum, external genitalia, and lower part of the anterior abdominal wall drain into the superficial inguinal nodes.

25–D. Space C in the diagram is the obturator foramen. The obturator nerve passes through the obturator foramen, where it divides into anterior and posterior branches.

26–A. Space D in the diagram is the lesser sciatic foramen, which transmits the pudendal nerve, internal pudendal vessels, and the obturator internus tendon.

27–E. Space A in the diagram is the greater sciatic foramen. The tendon of the obturator internus foramen. The tendon of the obturator internus muscle passes through the lesser sciatic foramen. The piriformis muscle and the superior gluteal, sciatic, and pudendal nerves pass through the greater sciatic foramen.

28-C. The external pudendal artery arises from the femoral artery inferior to the inguinal ligament and supplies the pubic area. The space deep to the inguinal ligament is separated by the iliopectineal arcus (ligament) into the lateral muscular lacunae, which transmits the iliopsoas muscle and the lateral femoral cutaneous nerve, and the medial vascular lacunae, which transmits the femoral nerve and vessels.

29–D. The pudendal nerve, internal pudendal vessels, and the nerve to the obturator internus muscle pass through both the greater (A) and lesser (D) sciatic foramina. The tendon of the obturator internus muscle runs only through the lesser sciatic foramen.

30–E. Urine will not extravasate into the thigh because Scarpa's fascia ends by firm attachment to the fascia lata of the thigh.

31–A. The apex of the urinary bladder is the anterior end of the bladder, or the junction of the superior and anteroinferior surfaces of the bladder behind the pubic symphysis. It cannot be palpated during vaginal examination.

32–E. The suspensory ligament of the ovary is a band of peritoneum that extends superiorly from the ovary to the pelvic wall, and is not contained in the broad ligament.

33–C. In the male, the pelvic part of the ureter lies lateral to the ductus deferens and enters the posterosuperior angle of the bladder, where it is situated anterior to the upper end of the seminal vesicle, and thus cannot be palpated during rectal examination. However, in the female the ureter can be palpated during vaginal examination because it runs near the uterine cervix and the lateral fornix of the vagina to enter the posterosuperior angle of the bladder.

34–D. The lateral wall of the ischiorectal fossa is formed by the obturator internus muscle.

35–D. The pelvic inlet (pelvic brim) is bounded by the promontory and the anterior border of the ala of the sacrum, the arcuate line of the ilium, the pectineal line, the pubic crest, and the superior margin of the pubic symphysis.

36–B. The skin of the urogenital triangle is innervated by the pudendal nerve, perineal branches of the posterior femoral cutaneous nerve, anterior scrotal or labial branches of the ilioinguinal nerve, and the genital branch of the genitofemoral nerve.

37–D. The sacrospinous ligament forms a boundary of the lesser sciatic foramen. The pubic arcuate ligament, tip of the coccyx, ischial tuberosities, and sacrotuberous ligament all form part of the boundary of the perineum.

38–B. The scrotum receives blood from the posterior scrotal branches of the internal pudendal arteries and the anterior scrotal branches of the external pudendal arteries. The lymph vessels from the scrotum drain into the superficial inguinal nodes, whereas the lymph vessels from the testis drain into the upper lumbar nodes.

39–A. The sacrogenital folds are peritoneal folds that extend posteriorly from the sides of the bladder, on either side of the rectum, to the sacrum, and thus they do not support the uterus. The uterus is supported by the pelvic diaphragm, the round ligament of the uterus, and the lateral cervical (cardinal), rectouterine, transverse cervical, pubocervical, and sacrocervical ligaments.

40–D. The levator ani is the major part of the pelvic diaphragm, which forms the pelvic floor and supports all of the pelvic organs. An obstetrician should avoid incising the levator ani and the external anal sphincter.

41–C. The uterine artery remains within the pelvic cavity. It does not leave the pelvic cavity to supply the perineal region.

42–C. The lumbosacral trunk is formed by part of the ventral ramus of the fourth lumbar nerve and the ventral ramus of the fifth lumbar nerve. This trunk contributes to the formation of the sacral plexus by joining the ventral ramus of the first sacral nerve in the pelvic cavity and does not leave the pelvic cavity.

43–D. The bulbourethral glands are found in the deep perineal space. The superficial perineal space contains the ischiocavernosus muscle, the superficial transverse perineal muscle, the greater vestibular glands, and the vestibular bulbs, which are covered with the bulbospongiosus.

44–D. Lymphatic vessels from the testis and epididymis ascend along the testicular vessels in the spermatic cord through the inguinal canal and continue upward in the abdomen to drain into the upper lumbar nodes.

45–E. The prostatic utricle is a minute pouch on the summit of the seminal colliculus.

46–D. The uterine artery does not cross the pelvic brim. It arises from the internal iliac artery and then runs medially in the base of the broad ligament to the junction of the cervix and body of the uterus.

47–E. Ejaculation occurs with the contraction of smooth muscle of the epididymal ducts and ductus deferens. During ejaculation, a sphincter at the neck of the bladder contracts (preventing sperm from entering the bladder and preventing urine from leaving it); the seminal vesicles, prostate gland, and bulbourethral glands contract to pump their secretions into the urethra; and semen is propelled through the ducts and out the external urethral opening.

48–C. Semen is a thick, yellowish white, viscid fluid containing spermatozoa; it is a mixture of the secretions of the testes, seminal vesicles, prostate, and bulbourethral glands. Sperm, or spermatozoa, are produced in the seminiferous tubules of the testis and matured in the head of the epididymis.

49–D. The sphincter urethrae is not enclosed in the pelvic fascia. The pelvic fascia is an inclusive term for the fascia in the pelvis that includes the parietal pelvic fascia lining the walls of the pelvis and the obturator internus muscle, and the visceral pelvic fascia investing the pelvic organs.

50-E. The pudendal nerve arises from the sacral plexus. It provides sensory innervation to the scrotum or labium majus. It leaves the pelvis through the greater sciatic foramen and enters the perineum through the lesser sciatic foramen near the inferior margin of the ischial spine.

51–D. The levator ani and coccygeus muscles form the pelvic diaphragm, which is a major uterine support.

52–A. The bulbospongiosus covers or is in close proximity to the major vestibular glands.

53–B. The ischiocavernosus lies on the surface of the crus of the penis or clitoris.

54–E. The obturator internus muscle forms the lateral wall of the ischiorectal fossa.

55–C. The bulbourethral glands are embedded in the sphincter urethrae.

56–B. The seminal vesicles are lobulated glandular structures and lie lateral to the ampullae of the ductus deferens, against the posterolateral aspect of the bladder.

57–D. The bulbourethral glands lie lateral to the membranous urethra in the deep perineal pouch.

58–C. The greater vestibular glands are found in the superficial perineal space (pouch).

59–A. The prostate gland opens its ducts into the prostatic sinus, which is a groove on each side of the urethral crest.

60–D. Ducts of the bulbourethral glands open into the bulb of the penis.

61–C. Ova are fertilized in the uterine tube, usually in the infundibulum or ampulla.

62–B. The uterus is supported by the cardinal ligaments.

63–C. The uterine tube opens into the peritoneal cavity.

64–A. The ovary is bounded by the external and internal iliac vessels. It is not enclosed by the peritoneum (broad ligament), but is attached along its posterior surface by the mesovarium.

65–E. The clitoris is attached to the pubic symphysis by the suspensory ligament.

7

Back

Vertebral Column

I. General Characteristics

- The vertebral column consists of 33 vertebrae (7 cervical, 12 thoracic, 5 lumbar, 5 fused sacral, and 4 fused coccygeal vertebrae).
- The **primary curvatures** are located in the thoracic and sacral regions, and the **secondary curvatures** are located in the cervical and lumbar regions.
- **Abnormal curvatures** may include the following:

A. **Kyphosis (hunchback),** an abnormal exaggerated thoracic curvature

B. **Lordosis (swayback or saddle back),** an abnormal accentuation of lumbar curvature

C. **Scoliosis,** a condition of lateral deviation due to unequal growth of the spinal column, pathologic erosion of vertebral bodies, or asymmetric paralysis or weakness of vertebral muscles

II. Typical Vertebra (Figure 7-1)

- consists of a **body** and a **vertebral arch** with several processes for muscular and articular attachments.

A. **Body**
- is a short cylinder, **supports weight,** and is separated and also bound together by the **intervertebral disks,** forming the **cartilaginous joints.**
- has **costal facets** on its sides, which articulate with the heads of the corresponding and subjacent ribs.

B. **Vertebral arch**
- consists of paired **pedicles** laterally and paired **laminae** posteriorly.
- gives rise to **seven processes:** one spinous, two transverse, and four articular.
- Failure of the vertebral arches to fuse results in **spina bifida,** which is classified as follows:

1. **Spina bifida occulta:** bony defect only

2. **Meningocele:** protrusion of the meninges through the unfused arch of the vertebra

3. **Meningomyelocele:** protrusion of the spinal cord as well as the meninges

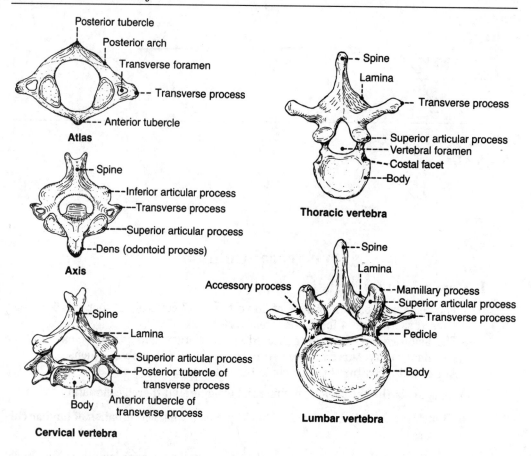

Figure 7-1. Typical cervical, thoracic, and lumbar vertebrae.

C. Vertebral processes

1. Spinous process

– projects posteriorly from the vertebral arch.

2. Transverse processes

– project on each side from the junction of the pedicle and the lamina and articulate with the tubercles of ribs 1 to 10 in the thoracic region.

3. Articular processes (facets)

– are two superior and two inferior projections from the junction of the laminae and pedicles.
– articulate with other articular processes of the arch above or below, forming **synovial joints.**

4. Costal facets (processes)

– arise on the sides of **thoracic vertebral bodies** anterior to the pedicles and articulate with the heads of the ribs.

5. Mamillary processes

– are tubercles on the superior articular processes of the **lumbar vertebrae.**

D. Foramina

1. Vertebral foramina

- are formed by the vertebral bodies and vertebral arches (pedicles and laminae). A series of vertebral foramina forms the **vertebral canal,** which contains the spinal cord, its meningeal coverings, and the associated vessels as well as the nerve roots (caudal to L2).
- transmit the **spinal cord** with its meningeal coverings and associated vessels.

2. Intervertebral foramina

- are located between the inferior and superior surfaces of the pedicles of adjacent vertebrae.
- transmit the **spinal nerves** and accompanying vessels as they exit the vertebral canal.

3. Transverse foramina

- are present in each **transverse process** of the cervical vertebra.
- transmit the **vertebral artery** (except for C7), **vertebral vein,** and **autonomic nerves.**

III. Intervertebral Disk

- lies between the bodies of two vertebrae from the axis to the sacrum.
- consists of a central mucoid substance (**nucleus pulposus**) with a surrounding fibrocartilaginous lamina (**anulus fibrosus**).
- permits **movements between the vertebrae** and **absorbs shocks.**

A. Nucleus pulposus

- is a remnant of the embryonic **notochord** and is situated in the central portion of the intervertebral disk.
- consists of **reticular and collagenous fibers** embedded in mucoid material.
- may protrude or extrude (herniate) through the anulus fibrosus, thereby compressing the roots of the spinal nerve.
- functions as a **shock-absorbing mechanism** by equalizing pressure.

B. Anulus fibrosus

- consists of concentric layers of fibrous tissue and fibrocartilage.
- **binds the vertebral column together,** retains the nucleus, and permits a small amount of movement.
- functions as a shock-absorbing mechanism.

IV. Regional Characteristics of Vertebrae (see Figure 7-1)

A. First cervical vertebra (atlas)

- **supports the skull.**
- has no body and no spinous process but consists of anterior and posterior arches and paired transverse processes.
- articulates superiorly with the **occipital condyles** of the skull to form the **atlanto–occipital joints** and inferiorly with the **axis** to form the **atlantoaxial joints.**

B. Second cervical vertebra (axis)

- has the **smallest transverse process** and is characterized by the presence of the dens (**odontoid process**).

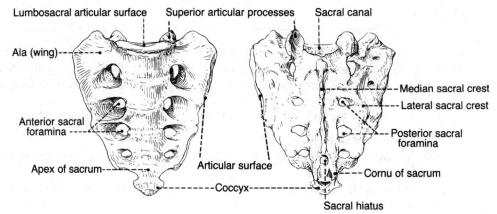

Figure 7-2. Sacrum.

- The **dens of the axis** projects superiorly from the body of the axis, articulates with the **anterior arch of the atlas** and thus forms the pivot around which the atlas rotates, and is supported by the cruciform, apical, and alar ligaments and the tectorial membrane.

C. Seventh cervical vertebra (C7)

- is called the **vertebra prominens** because of its long spinous process.
- is nearly horizontal (usually is not bifid) and forms a visible protrusion.
- gives attachment to the **ligamentum nuchae.**

D. Fifth lumbar vertebra (L5)

- has the **largest body** of the vertebrae. (Lumbar vertebrae are distinguished by their large bodies.)
- is characterized by a strong, massive transverse process and has **mamillary and accessory processes.**

E. Sacrum (Figure 7-2)

- is a large, triangular, wedge-shaped bone composed of **five fused sacral vertebrae.**
- has four pairs of foramina for the exit of the ventral and dorsal primary rami of the first four sacral nerves.
- forms the **posterior part of the pelvis** and provides the **strength and stability of the pelvis.**
- The **promontory** is the prominent anterior edge of the first sacral vertebra (S1).
- is characterized by the following structures:

1. The **ala** is formed by the fused transverse processes and fused costal processes.

2. The **median sacral crest** is formed by the fused spinous processes.

3. The **sacral hiatus** is formed by the failure of the laminae of vertebra S5 to meet.

4. The **sacral cornu or horn** is formed by the pedicles of the fifth sacral vertebra and is an important landmark for locating the sacral hiatus for the administration of **caudal (extradural) anesthesia.**

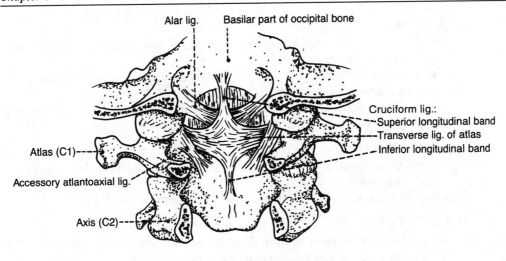

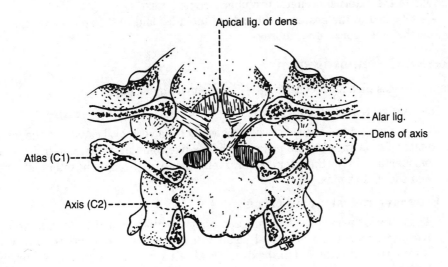

Figure 7-3. Ligaments of the atlas and the axis.

F. Coccyx

- is a wedge-shaped bone formed by the union of the **four coccygeal vertebrae** and provides attachment for the **coccygeus and levator ani muscles.**

V. Ligaments of the Vertebral Column (Figure 7-3)

A. Anterior longitudinal ligament

- runs from the skull to the sacrum on the anterior surface of the vertebral bodies and intervertebral disks.
- is narrowest at the upper end but widens as it descends.
- limits extension of the vertebral column, **supports the anulus fibrosus anteriorly,** and resists gravitational pull, **preventing hyperextension** of the vertebral column.

B. Posterior longitudinal ligament

- interconnects the vertebral bodies and intervertebral disks posteriorly and narrows as it descends.
- supports the posterior aspect of the vertebral column and also **supports the anulus fibrosus posteriorly.**
- limits flexion of the vertebral column and resists gravitational pull, **preventing hyperflexion** of the vertebral column.

C. Ligamentum flavum

- connects the laminae of two adjacent vertebrae and functions to **maintain the upright posture.**
- may be pierced during **lumbar (spinal) puncture.**

D. Ligamentum nuchae

- is a **triangular-shaped median fibrous septum** between the muscles on the two sides of the posterior aspect of the neck.
- is formed of **thickened supraspinous ligaments** that extend from vertebra C7 to the external occipital protuberance and crest.
- is attached to the posterior tubercle of the atlas and to the spinous processes of the other cervical vertebrae.

VI. Vertebral Venous System

- is a valveless plexiform consisting of interconnecting channels.

A. Internal vertebral venous plexus

- lies in the epidural space between the wall of the vertebral canal and the dura mater, forms anterior and posterior ladder-like configurations, and drains into segmental veins by the intervertebral veins that pass through the intervertebral and sacral foramina.

B. External vertebral venous plexus

- The **anterior part** consists of anastomosing longitudinal and transverse veins, lying in front of the vertebral column. These veins communicate with the **intervertebral veins** and the **basivertebral veins,** which lie within the vertebral bodies.
- The **posterior part** lies on the vertebral arch and communicates with the intervertebral veins and the internal plexus.
- also communicates above with the cranial dural sinus, below with the pelvic vein, and in the thoracic and abdominal regions with both the azygos and caval systems.
- is thought to be the **route of early metastasis of carcinoma** from the lung, breast, and prostate gland to bones and the central nervous system.

VII. Clinical Considerations

A. Herniated (slipped) disk

- is **protrusion of the nucleus pulposus** through the anulus fibrosus of the intervertebral disk into the intervertebral foramen, compressing the spinal nerve root.
- usually occurs posterolaterally where the anulus fibrosus is not reinforced by the posterior longitudinal ligament.

B. **Compression fracture**

   – is produced by **collapse of the vertebral bodies** resulting from trauma.
   – may result in **kyphosis** or **scoliosis** and may cause spinal nerve compression.

C. **Whiplash injury of the neck**

   – is produced by a force that drives the trunk forward while the head lags behind, causing **the head to hyperextend and the neck to hyperflex rapidly,** such as occurs in a rear-end automobile collision.
   – occurs frequently at the junction of vertebrae C4 and C5, and thus vertebrae C1–C4 act as the lash, and vertebrae C5–C7 act as the handle of the whip.
   – results in neck pain, stiff neck, and headache.
   – can be treated by supporting the head and neck using a cervical collar that is higher in the back than in the front; the collar keeps the cervical vertebral column in a flexed position.

Soft Tissues of the Back

I. Superficial Tissues

A. Superficial muscles of the back (Figure 7-4)

Muscle	Origin	Insertion	Nerve	Action
Trapezius	External occipital protuberance, superior nuchal line, ligamentum nuchae, spines of C7–T12	Spine of scapula, acromion, and lateral third of clavicle	Spinal accessory nerve, C3–C4	Adducts, rotates, elevates, and depresses scapula
Levator scapulae	Transverse processes of C1–C4	Medial border of scapula	Nerves to levator scapulae, C3–C4; dorsal scapular nerve	Elevates scapula
Rhomboid minor	Spines of C7–T1	Root of spine of scapula	Dorsal scapular nerve, C5	Adducts scapula
Rhomboid major	Spines of T2–T5	Medial border of scapula	Dorsal scapular nerve	Adducts scapula
Latissimus dorsi	Spines of T5–T12, thoracodorsal fascia, iliac crest, ribs 9–12	Floor of bicipital groove of humerus	Thoracodorsal nerve	Adducts, extends, and rotates arm medially
Serratus posterior–superior	Ligamentum nuchae, supraspinal ligament, and spines of C7–T3	Upper border of ribs 2–5	Intercostal nerve, T1–T4	Elevates ribs
Serratus posterior–inferior	Supraspinous ligament and spines of T11–L3	Lower border of ribs 9–12	Intercostal nerve, T9–12	Depresses ribs

B. Blood vessels with nerves

1. Occipital artery

– arises from the external carotid artery, runs deep to the sternocleidomastoid muscle, and lies on the obliquus capitis superior and the semispinalis capitis.
– pierces the trapezius, is accompanied by the **greater occipital nerve (C2)**, and supplies the scalp.
– gives off the **descending branch** which divides into the **superficial branch** that anastomoses with the transverse cervical artery and the **deep branch** that anastomoses with the deep cervical artery from the costocervical trunk.

2. Transverse cervical artery

– arises from the thyrocervical trunk of the subclavian artery.
– Its **superficial branch** accompanies the **spinal accessory nerve** on the deep surface of the trapezius.
– Its **deep branch** accompanies the **dorsal scapular nerve (C5)** deep to the levator scapulae and the rhomboids along the medial side of the scapula.

3. Vertebral artery

– arises from the subclavian artery, ascends through the transverse foramina of the upper six cervical vertebrae.
– winds behind the lateral mass of the atlas, runs in a groove on the superior surface of the posterior arch of the atlas, pierces the dura mater to enter the vertebral canal, and ascends into the cranial cavity through the foramen magnum.

C. Triangles and fascia

1. Triangle of auscultation (see Figure 7-4)

– is bounded by the upper border of the latissimus dorsi, the lateral border of the trapezius, and the medial border of the scapula.
– Its floor is formed by the **rhomboid major.**
– is the site where **breathing sounds** can be heard most clearly.

2. Lumbar triangle

– is formed by the iliac crest, latissimus dorsi, and posterior free border of the external oblique abdominis muscle.

3. Thoracolumbar (lumbodorsal) fascia

– invests the deep muscles of the back.
– Its **anterior layer** lies anterior to the erector spinae and attaches to the vertebral transverse process.
– Its **posterior layer** lies posterior to the erector spinae and attaches to the spinous processes.

II. Deep Tissues of the Back

A. Deep or intrinsic muscles of the back

1. Muscles of the superficial layer: spinotransverse group

– consist of the **splenius capitis** and the **splenius cervicis**.
– originate from the spinous processes and insert into the transverse processes. The **splenius capitis** inserts on the mastoid process and the superior nuchal line.

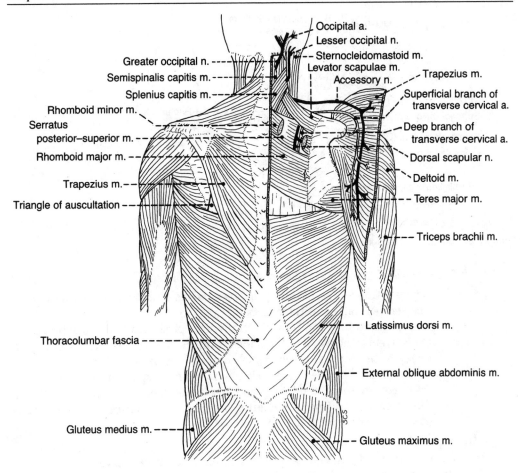

Figure 7-4. Superficial muscles of the back.

- are innervated by the dorsal primary rami of the middle and lower cervical spinal nerves.
- act to **rotate the head and neck** toward the same side and to **extend the head and the trunk.**

2. **Muscles of the intermediate layer: sacrospinalis group**
   - originate from the sacrum, ilium, ribs, and spinous processes of lumbar and lower thoracic vertebrae.
   - are innervated by the dorsal primary rami of the spinal nerves.
   - act to laterally **flex, extend, and rotate the vertebral column and head.**
   - consist of the **erector spinae (sacrospinalis),** which is divided into three columns:

   a. **Iliocostalis (lateral column)**
      - inserts on the ribs and cervical transverse processes.

   b. **Longissimus (medial column)**
      - inserts into the ribs, transverse processes, and mastoid process.

   c. **Spinalis (most medial column)**
      - arises from and inserts into the spinous processes.

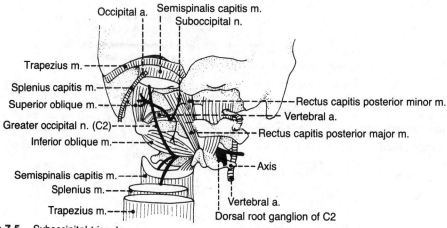

Figure 7-5. Suboccipital triangle.

3. Muscles of the deep layer: transversospinalis group

– originate from the transverse processes and insert into the spinous processes.
– The **semispinalis capitis** inserts into the skull.
– The **long rotators** run from the transverse processes to spinous processes two vertebrae above, and the **short rotators** run from the transverse processes to the spinous processes of adjacent vertebrae.
– are innervated by the dorsal primary rami of the spinal nerves.
– act to **rotate and extend the head, neck, and trunk.**
– consist of the **semispinalis capitis, cervicis, and thoracis;** the **multifidus** (deep to semispinalis); and the **rotators.**

B. Segmental muscles

– are innervated by the dorsal primary rami of the spinal nerves.
– consist of the following:

1. Interspinales

– run between adjacent spinous processes and aid in extension of the vertebral column.

2. Intertransversarii

– run between adjacent transverse processes and aid in lateral flexion of the vertebral column.

III. Suboccipital Area (Figure 7-5)

A. Suboccipital triangle

– is bound medially by the rectus capitis posterior major, laterally by the obliquus capitis superior muscle, and inferiorly by the obliquus capitis inferior muscle.
– Its roof is formed by the semispinalis capitis and longissimus capitis.
– Its floor is formed by the posterior arch of the atlas and posterior atlanto-occipital membrane.
– contains the vertebral artery and suboccipital nerve and vessels.

B. Suboccipital nerve

– is derived from the dorsal ramus of C1 and emerges between the vertebral artery above and the posterior arch of the atlas below.
– supplies the muscles of the suboccipital triangle and semispinalis capitis.

C. Suboccipital muscles

Muscle	Origin	Insertion	Nerve	Action
Rectus capitis posterior major	Spine of axis	Lateral portion of inferior nuchal line	Suboccipital	Extends, rotates, and flexes head laterally
Rectus capitis posterior minor	Posterior tubercle of atlas	Occipital bone below inferior nuchal line	Suboccipital	Extends and flexes head laterally
Obliquus capitis superior	Transverse process of atlas	Occipital bone above inferior nuchal line	Suboccipital	Extends, rotates, and flexes head laterally
Obliquus capitis inferior	Spine of axis	Transverse process of atlas	Suboccipital	Extends head and rotates it laterally

D. Joints

1. Atlanto–occipital joint

- is an **ellipsoidal synovial joint.**
- occurs between the superior articular facets of the atlas and the occipital condyles.
- is involved primarily in **flexion, extension, and lateral bending of the head.**

2. Atlantoaxial joints

- are **synovial joints.**
- consist of **two plane joints,** which are between the superior and inferior articular facets of the atlas and axis, and **one pivot joint,** which is the median joint between the dens of the axis and the anterior arch of the atlas.
- are involved in **rotation of the head.**

E. Components of the occipitoaxial ligament (see Figure 7-3)

1. Cruciform ligament

a. Transverse ligament

- runs between the lateral masses of the atlas, arching over the dens of the axis.

b. Longitudinal ligament

- extends from the dens of the axis to the anterior aspect of the foramen magnum and to the body of the axis.

2. Apical ligament

- extends from the apex of the dens to the anterior aspect of the foramen magnum (of the occipital bone).

3. Alar ligament

- extends from the apex of the dens to the tubercle of the medial side of the occipital condyle.

4. Tectorial membrane

– is an upward extension of the posterior longitudinal ligament from the body of the axis to the basilar part of the occipital bone anterior to the foramen magnum.

– covers the posterior surface of the dens and the apical, alar, and cruciform ligaments.

Spinal Cord and Associated Structures

I. Spinal Cord

– is cylindrical, occupies about the upper two-thirds of the **vertebral canal,** and is enveloped by the three **meninges.**

– has cervical and lumbar enlargements for nerve supply of the upper and lower limbs, respectively.

– **Gray matter** is located in the interior (in contrast to the cerebral hemispheres); the spinal cord is surrounded by **white matter.**

– has a conical end known as the **conus medullaris.**

– grows much more slowly than the bony vertebral column during fetal development, and thus its end gradually shifts to a higher level; ends at the level of L2 in the adult and at the level of L3 in the newborn.

– receives blood from the anterior spinal artery and two posterior spinal arteries as well as from branches of the vertebral, cervical, and posterior intercostal and lumbar arteries.

II. Spinal Nerves

– consist of 31 pairs of nerves (8 cervical, 12 thoracic, 5 lumbar, 5 sacral, and 1 coccygeal).

– are divided into the **dorsal primary rami,** which innervate the skin and deep muscles of the back; the **ventral primary rami,** which form the plexuses (C1–C4, cervical; C5–T1, brachial; L1–L4, lumbar; and L4–S4, sacral); and the **intercostal (T1–T11) and subcostal (T12) nerves.**

– are connected with the sympathetic chain ganglia by **rami communicantes.**

– are mixed nerves, containing all of the **general functional components** (i.e., general somatic afferent [GSA]; general somatic efferent [GSE]; general visceral afferent [GVA]; and general visceral efferent [GVE]).

– contain **sensory (GSA and GVA) fibers** with cell bodies in the dorsal root ganglion.

– contain **motor (GSE) fibers** with cell bodies in the anterior horn of the spinal cord.

– contain **preganglionic sympathetic (GVE) fibers** with cell bodies in the intermediolateral cell column in the lateral horn of the spinal cord (segments between T1 and L2).

– contain **preganglionic parasympathetic (GVE) fibers** with cell bodies in the intermediolateral cell column of the spinal cord segments between S2 and S4. These GVE fibers leave the sacral nerves via the pelvic splanchnic nerves.

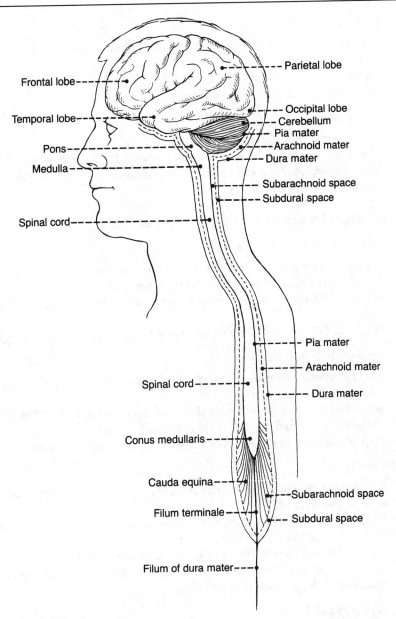

Figure 7-6. Meninges.

III. Meninges (Figure 7-6)

A. Pia mater

 – is the innermost meningeal layer; is closely applied to the spinal cord, and thus cannot be dissected from it. It also enmeshes blood vessels on the surfaces of the spinal cord.
 – has lateral extensions (**denticulate ligaments**) between nerve roots of spinal nerves and an inferior extension as the filum terminale.

B. Arachnoid layer

- is a filmy, transparent, spidery layer connected to the pia mater by web-like trabeculations.
- forms the **subarachnoid space,** the space between the arachnoid layer and the pia mater that is filled with **cerebrospinal fluid (CSF).**

C. Dura mater

- is the tough, fibrous, outermost layer of the meninges.
- The **subdural space** is internal to it (between the arachnoid and dura).
- The **epidural space** is external to it and contains the internal vertebral venous plexus and epidural fat.

IV. Structures Associated with the Spinal Cord

A. Cauda equina ("horse's tail")

- is formed by a great lash of dorsal and ventral roots of the lumbar and sacral spinal nerves that surround the **filum terminale.**
- is located within the **subarachnoid space,** below the level of the **conus medullaris.**
- is free to float in the CSF and therefore is not damaged during a spinal tap.

B. Denticulate ligaments

- are lateral extensions of the pia through the arachnoid, attaching to the inner surface of the dura mater.
- consist of 21 pairs of toothpick-like processes that attach to the dura mater between dorsal and ventral roots of the spinal nerves.
- help to **suspend the spinal cord** within the subarachnoid space.

C. Filum terminale

- is a prolongation of the pia mater from the tip (conus medullaris) of the spinal cord at the level of L2.
- lies in the midst of the cauda equina and ends at the level of S2 by attaching to the apex of the dural sac.
- blends with the dura at the apex of the dural sac, and then the dura continues downward as the **filum of the dura mater (coccygeal ligament),** which is attached to the dorsum of the coccyx.

D. Cerebrospinal fluid

- is contained in the subarachnoid space between the arachnoid and pia mater.
- is formed by **vascular choroid plexuses** in the ventricles of the brain.
- circulates through the ventricles, enters the subarachnoid space, and eventually filters into the venous system through arachnoid villi projecting into the dural venous sinuses, particularly the superior sagittal sinus.

V. Dermatome

- is an area of skin innervated by sensory fibers derived from a particular spinal nerve or segment of the spinal cord. Knowledge of the segmental innervation is useful clinically to produce a region of anesthesia or to determine which nerve has been damaged.

VI. Clinical Considerations

A. Lumbar puncture (spinal tap)

– is the tapping of the subarachnoid space in the lumbar region, usually between the laminae of vertebrae L3 and L4 or vertebrae L4 and L5.

– allows **measurement of CSF pressure** and withdrawal of some of the fluid for **bacteriologic and chemical examinations.**

– allows introduction of **anesthesia, drugs, or radiopaque material** into the subarachnoid space.

B. Caudal (epidural) anesthesia

– is used to **block the spinal nerves in the epidural space** by injection of local anesthetic agents via the sacral hiatus located between the sacral cornua.

Review Test

Directions: Each of the numbered items or incomplete statements in this section is followed by answers or by completions of the statement. Select the ONE lettered answer or completion that is BEST in each case.

1. Lumbar punctures or spinal taps are performed to withdraw cerebrospinal fluid (CSF), which is found

(A) in the epidural space
(B) in the subdural space
(C) between the pia mater and the spinal cord
(D) in the subarachnoid space
(E) between the arachnoid layer and dura mater

2. Which of the following statements concerning the dura mater is true?

(A) It adheres closely to the surface of the brain and spinal cord
(B) It is the innermost layer of the meninges
(C) It lies internal to the epidural space
(D) It forms the filum terminale
(E) It forms the denticulate ligaments

3. Which of the following ligaments is posterior to the spinal cord?

(A) Anterior longitudinal ligament
(B) Alar ligament
(C) Posterior longitudinal ligament
(D) Cruciform ligament
(E) Ligamentum nuchae

4. Which of the following statements concerning the anterior longitudinal ligament is true?

(A) It lies between the intervertebral disk and the dura
(B) It extends from the coccyx to the atlas
(C) It ends superiorly as the tectorial membrane
(D) It limits flexion of the vertebral column
(E) It is narrow at the upper end but widens as it descends

5. Which of the following structures is the primary site for absorption of CSF into the venous system?

(A) Choroid plexus
(B) Vertebral venous plexus
(C) Arachnoid villi
(D) Internal jugular vein
(E) Subarachnoid trabeculae

6. A patient is brought to the emergency department (ED) with multiple fractures of the transverse processes of the cervical and upper thoracic vertebrae. Which of the following muscles might be paralyzed?

(A) Trapezius
(B) Levator scapulae
(C) Rhomboid major
(D) Serratus posterior superior
(E) Rectus capitis posterior major

7. The body of vertebra T4 articulates with the

(A) head of the third rib
(B) neck of the fourth rib
(C) tubercle of the fourth rib
(D) head of the fifth rib
(E) tubercle of the fifth rib

Directions: Each of the numbered items or incomplete statements in this section is negatively phrased, as indicated by a capitalized word such as NOT, LEAST, or EXCEPT. Select the ONE lettered answer or completion that is BEST in each case.

8. All of the following statements concerning the suboccipital triangle are true EXCEPT

(A) the rectus capitis posterior major bounds the suboccipital triangle
(B) the rectus capitis posterior minor does not bound the suboccipital triangle
(C) the obliquus capitis superior and obliquus capitis inferior bound the suboccipital triangle
(D) the greater occipital nerve innervates the muscles of the suboccipital triangle
(E) within the triangle, the vertebral artery occupies the grooves on the superior surface of the posterior arch of the atlas

9. A patient is brought to the ED with multiple injuries, including lesions of the dorsal primary rami of the spinal nerves. These lesions could result in paralysis of all of the following muscles EXCEPT

(A) semispinalis capitis
(B) splenius
(C) serratus posterior superior
(D) iliocostalis
(E) spinalis

10. A stab near the superior angle of the scapula, injuring both the dorsal scapular nerve and the spinal accessory nerve, would result in paralysis or weakness of all of the following muscles EXCEPT

(A) trapezius
(B) rhomboid major
(C) rhomboid minor
(D) splenius cervicis
(E) levator scapulae

11. When CSF is withdrawn by lumbar puncture, the needle will penetrate all of the following structures EXCEPT

(A) the dura mater
(B) the subdural space
(C) the pia mater
(D) the subarachnoid space
(E) the arachnoid layer

12. All of the following statements concerning the spinal epidural space are true EXCEPT

(A) it may be entered via the sacral hiatus
(B) it is continuous from the sacrum to the base of the skull
(C) it contains a venous plexus
(D) it contains CSF
(E) it is the space between the dura mater and the wall of the vertebral canal

13. All of the following statements concerning the vertebral venous plexus are true EXCEPT

(A) it communicates with the cranial venous sinuses
(B) it is composed of thin-walled veins containing many valves
(C) it communicates with veins of the thorax, abdomen, and pelvis
(D) it is located in the epidural space
(E) it is thought to be the route of early metastasis of carcinoma from the lung, breast, and prostate gland to bones and the central nervous system

14. If the suboccipital nerve is injured, paralysis will occur in all of the following muscles EXCEPT

(A) rectus capitis posterior major
(B) semispinalis capitis
(C) rectus capitis anterior
(D) obliquus capitis superior
(E) obliquus capitis inferior

15. All of the following statements concerning the atlantoaxial joint are true EXCEPT

(A) it is a synovial joint
(B) it has two plane joints between the superior and inferior articular facets of the atlas and axis
(C) it has one pivot joint between the dens of the axis and the anterior arch of the atlas
(D) it permits flexion and extension of the head
(E) it is involved in rotation of the head

Directions: Each set of matching questions in this section consists of a list of four to twenty-six lettered options (some of which may be in figures) followed by several numbered items. For each numbered item, select the ONE lettered option that is most closely associated with it. To avoid spending too much time on matching sets with large numbers of options, it is generally advisable to begin each set by reading the list of options. Then, for each item in the set, try to generate the correct answer and locate it in the option list, rather than evaluating each option individually. Each lettered option may be selected once, more than once, or not at all.

Questions 16–20

Match each description with the most appropriate structure.

(A) Conus medullaris
(B) Dorsal root ganglion
(C) Cauda equina
(D) Internal vertebral venous plexus
(E) Subarachnoid space

16. Is found within the vertebral canal external to the dura mater of the spinal cord

17. Contains the CSF

18. Its inferior limit is at the level of the vertebra L2

19. Contains the cell bodies of sensory nerve fibers

20. Is formed by the roots of the lumbar and sacral nerves

Questions 21–25

Match each description below with the most appropriate component or condition of the spinal column.

(A) Whiplash injury
(B) Scoliosis
(C) Kyphosis
(D) Herniated intervertebral disk
(E) Lordosis

21. Is an abnormally exaggerated curvature of the thoracic vertebral column

22. Is produced by a force that drives the trunk forward while the head lags behind

23. Compresses the spinal nerve roots when the nucleus pulposus is protruded through the anulus fibrosus

24. An abnormal accentuation of the lumbar curvature

25. A lateral deviation due to unequal growth of the spinal column

Questions 26–30

Match each description below with the most appropriate ligament.

(A) Longitudinal ligament of the cruciform ligament
(B) Apical ligament
(C) Alar ligament
(D) Ligamentum flavum
(E) Denticulate ligament

26. Extends from the apex of the dens to the anterior aspect of the foramen magnum

27. Extends from the dens of the axis to the anterior aspect of the foramen magnum and the body of the axis

28. Extends from the apex of the dens to the medial side of the occipital bone

29. A lateral extension of the pia mater

30. Connects the laminae of adjacent vertebrae

Questions 31–35

Match each description below with the most appropriate muscle.

(A) Latissimus dorsi
(B) Rhomboid minor
(C) Semispinalis capitis
(D) Obliquus capitis inferior
(E) Trapezius

31. Is innervated by the dorsal scapular nerve

32. Is attached to the spinous and transverse processes of the vertebrae

33. Forms the boundaries of the triangle of auscultation and the lumbar triangle

34. Receives blood from the superficial or ascending branch of the transverse cervical artery

35. Forms the roof of the suboccipital triangle

Answers and Explanations

1–D. Cerebrospinal fluid (CSF) is found in the subarachnoid space, which is a wide interval between the arachnoid layer and the pia mater.

2–C. The dura mater is the tough, fibrous outermost layer of the meninges and is internal to the epidural space, which contains the internal vertebral venous plexus and the middle meningeal arteries in the cranial cavity. The dura mater is external to the subdural space, which contains a film of fluid to moisten its walls. The pia mater adheres closely to the surface of the brain and spinal cord; the filum terminale is a prolongation of the pia mater from the conus medullaris. The denticulate ligaments are lateral extensions of the pia mater through the arachnoid layer.

3–E. The ligamentum nuchae is a triangular-shaped median fibrous septum between the muscles on the two sides of the posterior aspect of the neck. It is an upward extension of the supraspinous ligament from vertebra C7 to the external occipital protuberance and crest.

4–E. The anterior longitudinal ligament extends from the base of the skull to the sacrum on the anterior surface of the vertebral bodies and intervertebral disks, and limits extension of the vertebral column. The tectorial membrane is an upward extension of the posterior longitudinal ligament from the body of the axis to the basilar part of the occipital bone.

5–C. Cerebrospinal fluid is produced by the choroid plexuses of the ventricles of the brain, circulates in the subarachnoid space, and is absorbed into the venous system primarily through the arachnoid villi projecting into the cranial dural venous sinuses, particularly the superior sagittal sinus.

6–B. The levator scapulae arises from the transverse processes of the upper cervical vertebrae and inserts on the medial border of the scapula. The other muscles of the back are attached to the spinous processes of the vertebrae.

7–D. The body of vertebra T4 articulates with the heads of the fourth and fifth ribs. The transverse process of vertebra T4 articulates with the tubercle of the fourth rib.

8–D. The suboccipital triangle is bounded by the rectus capitis posterior major and the obliquus capitis superior and inferior, which are innervated by the suboccipital nerve. The greater occipital nerve is a sensory nerve.

9–C. The serratus posterior superior is innervated by the anterior primary rami of the upper four thoracic nerves. The dorsal (posterior) primary rami of the spinal nerves innervate the deep muscles of the back.

10–D. The splenius capitis and cervicis muscles are the deep muscles of the back, which are innervated by dorsal primary rami of the spinal nerves.

11–C. The CSF is located in the subarachnoid space, between the arachnoid layer and pia mater.

12–D. The subarachnoid space contains CSF. The spinal epidural space is external to the dura mater and contains the internal vertebral venous plexus. It extends from the base of the skull to the sacrum and can be entered through the sacral hiatus for caudal (extradural) anesthesia.

13–B. The vertebral venous system consists of thin-walled veins; however, these veins do not have valves. The vertebral venous system lies in the epidural space and communicates with the cranial venous sinuses and paravertebral veins in the thorax, abdomen, and pelvis.

14–C. The suboccipital nerve (dorsal primary ramus of C1) supplies the muscles of the suboccipital area (e.g., the rectus capitis posterior major) and the semispinalis capitis. The rectus capitis anterior is innervated by the ventral primary rami of the first and second cervical nerves.

15–D. The atlantoaxial joints are synovial joints that consist of two plane joints and one pivot joint, and are involved primarily in rotation of the head.

16–D. The epidural space contains the internal vertebral venous plexus.

17–E. The CSF is found in the subarachnoid space between the pia mater and the arachnoid.

18–A. The conus medullaris is a conical end of the spinal cord at the level of vertebra L1 or L2.

19–B. The dorsal root ganglion contains the cell bodies of the visceral and somatic sensory nerve fibers.

20–C. The dorsal and ventral roots of the lumbar and sacral nerves form the cauda equina.

21–C. Kyphosis (hunchback or humpback) is an abnormally increased convexity in the thoracic curvature of the vertebral column.

22–A. Whiplash injury is a popular term for hyperextension–hyperflexion injury of the neck, such as occurs in a rear-end automobile collision.

23–D. Herniated (slipped) disk is a herniation of the nucleus pulposus through the anulus fibrosus of the intervertebral disk, compressing the spinal nerve root.

24–E. Lordosis (swayback) is an abnormal accentuation of the lumbar curvature.

25–B. Scoliosis is a lateral deviation due to unequal growth of the spinal column, pathologic erosion of vertebral bodies, or asymmetric paralysis or weakness of vertebral muscles.

26–B. The apical ligament of the dens extends from the tip of the dens to the anterior margin of the foramen magnum of the occipital bone.

27–A. The longitudinal ligament of the cruciform ligament extends from the body of the axis, over the dens, to the occipital bone (anterior margin of the foramen magnum).

28–C. The alar ligament extends from the side of the apex of the dens to the medial side of the occipital condyle (lateral margin of the foramen magnum).

29–E. The denticulate ligament is a lateral extension of the pia mater. It helps anchor the spinal cord within the vertebral canal.

30–D. The ligamentum flavum extends between the laminae of adjacent vertebrae.

31–B. The rhomboid minor muscle is innervated by the dorsal scapular nerve, which arises from the root (C5) of the brachial plexus.

32–D. The obliquus capitis inferior originates from the spine of the axis and inserts into the transverse process of the atlas.

33–A. The latissimus dorsi forms the boundaries of both the triangle of auscultation and the lumbar triangle.

34–E. The superficial branch of the transverse cervical artery ascends deep to the trapezius, and the deep or descending branch runs deep to the levator scapulae and rhomboid muscles to supply these muscles.

35–C. The roof of the suboccipital triangle is formed by the semispinalis capitis and longissimus capitis muscles.

8

Head and Neck

Structures of the Head and Neck

I. Major Divisions of the Neck (Figure 8-1)

A. Posterior triangle

- is bounded by the posterior border of the sternocleidomastoid (sternomastoid) muscle, the anterior border of the trapezius muscle, and the superior border of the clavicle.
- Its roof is formed by the **platysma** and the investing layer of the **deep cervical fascia.**
- Its floor is formed by the splenius capitis and levator scapulae muscles, and the anterior, middle, and posterior scalene muscles.
- contains the accessory nerve, cutaneous branches of the cervical plexus, external jugular vein, posterior (inferior) belly of the omohyoid, roots and trunks of the brachial plexus, and transverse cervical and suprascapular vessels.
- is further divided into the **occipital and subclavian (supraclavicular or omoclavicular) triangles** by the omohyoid posterior belly.

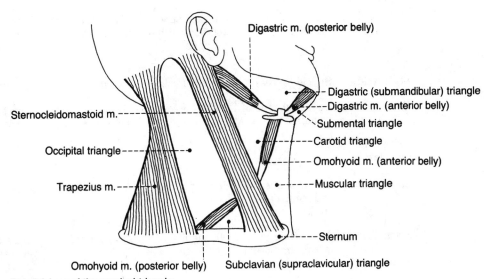

Sternocleidomastoid m.

Occipital triangle

Trapezius m.

Digastric m. (posterior belly)

Digastric (submandibular) triangle

Digastric m. (anterior belly)

Submental triangle

Carotid triangle

Omohyoid m. (anterior belly)

Muscular triangle

Sternum

Omohyoid m. (posterior belly)　　Subclavian (supraclavicular) triangle

Figure 8-1. Subdivisions of the cervical triangle.

B. Anterior triangle

– is bounded by the anterior border of the sternocleidomastoid, the anteromedian line of the neck, and the inferior border of the mandible.

– Its roof is formed by the platysma and the investing layer of the deep cervical fascia.

– is further divided by the omohyoid anterior belly and the digastric anterior and posterior bellies into the **digastric (submandibular), submental (suprahyoid), carotid, and muscular (inferior carotid) triangles.**

II. Muscles of the Neck (Figure 8-2)

Muscle	Origin	Insertion	Nerve	Action
Cervical muscles:				
Platysma	Superficial fascia over upper part of deltoid and pectoralis major	Mandible; skin and muscles over mandible and angle of mouth	Facial n.	Depresses lower jaw and lip and angle of mouth; wrinkles skin of neck
Sternocleido-mastoid	Manubrium sterni and medial one-third of clavicle	Mastoid process and lateral one-half of superior nuchal line	Spinal accessory n.; C2–C3 (sensory)	Singly turns face toward opposite side; together flex head, raise thorax
Suprahyoid muscles:				
Digastric	Anterior belly from digastric fossa of mandible; posterior belly from mastoid notch	Intermediate tendon attached to body of hyoid	Posterior belly by facial n.; anterior belly by mylohyoid n. of trigeminal n.	Elevates hyoid and tongue; depresses mandible
Mylohyoid	Mylohyoid line of mandible	Median raphe and body of hyoid bone	Mylohyoid n. of trigeminal n.	Elevates hyoid and tongue; depresses mandible
Stylohyoid	Styloid process	Body of hyoid	Facial n.	Elevates hyoid
Geniohyoid	Genial tubercle of mandible	Body of hyoid	C1 via hypoglossal n.	Elevates hyoid and tongue
Infrahyoid muscles:				
Sternohyoid	Manubrium sterni and medial end of clavicle	Body of hyoid	Ansa cervicalis	Depresses hyoid and larynx
Sternothyroid	Manubrium sterni; first costal cartilage	Oblique line of thyroid cartilage	Ansa cervicalis	Depresses thyroid cartilage and larynx

(Continued on next page)

Muscle	Origin	Insertion	Nerve	Action
Thyrohyoid	Oblique line of thyroid cartilage	Body and greater horn of hyoid	C1 via hypoglossal n.	Depresses and retracts hyoid and larynx
Omohyoid	Inferior belly from medial lip of suprascapular notch and suprascapular ligament; superior belly from intermediate tendon	Inferior belly to intermediate tendon; superior belly to body of hyoid	Ansa cervicalis	Depresses and retracts hyoid and larynx

III. Nerves of the Neck (Figures 8-3 and 8-4)

A. Accessory nerve

- is formed by the **union of cranial and spinal roots.**
- Its cranial roots arise from the medulla oblongata below the roots of the vagus.
- Its spinal roots arise from the cervical segment of the spinal cord between C1 and C3 (or C1 and C7) and unite to form a trunk that extends between the dorsal and ventral roots of the spinal nerves in the vertebral canal and passes through the foramen magnum.
- Both spinal and cranial portions traverse the **jugular foramen,** where they interchange fibers.
- Its cranial portion contains motor fibers that join the vagus nerve and innervate the soft palate, pharyngeal constrictors, and larynx.
- Its spinal portion innervates the sternocleidomastoid and trapezius muscles.
- lies on the levator scapulae in the posterior cervical triangle and then passes deep to the trapezius.

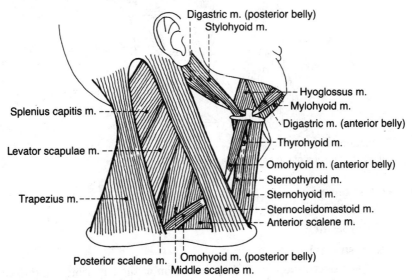

Figure 8-2. Muscles of the cervical triangle.

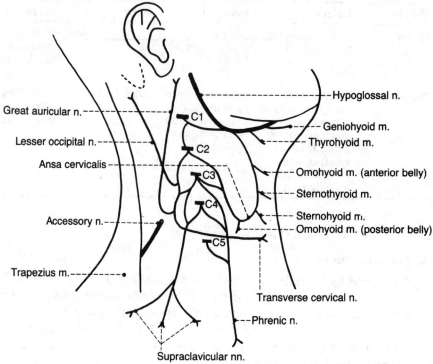

Figure 8-3. Cervical plexus.

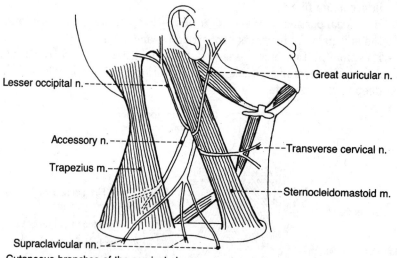

Figure 8-4. Cutaneous branches of the cervical plexus.

B. Cervical plexus

 – is formed by the **ventral primary rami of C1–C4.**

1. Motor branches

a. Ansa cervicalis

 – is a **nerve loop** formed by the union of the superior root (C1 or C1 and C2; **descendens hypoglossi**) and the inferior root (C2 and C3; **descendens cervicalis**).

– innervates the infrahyoid (or strap) muscles, such as the omohyoid, sterno-hyoid, and sternothyroid muscles, with the exception of the thyrohyoid muscle, which is innervated by C1 via the hypoglossal nerve.

b. Phrenic nerve (C3–C5)

– lies on the anterior surface of the anterior scalene muscle and passes into the thorax deep to the subclavian vein.

– passes between the mediastinal pleura and fibrous pericardium and supplies the diaphragm, pericardium, and mediastinal pleura.

c. Twigs to the longus capitis and cervicis, sternocleidomastoid, trapezius, levator scapulae, and scalene muscles

2. Cutaneous branches

a. Lesser occipital nerve (C2)

– ascends along the posterior border of the sternocleidomastoid to the scalp behind the auricle.

b. Great auricular nerve (C2–C3)

– ascends on the sternocleidomastoid to innervate the skin behind the auricle and on the parotid gland.

c. Transverse cervical nerve (C2–C3)

– turns around the posterior border of the sternocleidomastoid and innervates the skin of the anterior cervical triangle.

d. Supraclavicular nerve (C3–C4)

– emerges as a common trunk from under the sternocleidomastoid; then divides into **anterior, middle, and posterior branches** to the skin over the clavicle and the shoulder.

C. Brachial plexus (see Figure 2-9)

– is formed by the **union of the ventral primary rami of C5–T1** and passes between the anterior scalene and middle scalene muscles.

1. Its roots give rise to:

a. Dorsal scapular nerve (C5)

– emerges from behind the anterior scalene muscle and runs downward and backward through the middle scalene muscle and then deep to the trapezius.

– passes deep to or through the levator scapulae and descends along with the dorsal scapular artery on the deep surface of the rhomboid muscles along the medial border of the scapula, innervating the levator scapulae and rhomboid muscles.

b. Long thoracic nerve (C5–C7)

– pierces the middle scalene muscle, descends behind the brachial plexus, and enters the axilla to innervate the serratus anterior.

2. Its upper trunk gives rise to:

a. Suprascapular nerve (C5–C6)

– passes deep to the trapezius and joins the suprascapular artery in a course toward the shoulder.

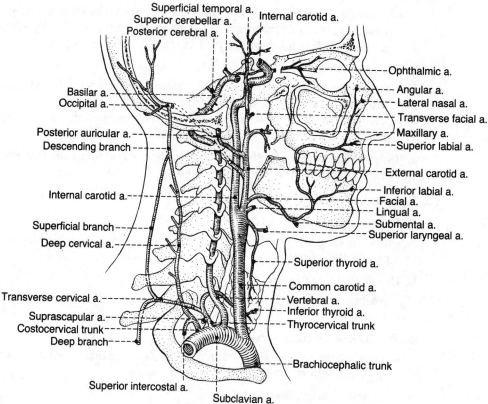

Figure 8-5. Subclavian and carotid arteries and their branches.

– passes through the **scapular notch** under the superior transverse scapular ligament and supplies the supraspinatus and infraspinatus muscles.

b. Nerve to the subclavius muscle (C5)

– descends in front of the plexus and behind the clavicle to innervate the subclavius.
– in many cases communicates with the phrenic nerve as the **accessory phrenic nerve.**

IV. Blood Vessels of the Neck (Figure 8-5)

A. Subclavian artery

– is a branch of the **brachiocephalic trunk** on the right but arises directly from the **arch of the aorta** on the left.
– is divided into three parts by the anterior scalene muscle: The first part passes from the origin of the vessel to the medial margin of the anterior scalene; the second part lies behind this muscle; and the third part passes from the lateral margin of the muscle to the outer border of the first rib.
– Its branches include the following:

1. Vertebral artery

– arises from the first part of the subclavian artery and ascends between the anterior scalene and longus coli muscles.

– ascends through the transverse foramina of vertebrae C1–C6, winds around the superior articular process of the atlas, and passes through the foramen magnum into the cranial cavity.

2. **Thyrocervical trunk**
   – is a short trunk from the first part of the subclavian artery that divides into the following arteries:

   a. **Inferior thyroid artery**
      – ascends in front of the anterior scalene muscle, turns medially behind the carotid sheath but in front of the vertebral vessels, and then arches downward to the lower pole of the thyroid gland.
      – gives rise to an **ascending cervical artery,** which ascends on the anterior scalene muscle medial to the phrenic nerve.

   b. **Transverse cervical artery**
      – runs laterally across the anterior scalene muscle, phrenic nerve, and trunks of the brachial plexus, passing deep to the trapezius.
      – divides into a superficial branch and a deep branch, which takes the place of the **dorsal (descending) scapular artery.** In the absence of the deep branch, the superficial branch is known as the **superficial cervical artery.**

   c. **Suprascapular artery**
      – passes in front of the anterior scalene muscle and the brachial plexus parallel to but below the transverse cervical artery.
      – passes superior to the superior transverse scapular ligament, whereas the suprascapular nerve passes inferior to it.

3. **Internal thoracic artery**
   – arises from the first part of the subclavian artery, descends through the thorax behind the upper six costal cartilages, and ends at the sixth intercostal space by dividing into the **superior epigastric and musculophrenic arteries.**

4. **Costocervical trunk**
   – arises from the posterior aspect of the second part of the subclavian artery behind the anterior scalene muscle and divides into the following arteries:

   a. **Deep cervical artery**
      – passes between the transverse process of vertebra C7 and the neck of the first rib, ascends between the semispinalis capitis and semispinalis cervicis muscles, and anastomoses with the deep branch of the descending branch of the occipital artery.

   b. **Superior intercostal artery**
      – descends behind the cervical pleura anterior to the necks of the first two ribs, and gives rise to the first two posterior intercostal arteries.

5. **Dorsal (descending) scapular artery**
   – arises from the third part of the subclavian artery or arises as the deep (descending) branch of the transverse cervical artery.

B. **Common carotid arteries**
   – The **right common carotid artery** begins at the bifurcation of the brachiocephalic artery, and the **left common carotid artery** arises from the aortic arch.

– ascends within the carotid sheath and divides at the level of the upper border of the thyroid cartilage into the **external and internal carotid arteries.**

1. Receptors

a. Carotid body

– lies at the bifurcation of the common carotid artery as an ovoid body.
– is a **chemoreceptor** that is stimulated by chemical changes (such as oxygen tension) in the circulating blood.
– is innervated by the **nerve to the carotid body,** which arises from the pharyngeal branch of the vagus nerve, and by the **carotid sinus branch** of the glossopharyngeal nerve.

b. Carotid sinus

– is a **spindle-shaped dilatation** located at the origin of the internal carotid artery and functions as a **pressoreceptor (baroreceptor),** stimulated by changes in blood pressure.
– is innervated primarily by the **carotid sinus branch** of the glossopharyngeal nerve but also by the nerve to the carotid body.

2. Internal carotid artery

– has no branches in the neck.
– ascends within the carotid sheath in company with the vagus nerve and the internal jugular vein.
– enters the cranium through the **carotid canal** in the petrous part of the temporal bone.
– in the middle cranial fossa, gives rise to the **ophthalmic artery** and the **anterior and middle cerebral arteries.**

3. External carotid artery

– extends from the level of the upper border of the thyroid cartilage to the neck of the mandible, where it ends in the substance of the parotid gland by dividing into the maxillary and superficial temporal arteries.
– has eight named branches:

a. Superior thyroid artery

– arises below the level of the greater horn of the hyoid bone.
– descends obliquely forward in the carotid triangle and passes deep to the infrahyoid muscles to reach the superior pole of the thyroid gland.
– gives rise to an infrahyoid, sternocleidomastoid, superior laryngeal, cricothyroid, and several glandular branches.

b. Lingual artery

– arises at the level of the tip of the greater horn of the hyoid bone and passes deep to the hyoglossus to reach the tongue.
– gives rise to suprahyoid, dorsal lingual, sublingual, and deep lingual branches.

c. Facial artery

– arises just above the lingual artery and ascends forward, deep to the posterior belly of the digastric and stylohyoid muscles.
– hooks around the lower border of the mandible at the anterior margin of the masseter to enter the face.

d. Ascending pharyngeal artery

– arises from the deep surface of the external carotid artery in the carotid triangle and ascends between the internal carotid artery and the wall of the pharynx.
– gives rise to pharyngeal, palatine, inferior tympanic, and meningeal branches.

e. Occipital artery

– arises from the posterior surface of the external carotid artery, just above the level of the hyoid bone.
– passes deep to the digastric posterior belly, occupies the groove on the mastoid process, and appears on the skin above the occipital triangle.
– gives rise to the following:

   (1) Sternocleidomastoid branch
   – descends inferiorly and posteriorly over the hypoglossal nerve and enters the substance of the muscle.
   – anastomoses with the sternocleidomastoid branch of the superior thyroid artery.

   (2) Descending branch
   – Its superficial branch anastomoses with the superficial branch of the transverse cervical artery.
   – Its deep branch anastomoses with the deep cervical artery of the costocervical trunk.

f. Posterior auricular artery

– arises from the posterior surface of the external carotid artery just above the digastric posterior belly.
– ascends superficial to the styloid process and deep to the parotid gland and ends between the mastoid process and the external acoustic meatus.
– gives rise to stylomastoid, auricular, and occipital branches.

g. Maxillary artery

– arises behind the neck of the mandible as the larger terminal branch of the external carotid artery.
– runs deep to the neck of the mandible and enters the infratemporal fossa.

h. Superficial temporal artery

– arises behind the neck of the mandible as the smaller terminal branch of the external carotid artery.
– gives rise to the **transverse facial artery,** which runs between the zygomatic arch above and the parotid duct below.
– ascends in front of the external acoustic meatus into the scalp, accompanying the auriculotemporal nerve and the superficial temporal vein.

C. Veins of the cervical triangle (Figure 8-6)

1. Retromandibular vein

– is formed by the superficial temporal and maxillary veins.
– divides into an anterior branch, which joins the facial vein to form the common facial vein, and a posterior branch, which joins the posterior auricular vein to form the external jugular vein.

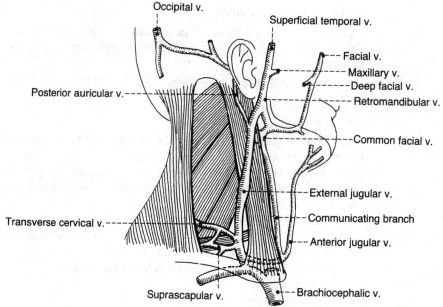

Figure 8-6. Veins of the cervical triangle.

2. External jugular vein

– is formed by the union of the **posterior auricular vein** and the posterior branch of the **retromandibular vein.**

– crosses the sternomastoid obliquely, under the platysma, and ends in the subclavian (or sometimes the internal jugular) vein.

– receives the suprascapular, transverse cervical, and anterior jugular veins.

3. Internal jugular vein (Figure 8-7)

– begins in the **jugular foramen** as a continuation of the sigmoid sinus, descends in the carotid sheath, and ends in the **brachiocephalic vein.**

– has the superior bulb at its beginning and the inferior bulb just above its termination.

– receives blood from the brain, face, and neck.

V. Lymphatics of the Head and Neck

A. Superficial lymph nodes of the head

– Lymph vessels from the face, scalp, and ear drain into the occipital, retroauricular, parotid, buccal (facial), submandibular, submental, and superficial cervical nodes, which in turn drain into the **deep cervical nodes** (including the jugulodigastric and juguloomohyoid nodes).

B. Deep lymph nodes of the head

– The middle ear drains into the retropharyngeal and upper deep cervical nodes; the nasal cavity and paranasal sinuses into the submandibular, retropharyngeal, and upper deep cervical; the tongue into the submental, submandibular, and upper and lower cervical; the larynx into the upper and lower deep cervical; the pharynx into the retropharyngeal and upper and lower deep cervical; and the thyroid gland into the lower deep cervical, prelaryngeal, pretracheal, and paratracheal.

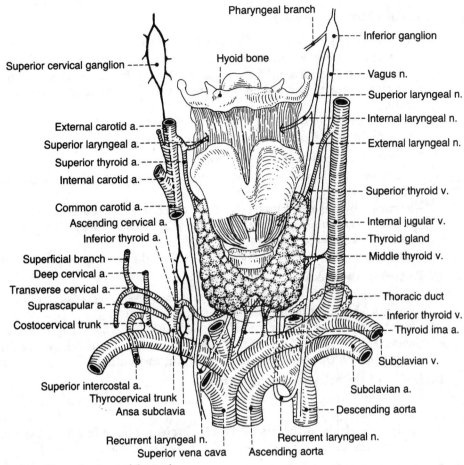

Figure 8-7. Deep structures of the neck.

C. Superficial cervical lymph nodes

– lie along the **external jugular vein** in the posterior triangle and along the **anterior jugular vein** in the anterior triangle, and drain into the deep cervical nodes.

D. Deep cervical lymph nodes

1. Superior deep cervical nodes

– lie along the **internal jugular vein** in the carotid triangle of the neck.
– receive afferent lymphatics from the back of the head and neck, tongue, palate, nasal cavity, larynx, pharynx, trachea, thyroid gland, and esophagus.
– Their efferent vessels join those of the inferior deep cervical nodes to form the **jugular trunk,** which empties into the thoracic duct on the left and into the junction of the internal jugular and subclavian veins on the right.

2. Inferior deep cervical nodes

– lie on the **internal jugular vein** near the subclavian vein.
– receive afferent lymphatics from the anterior jugular, transverse cervical, and apical axillary nodes.

VI. Clinical Considerations

A. Injury to the upper trunk of the brachial plexus

– may be caused by a **violent separation** of the head from the shoulder, such as occurs in a fall from a motorcycle. In this condition, the arm is in medial rotation owing to paralysis of the lateral rotators, resulting in **waiter's tip hand.**

– may be caused by stretching an infant's neck during a difficult delivery. This is referred to as **birth palsy** or obstetric paralysis.

B. Neurovascular compression syndrome

– produces symptoms of nerve compression of the brachial plexus and the subclavian vessels.

– is caused by abnormal insertion of the anterior and middle scalene muscles **(scalene syndrome)** and by the **cervical rib,** which is the cartilaginous accessory rib attached to vertebra C7.

– can be corrected by cutting the cervical rib or the anterior scalene muscle.

C. Torticollis (wryneck)

– is a **contracted state of the cervical muscles,** producing twisting of the neck with the chin pointing upward and to the other side.

– is due to injury to the sternocleidomastoid muscle or avulsion of the accessory nerve at the time of birth, and unilateral fibrosis in the muscle, which cannot lengthen with the growing neck **(congenital torticollis).**

D. Lesion of the external laryngeal nerve

– may occur during **thyroidectomy** because the nerve accompanies the superior thyroid artery.

– causes **paralysis of the cricothyroid muscle,** resulting in loss of the tension of the vocal cord, and weakness and hoarseness of voice.

E. Lesion of the accessory nerve in the neck

– denervates the trapezius, leading to atrophy of the muscle.

– causes a downward displacement or **drooping of the shoulder.**

F. Central venous line

– is inserted into the retroclavicular portion of the **subclavian vein.** The needle should be guided medially along the long axis of the clavicle to reach the posterior surface where the vein runs over the first rib.

Deep Neck and Prevertebral Region

I. Deep Structures of the Neck (see Figure 8-7)

A. Trachea

– begins at the inferior border of the cricoid cartilage (C6).

– has **16 to 20 incomplete hyaline cartilaginous rings** that open posteriorly to prevent the trachea from collapsing.

B. Thyroid gland

– is an endocrine gland that produces **thyroxine** and **thyrocalcitonin.**

– consists of right and left lobes connected by the **isthmus,** which crosses the second, third, and fourth tracheal rings. (The muscular band descending from the hyoid bone to the isthmus is called the **levator glandulae thyroideae.)**

– is supplied by the superior and inferior thyroid arteries (and the **arteria thyroidea ima**, an inconsistent branch from the brachiocephalic trunk).

– Venous drainage is via the superior and middle thyroid veins to the internal jugular vein and via the inferior thyroid vein to the brachiocephalic vein.

C. Parathyroid glands

– are endocrine glands that play a vital role in the regulation of calcium and phosphorus metabolism.

– consist usually of **four (can vary from two to six) small ovoid bodies** that lie against the dorsum of the thyroid under its sheath but with their own capsule.

– are supplied chiefly by the inferior thyroid artery.

D. Thyroid cartilage

– is a hyaline cartilage that forms a laryngeal prominence (**the Adam's apple**).

– Its superior horn is joined to the tip of the greater horn of the hyoid bone by the lateral thyroid ligament, and its inferior horn articulates with the cricoid cartilage.

E. Sympathetic trunk

– is covered by the prevertebral fascia, and runs behind the carotid sheath and in front of the longus colli and longus capitis muscles.

– contains preganglionic and postganglionic sympathetic fibers, cell bodies of the postganglionic sympathetic fibers, and visceral afferent fibers with cell bodies in the upper thoracic dorsal root ganglia.

– receives gray rami communicantes but no white rami communicantes in the cervical region.

– bears the following cervical ganglia:

1. Superior cervical ganglion

– lies in front of the transverse processes of vertebrae C1–C2, posterior to the internal carotid artery and anterior to the longus capitis.

– contains cell bodies of postganglionic sympathetic fibers that pass to the visceral structures of the head and neck.

– gives rise to the **internal carotid nerve** to form the internal carotid plexus; the **external carotid nerve** to form the external carotid plexus; the **pharyngeal branches** that join the pharyngeal branches of the glossopharyngeal and vagus nerves to form the pharyngeal plexus; and the **superior cervical cardiac nerve** to the heart.

2. Middle cervical ganglion

– lies at the level of the cricoid cartilage (vertebra C6).

– gives rise to a **middle cervical cardiac nerve,** which is the largest of the three cervical sympathetic cardiac nerves.

3. Inferior cervical ganglion

– fuses with the first thoracic ganglion to become the **cervicothoracic (stellate) ganglion.**

– lies in front of the neck of the first rib and the transverse process of vertebra C7 and behind the dome of the pleura.

– gives rise to the **inferior cervical cardiac nerve.**

F. Ansa subclavia

– is the cord connecting the middle and inferior cervical sympathetic ganglia, forming a loop around the first part of the subclavian artery.

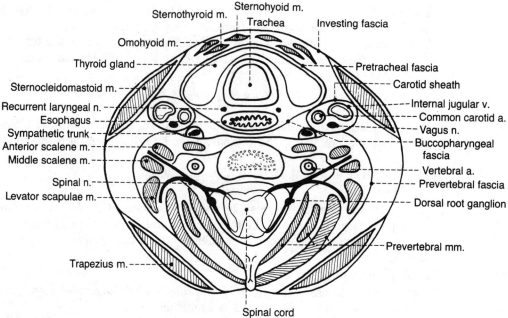

Figure 8-8. Cross-section of the neck.

II. Deep Cervical Fasciae (Figure 8-8)

A. Superficial (investing) layer of deep cervical fascia

– surrounds all of the **deeper parts of the neck.**
– splits to enclose the sternocleidomastoid and trapezius muscles.
– is attached superiorly along the mandible, mastoid process, external occipital protuberance, and superior nuchal line of the occipital bone.
– is attached inferiorly along the acromion and spine of the scapula, clavicle, and manubrium sterni.

B. Prevertebral layer of deep cervical fascia

– is cylindrical and encloses the **vertebral column** and its associated muscles.
– covers the scalene muscles and the deep muscles of the back.
– attaches to the external occipital protuberance and the basilar part of the occipital bone and becomes continuous with the **endothoracic fascia** and the **anterior longitudinal ligament** of the bodies of the vertebrae in the thorax.

C. Carotid sheath

– contains the **common and internal carotid arteries, internal jugular vein, and vagus nerve.**
– does not contain the sympathetic trunk, which lies posterior to the carotid sheath and anterior to the prevertebral fascia.
– blends with the prevertebral, pretracheal, and investing layers and also attaches to the base of the skull.

D. Pretracheal layer of deep cervical fascia

– invests the **larynx** and **trachea,** enclosing the **thyroid gland,** and contributing to the formation of the carotid sheath.
– attaches superiorly to the thyroid and cricoid cartilages and inferiorly to the pericardium.

E. **Buccopharyngeal fascia**

   – covers the **buccinator muscles** and the **pharynx.**
   – is attached to the pharyngeal tubercle and the pterygomandibular raphe.

F. **Pharyngobasilar fascia**

   – is the **fibrous coat in the wall of the pharynx,** situated between the mucous
   membrane and the pharyngeal constrictor muscles.

III. Prevertebral Muscles

Muscle	Origin	Insertion	Nerve	Action
Lateral vertebral:				
Anterior scalene	Transverse processes of CV3–CV6	Scalene tubercle on first rib	Lower cervical (C5–C8)	Elevates first rib; bends neck
Middle scalene	Transverse processes of CV2–CV7	Upper surface of first rib	Lower cervical (C5–C8)	Elevates first rib; bends neck
Posterior scalene	Transverse processes of CV4–CV6	Outer surface of second rib	Lower cervical (C6–C8)	Elevates second rib; bends neck
Anterior vertebral:				
Longus capitus	Transverse processes of CV3–CV6	Basilar part of occipital bone	C1–C4	Flexes and rotates head
Longus colli (L. cervicis)	Transverse processes and bodies of CV3–TV3	Anterior tubercle of atlas; bodies of CV2–CV4; transverse process of CV5–CV6	C2–C6	Flexes and rotates head
Rectus capitis anterior	Lateral mass of atlas	Basilar part of occipital bone	C1–C2	Flexes and rotates head
Rectus capitis lateralis	Transverse process of atlas	Jugular process of occipital bone	C1–C2	Flexes head laterally

IV. Clinical Considerations

A. **Goiter**

   – is a **pathologic enlargement of the thyroid gland,** causing a swelling in the
   front part of the neck.

B. **Tracheotomy (tracheostomy)**

   – is **opening into the trachea** by incising the third and fourth rings of the
   trachea, after making a vertical midline skin incision from the jugular notch
   of the manubrium sterni to the thyroid notch of the thyroid cartilage.

C. Cricothyrotomy

– is **incision through the skin and cricothyroid membrane** for relief of acute respiratory obstruction.
– is preferable to tracheostomy for nonsurgeons in emergency respiratory obstructions.

D. Parathyroidectomy

– may occur during a total thyroidectomy, and cause death if parathyroid hormone, calcium, or vitamin D is not provided.
– decreases the plasma calcium level, causing increased neuromuscular activity, such as muscular spasms and nervous hyperexcitability, called **tetany.**

Face and Scalp

I. Muscles of Facial Expression (Figure 8-9)

Muscle	Origin	Insertion	Nerve	Action
Occipitofrontalis	Superior nuchal line; upper orbital margin	Epicranial aponeurosis	Facial n.	Elevates eyebrows; wrinkles forehead (surprise)
Corrugator supercilii	Medial supraorbital margin	Skin of medial eyebrow	Facial n.	Draws eyebrows downward medially (anger, frowning)
Orbicularis oculi	Medial orbital margin; medial palpebral ligament; lacrimal bone	Skin and rim of orbit; tarsal plate; lateral palpebral raphe	Facial n.	Closes eyelids (squinting)
Procerus	Nasal bone and cartilage	Skin between eyebrows	Facial n.	Wrinkles skin over bones (sadness)
Nasalis	Maxilla lateral to incisive fossa	Ala of nose	Facial n.	Draws ala of nose toward septum
Depressor septi*	Incisive fossa of maxilla	Ala and nasal septum	Facial n.	Constricts nares
Orbicularis oris	Maxilla above incisor teeth	Skin of lip	Facial n.	Closes lips
Levator anguli oris	Canine fossa of maxilla	Angle of mouth	Facial n.	Elevates angle of mouth medially (disgust)
Levator labii superioris	Maxilla above infraorbital foramen	Skin of upper lip	Facial n.	Elevates upper lip; dilates nares (disgust)
Levator labii superioris alaeque nasi*	Frontal process of maxilla	Skin of upper lip	Facial n.	Elevates ala of nose and upper lip

(Continued on next page)

Muscle	Origin	Insertion	Nerve	Action
Zygomaticus major	Zygomatic arch	Angle of mouth	Facial n.	Draws angle of mouth backward and upward (smile)
Zygomaticus minor	Zygomatic arch	Angle of mouth	Facial n.	Elevates upper lip
Depressor labii inferioris	Mandible below mental foramen	Orbicularis oris and skin of lower lip	Facial n.	Depresses lower lip
Depressor anguli oris	Oblique line of mandible	Angle of mouth	Facial n.	Depresses angle of mouth
Risorius	Fascia over masseter	Angle of mouth	Facial n.	Retracts angle of mouth (false smile)
Buccinator	Mandible; pterygomandibular raphe; alveolar processes	Angle of mouth	Facial n.	Presses cheek to keep it taut
Mentalis	Incisive fossa of mandible	Skin of chin	Facial n.	Elevates and protrudes lower lip
Auricularis anterior, superior, and posterior*	Temporal fascia; epicranial aponeurosis; mastoid process	Anterior, superior, and posterior sides of auricle	Facial n.	Retract and elevate ear

* Indicates less important muscles.

II. Nerve Supply to the Face and Scalp (Figures 8-10 and 8-11)

A. Facial nerve (Figure 8-12)

– comes through the **stylomastoid foramen** and appears posterior to the parotid gland.

– enters the parotid gland to give rise to five terminal branches, which radiate forward in the face as **temporal, zygomatic, buccal, mandibular, and cervical branches.**

– innervates the **muscles of facial expression** and sends the **posterior auricular branch** to muscles of the auricle and the occipitalis muscle.

– also innervates the digastric posterior belly and stylohyoid muscles.

B. Trigeminal nerve

– provides sensory innervation to the **skin of the face.**

1. Ophthalmic division

– innervates the area above the upper eyelid and dorsum of the nose.

– Its branches are the **supraorbital, supratrochlear, infratrochlear, external nasal, and lacrimal nerves.**

2. Maxillary division

– innervates the face below the level of the eyes and above the upper lip.

– Its branches are the **zygomaticofacial, zygomaticotemporal, and infraorbital nerves.**

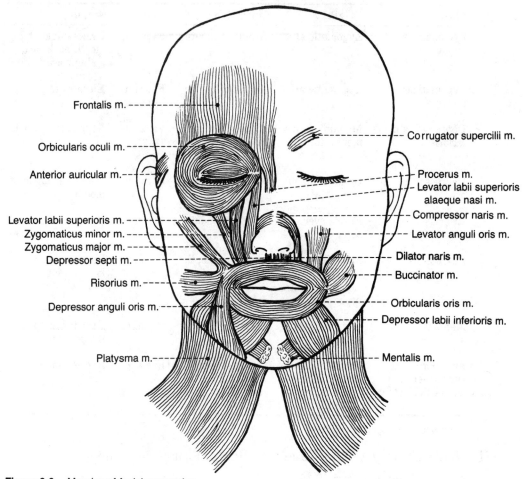

Frontalis m.

Orbicularis oculi m.

Anterior auricular m.

Levator labii superioris m.
Zygomaticus minor m.
Zygomaticus major m.
Depressor septi m.

Risorius m.

Depressor anguli oris m.

Platysma m.

Corrugator supercilii m.

Procerus m.
Levator labii superioris
alaeque nasi m.
Compressor naris m.

Levator anguli oris m.

Dilator naris m.

Buccinator m.

Orbicularis oris m.
Depressor labii inferioris m.

Mentalis m.

Figure 8-9. Muscles of facial expression.

3. Mandibular division

– innervates the face below the level of the lower lip.
– its branches are the **auriculotemporal, buccal, and mental nerves.**

III. Blood Vessels of the Face and Scalp (Figures 8-13 and 8-14)

A. Facial artery

– arises from the **external carotid artery** just above the upper border of the hyoid bone.
– passes deep to the mandible, winds around the lower border of the mandible, and runs upward and forward on the face.
– gives rise to the ascending palatine, tonsillar, glandular and submental branches in the neck and on the face the inferior labial, superior labial, and lateral nasal branches, and terminates as an angular artery that anastomoses with the palpebral and dorsal nasal branches of the ophthalmic artery to establish a communication between the external and internal carotid arteries.

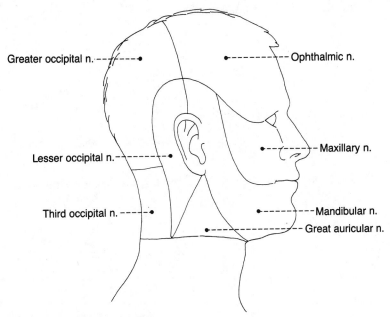

Figure 8-10. Sensory innervation of the face.

Greater occipital n.

Ophthalmic n.

Lesser occipital n.

Maxillary n.

Third occipital n.

Mandibular n.

Great auricular n.

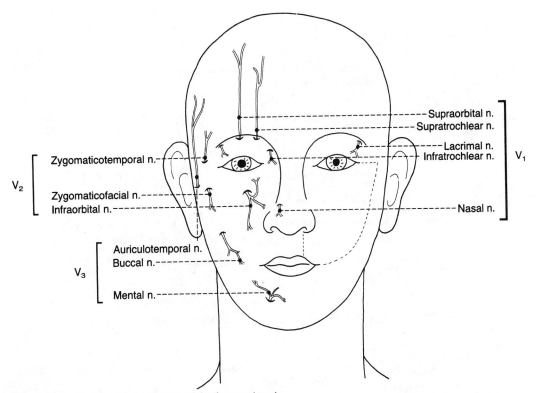

Figure 8-11. Cutaneous innervation of the face and scalp.

Supraorbital n.

Supratrochlear n.

Lacrimal n.

Infratrochlear n.

V_1

Zygomaticotemporal n.

V_2

Zygomaticofacial n.

Infraorbital n.

Nasal n.

Auriculotemporal n.

Buccal n.

V_3

Mental n.

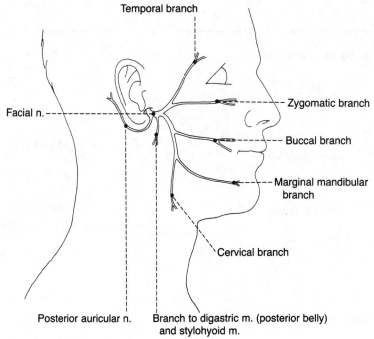

Temporal branch

Facial n.

Zygomatic branch

Buccal branch

Marginal mandibular branch

Cervical branch

Posterior auricular n.

Branch to digastric m. (posterior belly) and stylohyoid m.

Figure 8-12. Distribution of the facial nerve.

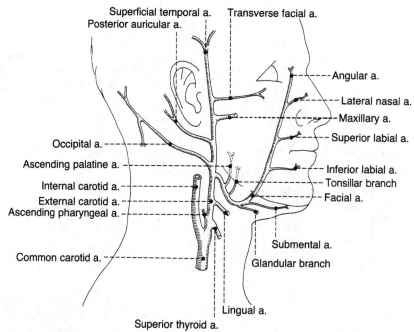

Superficial temporal a.
Posterior auricular a.

Transverse facial a.

Angular a.

Lateral nasal a.

Maxillary a.

Superior labial a.

Occipital a.

Ascending palatine a.

Internal carotid a.

External carotid a.

Ascending pharyngeal a.

Inferior labial a.

Tonsillar branch

Facial a.

Common carotid a.

Submental a.

Glandular branch

Lingual a.

Superior thyroid a.

Figure 8-13. Blood supply to the face and scalp.

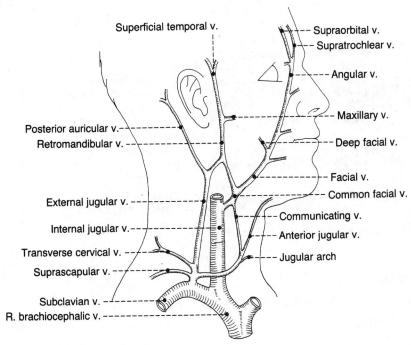

Figure 8-14. Veins of the head and neck.

B. Superficial temporal artery

- arises behind the neck of the mandible as the smaller terminal branch of the external carotid artery and ascends anterior to the external acoustic meatus into the scalp.
- accompanies the auriculotemporal nerve along its anterior surface.
- gives rise to the **transverse facial artery,** which passes forward across the masseter between the zygomatic arch above and the parotid duct below.
- also gives rise to zygomatico-orbital, middle temporal, anterior auricular, frontal, and parietal branches.

C. Facial vein

- begins as an angular vein by the confluence of the supraorbital and supratrochlear veins. The angular vein is continued at the lower margin of the orbital margin into the facial vein.
- receives tributaries corresponding to the branches of the facial artery and also receives the **infraorbital and deep facial veins.**
- drains either directly into the internal jugular vein or by joining the anterior branch of the retromandibular vein to form the **common facial vein,** which then enters the internal jugular vein.
- communicates with the superior ophthalmic vein and thus with the **cavernous sinus,** allowing a route of infection from the face to the cranial dural sinus.

D. Retromandibular vein

- is formed by the union of the superficial temporal and maxillary veins behind the mandible.
- divides into an **anterior branch,** which joins the facial vein to form the common facial vein, and a **posterior branch,** which joins the posterior auricular vein to form the external jugular vein.

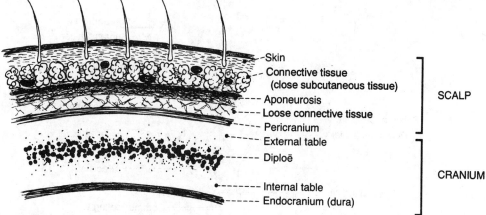

Figure 8-15. Layers of the scalp and cranium.

IV. Scalp

A. Layers of the scalp (Figure 8-15)

1. Skin

2. Connective tissue (close subcutaneous tissue)

– contains the larger blood vessels and nerves. Because of its toughness, the scalp gapes when cut and the blood vessels do not contract, which leads to severe bleeding.

3. Aponeurosis epicranialis (galea aponeurotica)

– is a **fibrous sheet** that covers the vault of the skull and unites the occipitalis and frontalis muscles.

4. Loose connective tissue

– forms the **subaponeurotic space** and contains the emissary veins.
– is termed a dangerous area because infection (blood and pus) can spread easily in it or from the scalp to the intracranial sinuses by way of the emissary veins.

5. Pericranium

– is the **periosteum** over the surface of the skull.

B. Innervation and blood supply of the scalp (Figure 8-16)

– is innervated by the supratrochlear, supraorbital, zygomaticotemporal, auriculotemporal, lesser occipital, greater occipital, and third occipital nerves.
– is supplied by the supratrochlear and supraorbital branches of the internal carotid and by the superficial temporal, posterior auricular, and occipital branches of the external carotid arteries.

V. Clinical Considerations

A. Bell's palsy (facial paralysis)

– is a **unilateral paralysis of the facial muscles** owing to a lesion of the facial nerve.

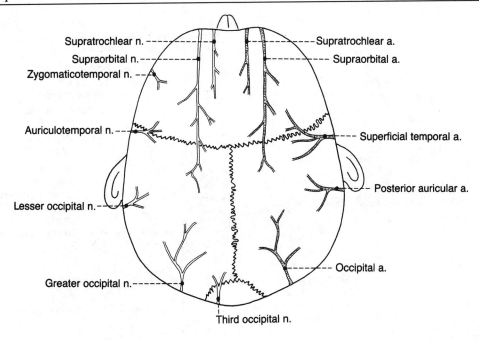

Figure 8-16. Nerves and arteries of the scalp.

- is marked by characteristic **distortions of the face** such as a sagging corner of the mouth and inability to smile, whistle, or blow; and drooping of the upper eyelid, eversion of the lower eyelid, and inability to close or blink the eye.
- causes decreased lacrimation (as a result of a lesion of the greater petrosal nerve), loss of taste in the anterior two-thirds of the tongue (chorda tympani), painful sensitivity to sounds (nerve to the stapedius), and deviation of the lower jaw and tongue (nerve to the digastric muscle).

B. Trigeminal neuralgia (tic douloureux)

- is marked by **paroxysmal pain** along the course of the trigeminal nerve.
- may be alleviated by sectioning the sensory root of the trigeminal nerve in the trigeminal (Meckel's) cave in the middle cranial fossa.

C. Danger area of the face

- is the **area of the face drained by the facial veins;** pustules (pimples) or other skin infections, particularly on the side of the nose and upper lip, may spread to the cavernous dural sinus through the ophthalmic veins and the pterygoid venous plexuses, which anastomose with the facial veins.

D. Corneal blink reflex

- is **closing of the eyes** caused by blowing on the cornea or touching it with a wisp of cotton wool; it is caused by bilateral contraction of the orbicularis oculi muscles.
- Its efferent limb (of the reflex arc) is the facial nerve; its afferent limb is the nasociliary nerve of the ophthalmic division of the trigeminal nerve.

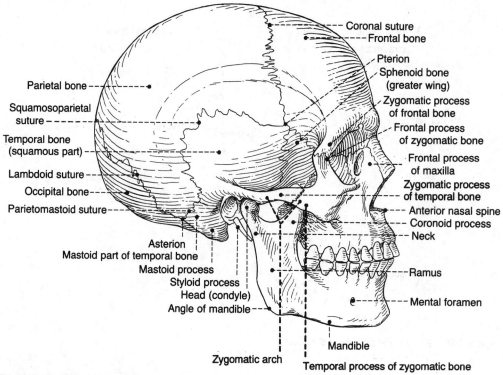

Figure 8-17. Lateral view of the skull.

Temporal and Infratemporal Fossae

I. Introduction

A. Infratemporal fossa (Figures 8-17 and 8-18)

– contains the lower portion of the temporalis muscle, the lateral and medial pterygoid muscles, the pterygoid plexus of veins, the mandibular nerve and its branches, the maxillary artery and its branches, the chorda tympani, and the otic ganglion.

– has the following boundaries:

1. Anterior: posterior surface of the maxilla

2. Posterior: styloid process

3. Medial: lateral pterygoid plate of the sphenoid bone

4. Lateral: ramus and coronoid process of the mandible

5. Roof: infratemporal surface of the greater wing of the sphenoid bone

B. Temporal fossa (see Figures 8-17 and 8-18)

– contains the temporalis muscle, the deep temporal nerves and vessels, the auriculotemporal nerve, and the superficial temporal vessels.

– has the following boundaries:

1. Anterior: zygomatic process of the frontal bone and the frontal process of the zygomatic bone

2. Posterior: temporal line

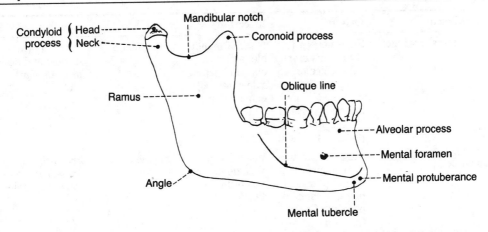

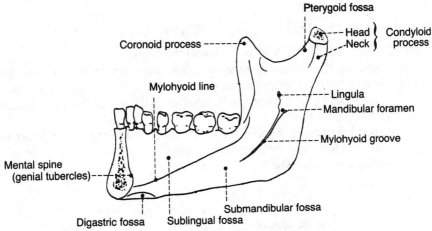

Figure 8-18. External (buccal) and internal (lingual) surfaces of the mandible.

3. **Superior:** temporal line

4. **Inferior:** zygomatic arch

5. **Floor:** parts of the frontal, parietal, temporal, and greater wing of the sphenoid bone

II. Muscles of Mastication (Figure 8-19)

Muscle	Origin	Insertion	Nerve	Action
Temporalis	Temporal fossa	Coronoid process and ramus of mandible	Trigeminal n.	Elevates and retracts mandible
Masseter	Lower border and medial surface of zygomatic arch	Lateral surface of coronoid process, ramus and angle of mandible	Trigeminal n.	Elevates mandible

(Continued on next page)

Muscle	Origin	Insertion	Nerve	Action
Lateral pterygoid	Superior head from infratemporal surface of sphenoid; inferior head from lateral surface of lateral pterygoid plate of sphenoid	Neck of mandible; articular disk and capsule of temporomandibular joint	Trigeminal n.	Protracts (protrudes) and depresses mandible
Medial pterygoid	Tuber of maxilla; medial surface of lateral pterygoid plate; pyramidal process of palatine bone	Medial surface of angle and ramus of mandible	Trigeminal n.	Protracts (protrudes) and elevates mandible

The jaws are opened by the lateral pterygoid muscle and are closed by the temporalis, masseter, and medial pterygoid muscles.

III. Nerves of the Infratemporal Region (see Figure 8-19)

A. Mandibular division of the trigeminal nerve

- passes through the **foramen ovale** and innervates the tensor veli palatini and tensor tympani muscles, muscles of mastication (temporalis, masseter, and lateral and medial pterygoid), anterior belly of the digastric muscle, and the mylohyoid muscle.
- provides sensory innervation to the lower teeth and to the lower part of the face below the lower lip and the mouth.
- gives rise to the following branches:

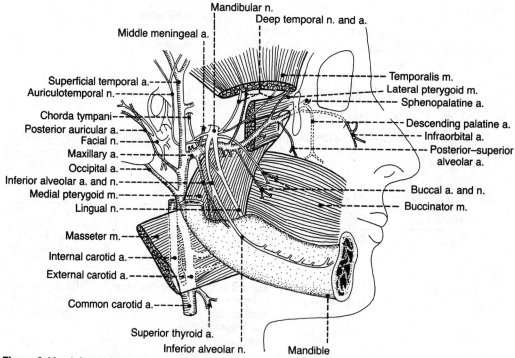

Figure 8-19. Infratemporal region.

1. Meningeal branch

   – accompanies the middle meningeal artery and enters the cranium through the foramen spinosum.

2. Masseteric, deep temporal, medial pterygoid, and lateral pterygoid nerves

   – innervate the corresponding muscles of mastication.

3. Buccal nerve

   – descends between the two heads of the lateral pterygoid muscle.
   – innervates skin and fascia on the buccinator muscle and penetrates this muscle to supply the mucous membrane of the cheek and gums.

4. Auriculotemporal nerve

   – arises from two roots that encircle the middle meningeal artery.
   – innervates sensory branches to the temporomandibular joint.
   – carries postganglionic parasympathetic and sympathetic fibers to the parotid gland.
   – Its terminal branches supply the skin of the auricle and the scalp.

5. Lingual nerve

   – descends deep to the lateral pterygoid muscle, where it joins the **chorda tympani,** which conveys the preganglionic parasympathetic (secretomotor) fibers to the submandibular ganglion and taste fibers from the anterior two-thirds of the tongue.
   – lies anterior to the inferior alveolar nerve on the medial pterygoid muscle, deep to the ramus of the mandible.
   – crosses lateral to the styloglossus and hyoglossus muscles, passes deep to the mylohyoid nerve, and descends across the submandibular duct.
   – supplies general sensation for the anterior two-thirds of the tongue.

6. Inferior alveolar nerve

   – passes deep to the lateral pterygoid muscle and then between the spheno-mandibular ligament and the ramus of the mandible.
   – enters the mandibular canal through the mandibular foramen.
   – gives rise to the following branches:

   a. The mylohyoid nerve to the mylohyoid and the anterior belly of the digastric muscle

   b. The inferior dental branch to the lower teeth

   c. The mental nerve to the skin over the chin

   d. The incisive branch to the canine and incisor teeth

B. Otic ganglion

   – lies in the infratemporal fossa, just below the foramen ovale between the mandibular nerve and the tensor veli palatini.
   – Preganglionic parasympathetic fibers that run in the glossopharyngeal nerve, tympanic plexus, and lesser petrosal nerve synapse in this ganglion.
   – Postganglionic fibers from the otic ganglion run in the **auriculotemporal nerve** to innervate the parotid gland.

IV. Blood Vessels of the Infratemporal Region (see Figure 8-19)

A. Maxillary artery

– arises from the external carotid artery at the posterior border of the ramus of the mandible.

– divides into three parts:

1. Mandibular part

– runs anteriorly between the neck of the mandible and the sphenomandibular ligament.

– gives rise to the following branches:

a. Deep auricular artery

– supplies the external acoustic meatus.

b. Anterior tympanic artery

– supplies the tympanic cavity and tympanic membrane.

c. Middle meningeal artery

– is embraced by two roots of the auriculotemporal nerve and enters the middle cranial fossa through the foramen spinosum.

– runs between the dura mater and the periosteum.

– Damage to the artery results in **epidural hematoma**.

d. Accessory meningeal artery

– passes through the foramen ovale.

e. Inferior alveolar artery

– follows the inferior alveolar nerve between the sphenomandibular ligament and the ramus of the mandible.

– enters the mandibular canal through the mandibular foramen and supplies the tissues of the chin and lower teeth.

2. Pterygoid part

– runs anteriorly deep to the temporalis and lies superficial (or deep) to the lateral pterygoid muscle.

– Its branches include the anterior and posterior deep temporal, pterygoid, masseteric, and buccal arteries, which supply chiefly the muscles of mastication.

3. Pterygopalatine part

– runs between the two heads of the lateral pterygoid muscle and then through the pterygomaxillary fissure into the pterygopalatine fossa.

– Its branches include the following arteries:

a. Posterior–superior alveolar arteries

– run downward on the posterior surface of the maxilla and supplies the molar and premolar teeth and the maxillary sinus.

b. Infraorbital artery

– runs upward and forward to enter the orbit through the inferior orbital fissure.

– traverses the infraorbital groove and canal and emerges on the face through the infraorbital foramen.

– divides into branches to supply the lower eyelid, lacrimal sac, upper lip, and cheek.

- gives rise to **anterior and middle superior alveolar branches** to the upper canine and incisor teeth and the maxillary sinus.

c. Descending palatine artery

- descends in the pterygopalatine fossa and the palatine canal.
- gives rise to the **greater and lesser palatine arteries,** which pass through the greater and lesser palatine foramina, respectively, and supplies the soft and hard palates.
- The greater palatine artery sends a branch to anastomose with the terminal (nasopalatine) branch of the sphenopalatine artery in the incisive canal or on the nasal septum.

d. Artery of the pterygoid canal

- passes through the pterygoid canal and supplies the upper part of the pharynx, auditory tube, and tympanic cavity.

e. Pharyngeal artery

- supplies the roof of the nose and pharynx, sphenoid sinus, and auditory tube.

f. Sphenopalatine artery

- is the terminal branch of the maxillary artery.
- enters the nasal cavity through the sphenopalatine foramen in company with the nasopalatine branch of the maxillary nerve.
- is the principal artery to the nasal cavity, supplying the conchae, meatus, and paranasal sinuses.
- Damage to this artery results in **epistaxis.**

B. Pterygoid venous plexus (Figure 8-20)

- lies on the lateral surface of the medial pterygoid muscle.
- communicates with the cavernous sinus via emissary veins and the inferior ophthalmic vein.
- communicates with the facial vein via the deep facial vein.

C. Retromandibular vein

- is formed by the superficial temporal vein and the maxillary vein.
- divides into an anterior branch, which joins the facial vein to form the common facial vein, and a posterior branch, which joins the posterior auricular vein to form the external jugular vein.

V. Parotid Gland

- occupies the **retromandibular space** between the ramus of the mandible and the mastoid process.
- is invested with a dense fibrous capsule derived from the investing layer of the deep cervical fascia.
- is separated from the submandibular gland by the **stylomandibular ligament,** which extends from the styloid process to the angle of the mandible. (Therefore, pus does not readily exchange between these two glands.)
- is innervated by parasympathetic (secretomotor) fibers of the glossopharyngeal nerve by way of the **lesser petrosal nerve, otic ganglion,** and **auriculotemporal nerve.**
- **Stensen's duct,** the duct of the parotid, crosses the masseter, pierces the buccinator muscle, and opens into the oral cavity opposite the second upper molar tooth.

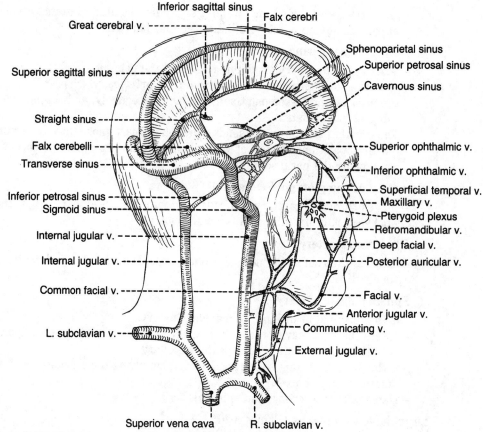

Figure 8-20. Cranial venous sinuses and veins of the head and neck.

– Complete surgical removal of the parotid may damage the facial nerve.

VI. Joints and Ligaments of the Infratemporal Region

A. Temporomandibular joint

– is a **synovial joint** between the articular tubercle and the mandibular fossa of the temporal bone above and the head of the mandible below.

– combines a hinge and a gliding joint and has two (superior and inferior) synovial cavities, divided by an **articular disk.**

– has an articular capsule that extends from the articular tubercle and the margins of the mandibular fossa to the neck of the mandible.

– is reinforced by the **lateral (temporomandibular) ligament,** which extends from the tubercle on the zygoma to the neck of the mandible, and the **sphenomandibular ligament,** which extends from the spine of the sphenoid bone to the lingula of the mandible.

– is innervated by the auriculotemporal, masseteric, and deep temporal branches of the mandibular division of the trigeminal nerve.

– is supplied by the superficial temporal, maxillary (middle meningeal and anterior tympanic branches), and ascending pharyngeal arteries.

B. Pterygomandibular raphe

 – is a **ligamentous band** (or a tendinous inscription) between the buccinator muscle and the superior pharyngeal constrictor.
 – extends between the pterygoid hamulus superiorly and the posterior end of the mylohyoid line of the mandible inferiorly.

C. Stylomandibular ligament

 – extends from the styloid process to the posterior border of the ramus of the mandible, near the angle of the mandible, separating the parotid from the submandibular gland.

VII. Clinical Considerations

A. Mumps (epidermic parotitis)

 – is an acute infectious and contagious disease caused by a viral infection; it irritates the auricolotemporal nerve, causing severe pain because of **inflammation and swelling of the parotid gland and stretching of its capsule;** and is also characterized by **inflammation of salivary glands.**
 – may be accompanied by inflammation of the testes or ovary, causing **sterility.**

B. Frey's syndrome

 – produces **flushing and sweating instead of salivation in response to taste of food,** following injury of the auriculotemporal nerve, which contains parasympathetic secretomotor fibers to the parotid gland and sympathetic fibers to the sweat glands. (When the nerve is severed, the fibers can regenerate along each others pathways and innervate the wrong gland.)
 – can occur after parotid surgery; may be treated by cutting the tympanic plexus in the ear.

C. Dislocation of the temporomandibular joint

 – occurs anteriorly by placing the heads of the mandible anterior to the articular tubercle during yawning and laughing.

D. Rupture of the middle meningeal artery

 – may be caused by **fracture of the squamous part of the temporal bone,** as it runs through the foramen spinosum and just deep to the inner surface of the temporal bone.
 – causes epidural hematoma with increased intracranial pressure.

Skull and Cranial Cavity

I. Skull (see Figure 8-17; Figures 8-21 and 8-22)

 – is the skeleton of the head, which may be divided into two types: 8 **cranial bones** for enclosing the brain (unpaired frontal, occipital, ethmoid, and sphenoid bones, and paired parietal and temporal bones); and 14 **facial bones** (paired lacrimal, nasal, palatine, inferior turbinate, maxillary, and zygomatic bones, and unpaired vomer and mandible).

A. Cranium

 – is sometimes restricted to the skull without the mandible.

B. Calvaria

 – is the **skullcap,** which is the vault of the skull without the facial bones; it consists of the superior portions of the frontal, parietal, and occipital bones.
 – Its highest point on the sagittal suture is the **vertex.**

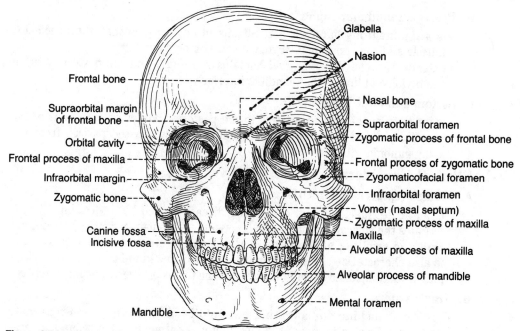

Figure 8-21. Anterior view of the skull.

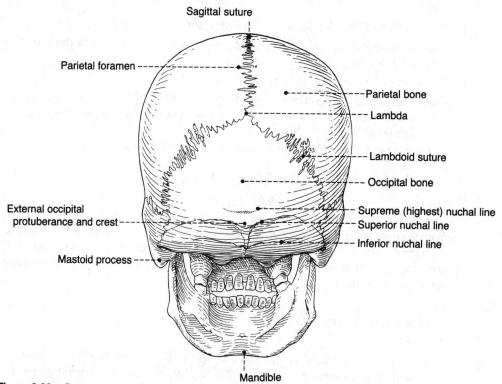

Figure 8-22. Posterior view of the skull.

II. Bones of the Cranium

A. Frontal bone
– underlies the forehead and the superior margin and roof of the orbit, and has a smooth median prominence called the **glabella.**

B. Parietal bone
– forms part of the superior and lateral surface of the skull.

C. Temporal bone
– consists of the **squamous part** external to the lateral surface of the temporal lobe of the brain; the **petrous part,** which encloses the internal and middle ears; the **mastoid part,** which contains mastoid air cells; and the **tympanic part,** which houses the external auditory meatus and the tympanic cavity.

D. Occipital bone
– consists of **squamous, basilar, and two lateral condylar parts.**
– encloses the foramen magnum and forms the cerebral and cerebellar fossae.

E. Sphenoid bone
– consists of the body (which houses the sphenoid sinus), the greater and lesser wings, and the **pterygoid process.**

F. Ethmoid bone
– is located between the orbits and consists of the **cribriform plate, perpendicular plate,** and two lateral masses enclosing ethmoid air cells.

III. Sutures of the Skull
– are the immovable fibrous joints between the bones of the skull.

A. Coronal suture lies between the frontal bone and the two parietal bones.

B. Sagittal suture lies between the two parietal bones.

C. Squamous (squamoparietal) suture lies between the parietal bone and the squamous part of the temporal bone.

D. Lambdoid suture lies between the two parietal bones and the occipital bone.

E. Junctions of the cranial sutures
1. **Lambda** is the intersection of the lambdoid and sagittal sutures.
2. **Bregma** is the intersection of the sagittal and coronal sutures.
3. **Pterion** is a craniometric point at the junction of the frontal, parietal, and temporal bones and the great wing of the sphenoid bone.
4. **Asterion** is a craniometric point at the junction of the parietal, occipital, and temporal (mastoid part) bones.
5. **Nasion** is a point on the middle of the nasofrontal suture (intersection of the frontal and two nasal bones).

IV. Foramina in the Skull (Figures 8-23 and 8-24)
– include the following, which are presented here with the structures that pass through them:

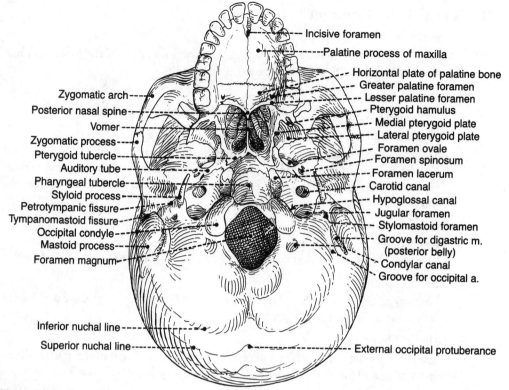

Figure 8-23. Base of the skull.

Incisive foramen
Palatine process of maxilla
Horizontal plate of palatine bone
Greater palatine foramen
Lesser palatine foramen
Pterygoid hamulus
Medial pterygoid plate
Lateral pterygoid plate
Foramen ovale
Foramen spinosum
Foramen lacerum
Carotid canal
Hypoglossal canal
Jugular foramen
Stylomastoid foramen
Groove for digastric m. (posterior belly)
Condylar canal
Groove for occipital a.

Zygomatic arch
Posterior nasal spine
Vomer
Zygomatic process
Pterygoid tubercle
Auditory tube
Pharyngeal tubercle
Styloid process
Petrotympanic fissure
Tympanomastoid fissure
Occipital condyle
Mastoid process
Foramen magnum

Inferior nuchal line
Superior nuchal line
External occipital protuberance

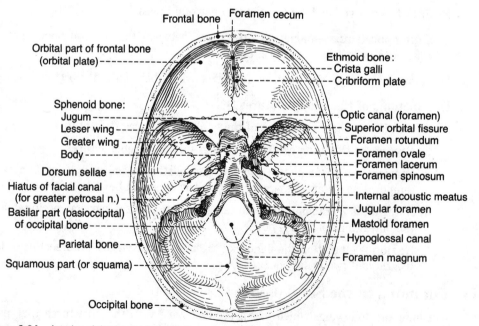

Figure 8-24. Interior of the base of the skull.

Frontal bone
Foramen cecum

Orbital part of frontal bone (orbital plate)

Ethmoid bone:
Crista galli
Cribriform plate

Sphenoid bone:
Jugum
Lesser wing
Greater wing
Body
Dorsum sellae
Hiatus of facial canal (for greater petrosal n.)
Basilar part (basioccipital) of occipital bone
Parietal bone
Squamous part (or squama)

Optic canal (foramen)
Superior orbital fissure
Foramen rotundum
Foramen ovale
Foramen lacerum
Foramen spinosum
Internal acoustic meatus
Jugular foramen
Mastoid foramen
Hypoglossal canal
Foramen magnum

Occipital bone

A. Anterior cranial fossa

1. **Cribriform plate:** olfactory nerves

2. **Foramen cecum:** occasional small emissary vein from nasal mucosa to superior sagittal sinus

3. **Anterior and posterior ethmoidal foramina:** anterior and posterior ethmoidal nerves, arteries, and veins

B. Middle cranial fossa

1. **Optic canal:** optic nerve, ophthalmic artery, and central vein of the retina

2. **Superior orbital fissure:** oculomotor, trochlear, and abducens nerves; ophthalmic division of trigeminal nerve; and ophthalmic veins

3. **Foramen rotundum:** maxillary division of trigeminal nerve

4. **Foramen ovale:** mandibular division of trigeminal nerve, accessory meningeal artery, and occasionally lesser petrosal nerve

5. **Foramen spinosum:** middle meningeal artery

6. **Foramen lacerum:** internal carotid artery and greater and deep petrosal nerves en route to the pterygoid canal

7. **Carotid canal:** internal carotid artery and sympathetic nerves (carotid plexus)

C. Posterior cranial fossa

1. **Internal auditory meatus:** facial and vestibulocochlear nerves and labyrinthine artery

2. **Jugular foramen:** glossopharyngeal, vagus, and spinal accessory nerves and beginning of internal jugular vein

3. **Hypoglossal canal:** hypoglossal nerve and meningeal artery

4. **Foramen magnum:** spinal cord, spinal accessory nerve, vertebral arteries, venous plexus of vertebral canal, and anterior and posterior spinal arteries

5. **Condyloid foramen:** condyloid emissary vein

6. **Mastoid foramen:** branch of occipital artery to dura mater and mastoid emissary vein

D. Foramina in the front of the skull (see Figure 8-21)

1. **Zygomaticofacial foramen:** zygomaticofacial nerve

2. **Supraorbital notch:** supraorbital nerve and vessels

3. **Infraorbital foramen:** infraorbital nerve and vessels

4. **Mental foramen:** mental nerve and vessels

E. Foramina in the base of the skull (see Figure 8-24)

1. **Petrotympanic fissure:** chorda tympani

2. **Stylomastoid foramen:** facial nerve

3. **Incisive canal:** nasopalatine nerve

4. **Greater palatine canal:** greater palatine nerve and vessels

5. **Lesser palatine canal:** lesser palatine nerve and vessels

V. Structures in the Cranial Fossae (see Figure 8-24)

A. Foramen cecum
- is a small pit in front of the **crista galli** between the ethmoid and frontal bones.
- may transmit an emissary vein from the nasal mucosa to the superior sagittal sinus.

B. Crista galli
- is the triangular midline process of the ethmoid bone extending upward from the cribriform plate.
- gives attachment to the **falx cerebri.**

C. Cribriform plate of the ethmoid bone
- supports the olfactory bulb and transmits olfactory nerve fibers from the olfactory mucosa to the olfactory bulb.

D. Anterior clinoid processes
- are two anterior processes of the lesser wing of the sphenoid bone.
- give attachment to the free border of the tentorium cerebelli.

E. Posterior clinoid processes
- are two tubercles from each side of the **dorsum sellae.**
- give attachment to the attached border of the tentorium cerebelli.

F. Lesser wing of the sphenoid bone
- forms the anterior boundary of the middle cranial fossa.
- forms the **sphenoidal ridge** separating the anterior from the middle cranial fossa.
- forms the boundary of the **superior orbital fissure** (the space between the lesser and greater wings).

G. Greater wing of the sphenoid bone
- forms the anterior wall and the floor of the middle cranial fossa.
- presents two openings, the **foramen ovale** and the **foramen spinosum.**

H. Sella turcica (Turk's saddle) of the sphenoid bone
- is bounded anteriorly by the tuberculum sellae and posteriorly by the dorsum sellae.
- Its deep central depression is the **hypophyseal fossa,** which accommodates the pituitary gland or the hypophysis.
- lies directly above the sphenoid sinus located within the body of the sphenoid bone; its dural roof is formed by the **diaphragma sellae.**

I. Jugum sphenoidale
- forms the roof for the **sphenoidal air sinus.**

VI. Meninges of the Brain

A. Pia mater
- is a delicate investment that is closely applied to the brain and dips into fissures and sulci.
- enmeshes blood vessels on the surfaces of the brain.

B. Arachnoid layer
- is a filmy, transparent, spidery layer that is connected to the pia mater by web-like trabeculations.

- is separated from the pia mater by the **subarachnoid space,** which is filled with cerebrospinal fluid (CSF); it may contain blood after hemorrhage of a cerebral artery.
- projects into the venous sinuses to form **arachnoid villi,** which serve as sites where CSF diffuses into the venous blood.

1. **Cerebrospinal fluid**
   - is formed by **vascular choroid plexuses** in the ventricle of the brain and is contained in the subarachnoid space.
   - circulates through the ventricles, enters the subarachnoid space, and eventually filters into the venous system.

2. **Arachnoid granulations**
   - are tuft-like collections of highly folded arachnoid that project into the superior sagittal sinus and the lateral lacunae, which are lateral extensions of the superior sagittal sinus.
   - absorb the CSF into the dural sinuses and often produce erosion or pitting of the inner surface of the calvaria.

C. **Dura mater**
   - is the tough, fibrous, outermost layer of the meninges external to the **subdural space,** the space between the arachnoid and the dura.
   - lies internal to the **epidural space,** a potential space that contains the middle meningeal arteries in the cranial cavity.
   - forms the **dural venous sinuses,** spaces between the periosteal and meningeal layers or between duplications of the meningeal layers.

   1. **Innervation of the dura mater**

      a. **Anterior and posterior ethmoidal branches** of the ophthalmic division of the trigeminal nerve in the anterior cranial fossa

      b. **Meningeal branches** of the maxillary and mandibular divisions of the trigeminal nerve in the middle cranial fossa

      c. **Meningeal branches** of the vagus and hypoglossal nerves in the posterior cranial fossa

   2. **Projections of the dura mater** (see Figure 8-20)

      a. **Falx cerebri**
         - is the sickle-shaped double layer of the dura mater, lying between the cerebral hemispheres.
         - is attached anteriorly to the crista galli and posteriorly to the tentorium cerebelli.
         - Its inferior concave border is free and contains the **inferior sagittal sinus,** and its upper convex margin encloses the **superior sagittal sinus.**

      b. **Falx cerebelli**
         - is a small sickle-shaped projection between the cerebellar hemispheres.
         - is attached to the posterior and inferior parts of the tentorium.
         - contains the **occipital sinus** in its posterior border.

      c. **Tentorium cerebelli**
         - is a crescentic fold of dura mater that supports the occipital lobes of the cerebral hemispheres and covers the cerebellum.

- Its internal concave border is free and bounds the **tentorial notch,** whereas its external convex border encloses the **transverse sinus** posteriorly and the **superior petrosal sinus** anteriorly.
- The free border is anchored to the anterior clinoid process, whereas the attached border is attached to the posterior clinoid process.

d. Diaphragma sellae

- is a circular, horizontal fold of dura that forms the roof of the sella turcica, covering the pituitary gland or the hypophysis.
- has a central aperture for the hypophyseal stalk or infundibulum.

VII. Cranial Venous Channels (see Figure 8-20)

A. Superior sagittal sinus

- lies in the midline along the convex border of the falx cerebri.
- begins at the crista galli and receives the cerebral, diploic, meningeal, and parietal emissary veins.

B. Inferior sagittal sinus

- lies in the free edge of the falx cerebri and is joined by the **great cerebral vein of Galen** to form the straight sinus.

C. Straight sinus

- runs along the line of attachment of the falx cerebri to the tentorium cerebelli.

D. Transverse sinus

- runs laterally from the confluence of sinuses along the edge of the tentorium cerebelli.

E. Sigmoid sinus

- is a continuation of the transverse sinus; arches downward and medially in an S-shaped groove on the mastoid part of the temporal bone.
- enters the superior bulb of the internal jugular vein.

F. Cavernous sinuses

- are located on each side of the sella turcica and the body of the sphenoid bone and lie between the meningeal and periosteal layers of the dura mater.
- The internal carotid artery and the abducens nerve pass through these sinuses.
- The oculomotor, trochlear, ophthalmic, and maxillary nerves pass forward in the lateral wall of these sinuses.
- communicate with the pterygoid venous plexus by emissary veins and receive the superior ophthalmic vein.

G. Superior petrosal sinus

- lies in the margin of the tentorium cerebelli, running from the posterior end of the cavernous sinus to the transverse sinus.

H. Inferior petrosal sinus

- drains the cavernous sinus into the bulb of the internal jugular vein.
- runs in a groove between the petrous part of the temporal bone and the basilar part of the occipital bone.

I. Sphenoparietal sinus

- lies along the posterior edge of the lesser wing of the sphenoid bone and drains into the cavernous sinus.

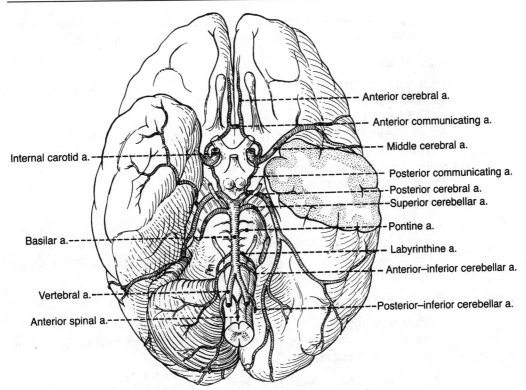

Figure 8-25. Arterial circle on the inferior surface of the brain.

J. Occipital sinus
– lies in the falx cerebelli and drains into the confluence of sinuses.

K. Basilar plexus
– consists of interconnecting venous channels on the basilar part of the occipital bone and connects the two inferior petrosal sinuses.
– communicates with the internal vertebral venous plexus.

L. Diploic veins
– lie in the **diploë** of the skull and are connected with the cranial dura sinuses by the emissary veins.

M. Emissary veins
– are small veins connecting the venous sinuses of the dura with the diploic veins and the veins of the scalp.

VIII. Blood Supply of the Brain (Figure 8-25)

A. Internal carotid artery
– enters the carotid canal in the petrous portion of the temporal bone.
– is separated from the tympanic cavity by a thin bony structure.
– lies within the cavernous sinus and gives rise to small twigs to the wall of the cavernous sinus, to the hypophysis, and to the semilunar ganglion of the trigeminal nerve.
– pierces the dural roof of the cavernous sinus between the anterior and middle clinoid processes.

– gives rise to the **superior hypophyseal, ophthalmic, posterior communicating, anterior choroid, anterior cerebral, and middle cerebral arteries.**

B. Vertebral arteries

– arise from the first part of the subclavian artery and ascend through the transverse foramina of the vertebrae C1–C6.
– curve posteriorly behind the lateral mass of the atlas, pierce the dura mater into the vertebral canal, and then enter the cranial cavity through the foramen magnum.
– join to form the **basilar artery.**
– give rise to the following:

1. Anterior spinal artery

– arises as two roots from the vertebral arteries shortly before the junction of the vertebral arteries.
– descends in front of the medulla, and the two roots unite to form a single median trunk at the level of the foramen magnum.

2. Posterior spinal artery

– arises from the vertebral artery or the posterior–inferior cerebellar artery and descends on the side of the medulla; the right and left roots unite at the lower cervical region.

3. Posterior–inferior cerebellar artery

– is the largest branch of the vertebral artery, distributes to the posterior–inferior surface of the cerebellum, and gives rise to the posterior spinal artery.

C. Basilar artery

– is formed by the union of the two vertebral arteries at the lower border of the pons.
– gives rise to the **pontine, anterior–inferior cerebellar, labyrinthine, and superior cerebellar arteries.**
– ends near the upper border of the pons by dividing into the right and left **posterior cerebral arteries.**

D. Circle of Willis (circulus arteriosus) [Figure 8-26]

– is formed by the posterior cerebral, posterior communicating, internal carotid, anterior cerebral, and anterior communicating arteries.
– forms an important means of **collateral circulation** in the event of obstruction.

IX. Clinical Considerations

A. Epidural hematoma

– is due to rupture of the middle meningeal artery.

B. Subdural hematoma

– is due to rupture of cerebral veins as they pass from the brain surface into one of the venous sinuses.

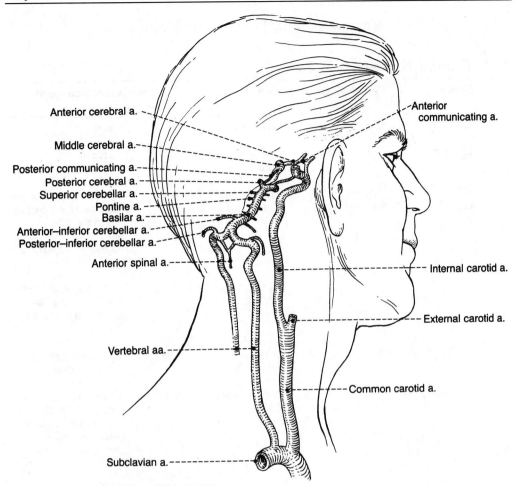

Figure 8-26. Formation of the circle of Willis.

C. Subarachnoid hemorrhage

– is due to rupture of cerebral arteries.

D. Pial hemorrhage

– is due to damage to the small vessels of the pia and brain tissue.

E. Cavernous sinus thrombophlebitis

– is an **infectious inflammation of the cavernous sinus** with secondary thrombus formation.

– is associated with significant morbidity and mortality because of the formation of **meningitis** (inflammation of the meninges).

– may produce papilledema (edema of the optic disk probably due to raised intracranial pressure), exophthalmos (protrusion of the eyeball), and ophthalmoplegia (paralysis of the eye muscles).

Nerves of the Head and Neck

I. Cranial Nerves (Figure 8-27)

Nerve	Cranial Exit	Cell Bodies	Compo-nents	Chief Functions
I: Olfactory	Cribriform plate	Nasal mucosa	SVA	Smell
II: Optic	Optic canal	Ganglion cells of retina	SSA	Vision
III: Oculo-motor	Superior orbital fissure	Nucleus CN III (midbrain)	GSE	Eye movements (superior, inferior, and medial recti, inferior oblique, and levator palpebrae superioris mm.)
		Edinger-Westphal nucleus (mid-brain)	GVE	Constriction of pupil (sphincter pupillae m.) and accommodation (ciliary m.)
IV: Trochlear	Superior orbital fissure	Nucleus CN IV (midbrain)	GSE	Eye movements (superior oblique m.)
V: Trigeminal	Superior orbital fissure; foramen rotundum and foramen ovale	Motor nucleus CN V (pons)	SVE	Muscles of mastication, (mylohyoid, anterior belly of digastric, tensor veli palatini, and tensor tympani mm.)
		Trigeminal ganglion	GSA	Sensation in head (skin and mucous membranes of face and head)
VI: Abducens	Superior orbital fissure	Nucleus CN VI (pons)	GSE	Eye movement (lateral rectus m.)
VII: Facial	Stylomastoid foramen	Motor nucleus CN VII (pons)	SVE	Muscle of facial expression (posterior belly of digastric, stylohyoid, and stapedius mm.)
		Salivatory nucleus (pons)	GVE	Lacrimal and salivary secretion
		Geniculate ganglion	SVA	Taste from anterior two-thirds of tongue and palate
		Geniculate ganglion	GVA	Sensation from palate

(Continued on next page)

Nerve	Cranial Exit	Cell Bodies	Compo-nents	Chief Functions
		Geniculate ganglion	GSA	Sensation from external acoustic meatus
VIII: Vestibulocochlear	Does not leave skull	Vestibular ganglion	SSA	Equilibrium
		Spiral ganglion	SSA	Hearing
IX: Glosso-pharyngeal	Jugular foramen	Nucleus ambiguus (medulla)	SVE	Elevation of pharynx (stylopharyngeus m.)
		Dorsal nucleus (medulla)	GVE	Secretion of saliva (parotid gland)
		Inferior ganglion	GVA	Sensation in carotid sinus and body, tongue, and pharynx
		Inferior ganglion	SVA	Taste from posterior one-third of tongue
		Inferior ganglion	GSA	Sensation in external and middle ear
X: Vagus	Jugular foramen	Nucleus ambiguus (medulla)	SVE	Muscles of movements of pharynx, larynx, and palate
		Dorsal nucleus (medulla)	GVE	Involuntary muscle and gland control in thoracic and abdominal viscerae
		Inferior ganglion	GVA	Sensation in pharynx, larynx, and other viscerae
		Inferior ganglion	SVA	Taste from root of tongue and epiglottis
		Superior ganglion	GSA	Sensation in external ear and external acoustic meatus
XI: Accessory	Jugular foramen	Spinal cord (cervical)	SVE	Movement of head and shoulder (sternocleidomastoid and trapezius mm.)
XII: Hypo-glossal	Hypoglossal canal	Nucleus CN XII (medulla)	GSE	Muscles of movements of tongue

A. Olfactory nerves (CN I)

- consist of about 20 bundles of unmyelinated special visceral afferent (SVA) fibers that arise from olfactory neurons in the olfactory area, the upper one-third of the nasal mucosa.

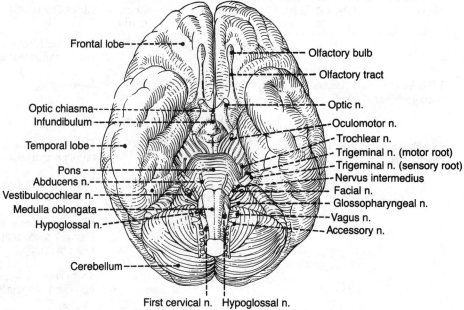

Figure 8-27. Cranial nerves on the base of the brain.

– pass through the foramina in the cribriform plate of the ethmoid bone and enter the **olfactory bulb,** where they synapse.

B. Optic nerve (CN II)

– is formed by the axons of **ganglion cells of the retina,** which converge at the optic disk.

– carries special somatic afferent (SSA) fibers from the retina to the brain.

– leaves the orbit through the optic canal and forms the **optic chiasma,** where fibers from the nasal side of either retina cross over to the opposite side of the brain.

C. Oculomotor nerve (CN III)

– enters the orbit through the superior orbital fissure within the tendinous ring.

– supplies general somatic efferent (GSE) fibers to the extraocular muscles (i.e., medial, superior, and inferior recti; inferior oblique; and levator palpebrae superioris).

– contains preganglionic parasympathetic general visceral efferent (GVE) fibers with cell bodies located in the Edinger-Westphal nucleus, and postganglionic fibers derived from the **ciliary ganglion** that run in the **short ciliary nerves** to supply the sphincter pupillae and the ciliary muscle.

D. Trochlear nerve (CN IV)

– passes through the lateral wall of the cavernous sinus during its course.

– enters the orbit by passing through the superior orbital fissure and supplies GSE fibers to the superior oblique muscle.

– is the **smallest cranial nerve** and the only cranial nerve that emerges from the dorsal aspect of the brainstem.

E. Trigeminal nerve (CN V)

– is the **first branchiomeric nerve** and supplies the first branchial arch.
– provides general somatic afferent (GSA) sensory fibers to the face, scalp, auricle, external auditory meatus, nose, paranasal sinuses, mouth (except the posterior one-third of the tongue), parts of the nasopharynx, auditory tube, and cranial dura mater.
– has a ganglion (**semilunar or trigeminal ganglion**) that consists of cell bodies of GSA fibers and occupies the **trigeminal impression** on the petrous portion of the temporal bone.
– has the following divisions:

1. Ophthalmic division

– runs in the dura of the lateral wall of the cavernous sinus and enters the orbit through the **supraorbital fissure.**
– provides sensory innervation to the eyeball, tip of the nose, and skin of the face above the eye.
– mediates the **afferent limb of the corneal reflex** by way of the nasociliary branch.

2. Maxillary division

– passes through the lateral wall of the cavernous sinus and through the **foramen rotundum.**
– provides sensory innervation to the midface (below the eye but above the upper lip), palate, paranasal sinuses, and maxillary teeth, with cell bodies in the trigeminal ganglion.
– mediates the **afferent limb of the sneeze reflex.**

3. Mandibular division

– passes through the **foramen ovale** and supplies special visceral efferent (SVE) fibers to the tensor veli palatini, tensor tympani, muscles of mastication (temporalis, masseter, and lateral and medial pterygoid), and the anterior belly of the digastric and mylohyoid muscles.
– provides sensory innervation to the lower part of the face (below the lower lip and mouth), scalp, jaw, mandibular teeth, and anterior two-thirds of the tongue.
– mediates the **afferent limb of the jaw jerk reflex.**

F. Abducens nerve (CN VI)

– pierces the dura on the dorsum sellae of the sphenoid bone.
– passes through the cavernous sinus, enters the orbit through the supraorbital fissure, and supplies GSE fibers to the lateral rectus.

G. Facial Nerve (CN VII) (Figure 8-28)

– consists of a larger root, which contains SVE fibers to innervate the muscles of facial expression, and a smaller root, termed the nervus intermedius, which contains SVA fibers from the anterior two-thirds of the tongue; preganglionic parasympathetic GVE fibers for the lacrimal, submandibular, sublingual, nasal, and palatine glands; general visceral afferent (GVA) fibers from the palate; and GSA fibers from the external acoustic meatus and the auricle.
– enters the **internal acoustic meatus,** the **facial canal** in the temporal bone, and emerges from the **stylomastoid foramen.**

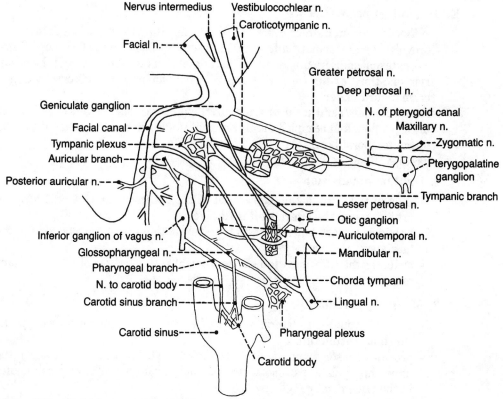

Figure 8-28. Facial nerve and its connections with other nerves.

- has a sensory ganglion, the **geniculate ganglion,** which lies at the knee-shaped bend or genu (L., "knee"). It gives rise to the **greater petrosal nerve,** which carries preganglionic parasympathetic fibers to the **pterygopalatine ganglion** (for the lacrimal, nasal, and palatine glands); a branch to the tympanic plexus; and a branch to the sympathetic plexus on the middle meningeal artery.
- gives rise to the following branches:

1. **Greater petrosal nerve**
   - contains preganglionic parasympathetic GVE fibers and joins the **deep petrosal nerve** (containing postganglionic sympathetic fibers) to form the **nerve of the pterygoid canal** (vidian nerve).
   - also contains SVA (taste) and GVA fibers, which pass from the palate through the pterygopalatine ganglion, the nerve of the pterygoid canal, and the greater petrosal nerve to the geniculate ganglion (where cell bodies are located).

2. **Communicating branch**
   - joins the lesser petrosal nerve.

3. **Stapedial nerve**
   - supplies motor (SVE) fibers to the stapedius.

4. **Chorda tympani**
   - arises in the descending part of the facial canal, crosses the medial aspect of the tympanic membrane, passing between the handle of the malleus and the long process of the incus.

– exits the skull through the **petrotympanic fissure** and joins the lingual nerve in the infratemporal fossa.

– contains preganglionic parasympathetic GVE fibers that synapse on postganglionic cell bodies in the **submandibular ganglion.** Their postganglionic fibers innervate the submandibular, sublingual, and lingual glands.

– also contains taste (SVA) fibers from the anterior two-thirds of the tongue and the soft palate, with cell bodies located in the geniculate ganglion.

– may communicate with the otic ganglion below the base of the skull.

5. **Muscular branches**

– supply motor (SVE) fibers to the stylohyoid and the posterior belly of the digastric muscle.

6. **Fine branch**

– joins the auricular branch of the vagus nerve and the glossopharyngeal nerve to supply GSA fibers to the external ear.

7. **Posterior auricular nerve**

– runs behind the auricle with the posterior auricular artery.

– supplies SVE fibers to the muscles of the auricle and the occipital belly of the occipitofrontalis muscle.

8. **Terminal branches**

– arise in the parotid gland and radiate onto the face.

– include the temporal, zygomatic, buccal, marginal mandibular, and cervical branches, which supply motor (SVE) fibers to the muscles of facial expression.

H. Vestibulocochlear (acoustic) nerve (CN VIII)

– enters the internal acoustic meatus and remains within the temporal bone to supply SSA fibers to the cochlea, the ampullae of the semicircular ducts, and the utricle and saccule.

– is split into a **cochlear portion** for hearing and a **vestibular portion** for equilibrium.

I. Glossopharyngeal nerve (CN IX) [see Figure 8-28]

– passes through the **jugular foramen** and gives rise to the following branches:

1. **Tympanic nerve**

– forms the **tympanic plexus** on the medial wall of the middle ear with sympathetic fibers from the internal carotid plexus (caroticotympanic nerves) and a branch from the geniculate ganglion of the facial nerve. The plexus supplies GVA fibers to the tympanic cavity, the mastoid antrum and air cells, and the auditory tube.

– continues beyond the plexus as the **lesser petrosal nerve,** which transmits preganglionic parasympathetic GVE fibers to the otic ganglion.

2. **Communicating branch**

– joins the auricular branch of the vagus nerve and provides GSA fibers.

3. **Pharyngeal branch**

– supplies GVA fibers to the pharynx and forms the **pharyngeal plexus** on the middle constrictor muscle along with the pharyngeal branch of the vagus nerve and branches from the sympathetic trunk.

4. Carotid sinus branch

   – supplies GVA fibers to the **carotid sinus** and the **carotid body.**

5. Tonsillar branches

   – supply GVA fibers to the palatine tonsil and the soft palate.

6. Motor branch

   – supplies SVE fibers to the stylopharyngeus.

7. Lingual branch

   – supplies GVA and SVA fibers to the posterior one-third of the tongue and SVA fibers to the vallate papillae.

J. Vagus nerve (CN X)

– supplies the fourth to sixth branchial arches during development.

– passes through the jugular foramen.

– provides motor (GVE) innervation to smooth muscle and cardiac muscle, secretory innervation to all glands, and afferent (GVA) fibers from all mucous membranes in the thoracic and abdominal visceral organs (except for the descending colon, sigmoid colon, rectum, and other pelvic organs).

– provides branchiomotor (SVE) innervation to all muscles of the larynx, pharynx (except the stylopharyngeus), and palate (except the tensor veli palatini).

– gives rise to the following branches:

1. Meningeal branch

   – arises from the superior ganglion and supplies the dura mater of the posterior cranial fossa.

2. Auricular branch

   – is joined by a branch from the glossopharyngeal nerve and the facial nerve and supplies GSA fibers to the external acoustic meatus.

3. Pharyngeal branch

   – supplies motor (SVE) fibers to all muscles of the pharynx, except the stylopharyngeus, by way of the pharyngeal plexus and all muscles of the palate except the tensor veli palatini.

   – gives rise to the **nerve to the carotid body,** which supplies the carotid body and the carotid sinus.

4. Superior, middle, and inferior cardiac branches

   – pass to the cardiac plexuses.

5. Superior laryngeal nerve

   – divides into internal and external branches:

   a. Internal laryngeal nerve

      – provides sensory (GVA) fibers to the larynx above the vocal cord.

      – supplies SVA fibers to the taste buds on the root of the tongue near the epiglottis.

   b. External laryngeal nerve

      – supplies motor (SVE) fibers to the cricothyroid and inferior pharyngeal constrictor muscles.

6. Recurrent laryngeal nerve

- hooks around the subclavian artery on the right and around the arch of the aorta lateral to the ligamentum arteriosum on the left.
- ascends in the groove between the trachea and the esophagus.
- provides sensory (GVA) fibers to the larynx below the vocal cord and motor fibers to all muscles of the larynx except the cricothyroid muscle.
- becomes the **inferior laryngeal nerve** at the lower border of the cricoid cartilage.

K. Accessory nerve (CN XI)

- passes through the jugular foramen.
- Its spinal roots unite to form the trunk that ascends between dorsal and ventral roots of the spinal nerves and passes through the foramen magnum.
- provides branchiomotor (SVE) fibers to the sternocleidomastoid and trapezius muscles.
- Its cranial portion contains motor fibers that join the vagus nerve and is best regarded as a part of the vagus nerve.

L. Hypoglossal nerve (CN XII)

- passes through the **hypoglossal canal.**
- loops around the occipital artery and passes between the external carotid and internal jugular vessels.
- passes above the hyoid bone on the lateral surface of the hyoglossus deep to the mylohyoid muscle.
- supplies GSE fibers to all of the intrinsic and extrinsic muscles of the tongue except the palatoglossus, which is supplied by the vagus nerve.

II. Parasympathetic Ganglia and Associated Autonomic Nerves (Figure 8-29)

Ganglion	Location	Parasympathetic Fibers	Sympathetic Fibers	Chief Distribution
Ciliary	Lateral to optic n.	Oculomotor n. and its inferior division	Internal carotid plexus	Ciliary muscle, and sphincter pupillae (parasympathetic); dilator pupillae and tarsal mm. (sympathetic)
Pterygopalatine	In pterygopalatine fossa	Facial n., greater petrosal n., and n. of pterygoid canal	Internal carotid plexus	Lacrimal gland and glands in palate and nose
Submandibular	On hyoglossus	Facial n., chorda tympani, and lingual n.	Plexus on facial a.	Submandibular and sublingual glands
Otic	Below foramen ovale	Glossopharyngeal n., its tympanic branch, and lesser petrosal n.	Plexus on middle meningeal a.	Parotid gland

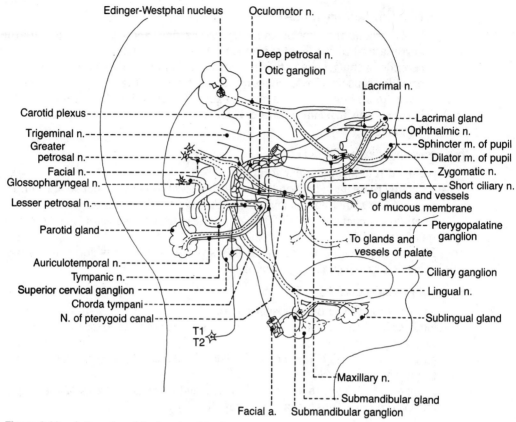

Figure 8-29. Autonomics of the head and neck.

A. Ciliary ganglion

– is situated behind the eyeball, between the optic nerve and the lateral rectus muscle.

– receives preganglionic parasympathetic fibers (with cell bodies in the Edinger-Westphal nucleus of CN III in the mesencephalon), which run in the inferior division of the oculomotor nerve.

– sends its postganglionic parasympathetic fibers to the sphincter pupillae and the ciliary muscle via the **short ciliary nerves.**

– receives postganglionic sympathetic fibers (derived from the superior cervical ganglion) that reach the dilator pupillae by way of the sympathetic plexus on the internal carotid artery, the long ciliary nerve and/or the ciliary ganglion (without synapsing), and the short ciliary nerves.

B. Pterygopalatine ganglion

– lies in the pterygopalatine fossa just below the maxillary nerve, lateral to the sphenopalatine foramen and anterior to the pterygoid canal.

– receives preganglionic parasympathetic fibers from the facial nerve by way of the greater petrosal nerve and the nerve of the pterygoid canal.

– sends postganglionic parasympathetic fibers to the nasal and palatine glands and to the lacrimal glands by way of the maxillary, zygomatic, and lacrimal nerves.

– also receives postganglionic sympathetic fibers (derived from the superior cervical ganglion) by way of the plexus on the internal carotid artery, the deep petrosal nerve, and the nerve of the pterygoid canal. The fibers merely pass through the ganglion and are distributed with the postganglionic parasympathetic fibers.

1. **Greater petrosal nerve**

   – arises from the **facial nerve** adjacent to the geniculate ganglion.
   – emerges at the hiatus of the canal for the greater petrosal nerve in the middle cranial fossa.
   – contains preganglionic parasympathetic fibers and joins the deep petrosal nerve (containing postganglionic sympathetic fibers) to form the **nerve of the pterygoid canal** (vidian nerve).
   – also contains taste fibers, which pass from the palate nonstop through the pterygopalatine ganglion, the nerve of the pterygoid canal, and the greater petrosal nerve to the geniculate ganglion (where cell bodies are found).

2. **Deep petrosal nerve**

   – arises from the plexus on the internal carotid artery.
   – contains postganglionic sympathetic fibers with cell bodies located in the superior cervical ganglion. These fibers run in the nerve of the pterygoid canal, pass through the pterygopalatine ganglion without synapsing, and then join the postganglionic parasympathetic fibers in supplying the lacrimal gland and the nasal and oral mucosa.

3. **Nerve of the pterygoid canal (vidian nerve)**

   – consists of preganglionic parasympathetic fibers from the greater petrosal nerve and postganglionic sympathetic fibers from the deep petrosal nerve.
   – passes through the pterygoid canal and ends in the pterygopalatine ganglion, which is slung from the maxillary nerve. The postganglionic parasympathetic fibers have cell bodies located in the pterygopalatine ganglion, and the postganglionic sympathetic fibers are distributed to the lacrimal, nasal, and palatine glands.
   – also contains SVA and GVA fibers from the palate.

C. **Submandibular ganglion**

   – lies on the lateral surface of the hyoglossus muscle but deep to the mylohyoid muscle and is suspended from the lingual nerve.
   – receives preganglionic parasympathetic (secretomotor) fibers that run in the facial nerve, chorda tympani, and lingual nerve.
   – sends postganglionic fibers to supply the submandibular gland mostly, although some join the lingual nerve to reach the sublingual and lingual glands.

D. **Otic ganglion**

   – lies in the infratemporal fossa, just below the foramen ovale, between the mandibular nerve and the tensor veli palatini.
   – receives preganglionic parasympathetic fibers that run in the glossopharyngeal nerve, tympanic plexus, and lesser petrosal nerve and synapse in the otic ganglion.
   – sends postganglionic fibers that run in the auriculotemporal nerve and supply the parotid gland.

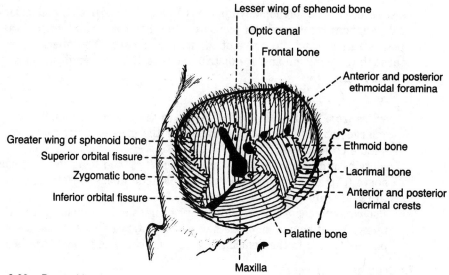

Figure 8-30. Bony orbit.

1. Tympanic nerve

– contains preganglionic parasympathetic (secretomotor) fibers for the parotid gland.
– arises from the inferior ganglion of the glossopharyngeal nerve.
– passes through a small canal between the jugular foramen and the carotid canal into the tympanic cavity.
– enters the tympanic plexus on the promontory of the medial wall of the tympanic cavity.

2. Lesser petrosal nerve

– is a continuation of the tympanic nerve beyond the tympanic plexus.
– runs just lateral to the greater petrosal nerve and leaves the middle cranial fossa through either the foramen ovale or the fissure between the petrous bone and the great wing of the sphenoid to enter the otic ganglion.
– contains preganglionic parasympathetic (secretomotor) fibers that run in the glossopharyngeal and tympanic nerves before synapsing in the otic ganglion. (The postganglionic fibers arising from the ganglion are passed to the parotid gland by the auriculotemporal nerves.)
– also transmits postganglionic sympathetic fibers to the parotid gland.

Orbit

I. Bony Orbit (Figure 8-30)

A. Orbital margin

– is formed by the frontal, maxillary, and zygomatic bones.

B. Walls of the orbit

1. Superior wall or roof: orbital part of frontal bone and lesser wing of sphenoid bone

2. **Lateral wall:** zygomatic bone (frontal process) and greater wing of sphenoid bone

3. **Inferior wall or floor:** maxilla (orbital surface), zygomatic, and palatine bones

4. **Medial wall:** ethmoid bone (orbital plate), frontal lacrimal bone, and body of sphenoid bone

C. Fissures, canals, and foramina

1. Superior orbital fissure
- communicates with the middle cranial fossa and is bounded by the greater and lesser wings of the sphenoid.
- transmits the oculomotor, trochlear, abducens, and ophthalmic nerves (three branches), as well as the ophthalmic veins.

2. Inferior orbital fissure
- communicates with the infratemporal and pterygopalatine fossae.
- is bounded by the greater wing of the sphenoid (above) and the maxillary and palatine bones (below).
- transmits the maxillary nerve and its zygomatic branch and the infraorbital vessels.

3. Optic canal
- connects the orbit with the middle cranial fossa.
- is formed by the two roots of the lesser wing of the sphenoid and is situated in the posterior part of the roof of the orbit.
- transmits the optic nerve and ophthalmic artery.

4. Infraorbital groove and infraorbital foramen
- transmit the infraorbital nerve and vessels.

5. Supraorbital notch or foramen
- transmits the supraorbital nerve and vessels.

6. Anterior and posterior ethmoidal foramina
- transmit the anterior and posterior ethmoidal nerves and vessels, respectively.

7. Nasolacrimal canal
- is formed by the maxilla, lacrimal bone, and inferior nasal concha.
- transmits the nasolacrimal duct from the lacrimal sac to the inferior nasal meatus.

II. Nerves of the Orbit (Figures 8-31 and 8-32)

A. Ophthalmic nerve
- enters the orbit through the superior orbital fissure and divides into three branches:

1. Lacrimal nerve
- enters the orbit through the superior orbital fissure.
- enters the lacrimal gland, giving rise to branches to the lacrimal gland, the conjunctiva, and the skin of the upper eyelid.

2. Frontal nerve
- enters the orbit through the superior orbital fissure.
- runs superior to the levator palpebrae superioris.

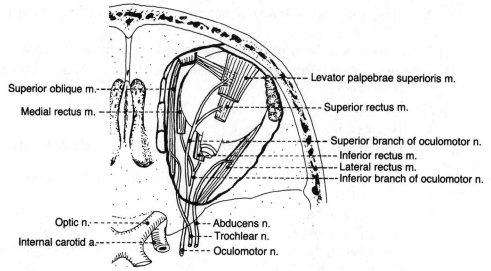

Figure 8-31. Motor nerves of the orbit.

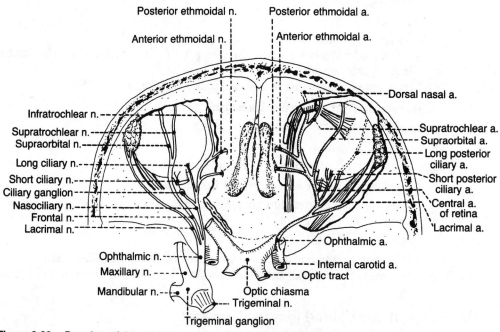

Figure 8-32. Branches of the ophthalmic nerve and ophthalmic artery.

- divides into the **supraorbital nerve,** which passes through the **supraorbital notch or foramen** and supplies the scalp, forehead, frontal sinus, and upper eyelid; and the **supratrochlear nerve,** which passes through the trochlea and supplies the scalp, forehead, and upper eyelid.

3. **Nasociliary nerve**

- is the sensory nerve to the eye.
- enters the orbit through the superior orbital fissure, within the common tendinous ring.
- gives rise to the following:

      a. A communicating branch to the ciliary ganglion

      b. Short ciliary nerves, which carry postganglionic parasympathetic and sympathetic fibers to the ciliary body and iris

      c. Long ciliary nerves, which transmit postganglionic sympathetic fibers to the dilator pupillae and afferent fibers from the iris and cornea

      d. The posterior ethmoidal nerve, which passes through the posterior ethmoidal foramen to the sphenoidal and posterior ethmoidal sinuses

      e. The anterior ethmoidal nerve, which passes through the anterior ethmoidal foramen to supply the anterior ethmoidal air cells. It divides into **internal nasal branches,** which supply the septum and lateral walls of the nasal cavity, and **external nasal branches,** which supply the skin of the tip of the nose.

      f. The infratrochlear nerve to the eyelids, conjunctiva, skin of the nose, and lacrimal sac

B. Optic nerve

    – consists of the axons of the ganglion cells of the retina.
    – leaves the orbit by passing through the **optic canal** and carries afferent optic fibers from the retina to the brain.
    – joins the optic nerve from the corresponding eye to form the optic chiasma.

C. Oculomotor nerve

    – leaves the cranium through the superior orbital fissure.
    – divides into a **superior division,** which innervates the superior rectus and levator palpebrae superioris muscles, and an **inferior division,** which innervates the medial rectus, inferior rectus, and inferior oblique muscles.
    – Its inferior division also carries preganglionic parasympathetic fibers (with cell bodies located in the Edinger-Westphal nucleus) to the **ciliary ganglion.**

D. Trochlear nerve

    – passes through the lateral wall of the cavernous sinus during its course.
    – enters the orbit by passing through the superior orbital fissure and innervates the superior oblique muscle.

E. Abducens nerve

    – enters the orbit through the superior orbital fissure and supplies the lateral rectus muscle.

F. Ciliary ganglion

    – is a parasympathetic ganglion situated behind the eyeball, between the optic nerve and the lateral rectus muscle.

III. Blood Vessels of the Orbit (see Figure 8-32)

A. Ophthalmic artery

    – is a branch of the internal carotid artery and enters the orbit through the **optic canal** beneath the optic nerve.
    – gives rise to the **ocular and orbital vessels,** which include the following:

1. **Central artery of the retina**
   - is the **most important branch** of the ophthalmic artery.
   - travels in the optic nerve; it divides into superior and inferior branches at the optic disk, and each of those further divides into temporal and nasal branches.
   - is an end artery that does not anastomose with other arteries, and thus its **occlusion results in blindness.**

2. **Long posterior ciliary arteries**
   - pierce the sclera and supply the ciliary body and the iris.

3. **Short posterior ciliary arteries**
   - pierce the sclera and supply the choroid.

4. **Lacrimal artery**
   - passes along the superior border of the lateral rectus and supplies the lacrimal gland, conjunctiva, and eyelids.
   - gives rise to two **lateral palpebral arteries,** which contribute to arcades in the upper and lower eyelids.

5. **Medial palpebral arteries**
   - contribute to arcades in the upper and lower eyelids.

6. **Muscular branches**
   - supply orbital muscles and give off the anterior ciliary arteries, which supply the iris.

7. **Supraorbital artery**
   - passes through the supraorbital notch (or foramen) and supplies the forehead and the scalp.

8. **Posterior ethmoidal artery**
   - passes through the posterior ethmoidal foramen to the posterior ethmoidal air cells.

9. **Anterior ethmoidal artery**
   - passes through the anterior ethmoidal foramen to the anterior and middle ethmoidal air cells, frontal sinus, nasal cavity, and external nose.

10. **Supratrochlear artery**
    - passes to the supraorbital margin and supplies the forehead and the scalp.

11. **Dorsal nasal artery**
    - supplies the side of the nose and the lacrimal sac.

B. **Ophthalmic veins** (Figure 8-33)

1. **Superior ophthalmic vein**
   - is formed by the union of the supraorbital, supratrochlear, and angular veins.
   - receives branches corresponding to most of those of the ophthalmic artery and, in addition, receives the inferior ophthalmic vein before draining into the cavernous sinus.

2. **Inferior ophthalmic vein**
   - begins by the union of small veins in the floor of the orbit.
   - communicates with the pterygoid venous plexus and often with the infraorbital vein and terminates directly or indirectly in the cavernous sinus.

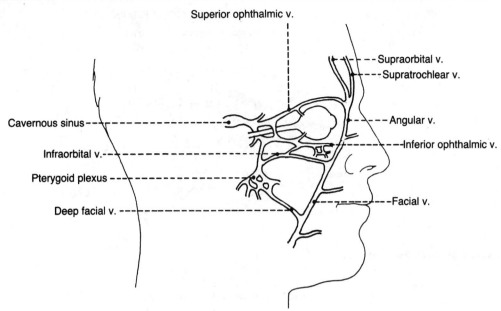

Figure 8-33. Ophthalmic veins.

IV. Muscles of Eye Movement (Figures 8-34 and 8-35)

Muscle	Origin	Insertion	Nerve	Actions
Superior rectus	Common tendinous ring	Sclera just behind cornea	Oculomotor n.	Elevates eyeball
Inferior rectus	Common tendinous ring	Sclera just behind cornea	Oculomotor n.	Depresses eyeball
Medial rectus	Common tendinous ring	Sclera just behind cornea	Oculomotor n.	Adducts eyeball
Lateral rectus	Common tendinous ring	Sclera just behind cornea	Abducens n.	Abducts eyeball
Levator palpebrae superioris	Lesser wing of sphenoid above and anterior to optic canal	Tarsal plate and skin of upper eyelid	Oculomotor n. Sympathetic n.	Elevates upper eyelid
Superior oblique	Body of sphenoid bone above optic canal	Sclera beneath superior rectus	Trochlear n.	Rotates downward and medially; depresses adducted eye
Inferior oblique	Floor of orbit lateral to lacrimal groove	Sclera beneath lateral rectus	Oculomotor n.	Rotates upward and laterally; elevates adducted eye

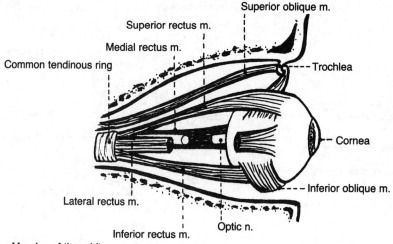

Figure 8-34. Muscles of the orbit.

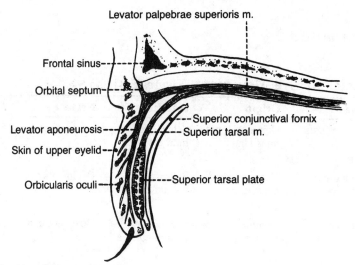

Figure 8-35. Structure of the upper eyelid.

A. **Innervation of muscles of the eyeball**

– can be summarized as **SO$_4$, LR$_6$,** and **Remainder$_3$,** which means that the superior oblique muscle is innervated by the trochlear nerve, the lateral rectus by the abducens nerve, and the remainder of these muscles by the oculomotor nerve.

B. **Movements of the eye**

1. **Intorsion**

– is an **inward (medial) rotation** of the upper pole of the vertical corneal meridians, caused by the superior oblique and superior rectus muscles.

2. **Extorsion**

– is an **outward (lateral) rotation** of the upper pole of the vertical corneal meridians, caused by the inferior oblique and inferior rectus muscles.

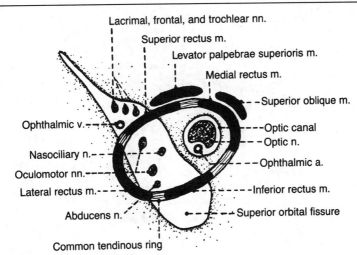

Figure 8-36. Common tendinous ring.

C. **Common tendinous ring** (Figure 8-36)
   – is a **fibrous ring** that surrounds the optic canal and the medial part of the superior orbital fissure.
   – gives origin to the four rectus muscles of the eye and transmits the following structures:
   1. **The oculomotor, nasociliary, and abducens nerves** enter the orbit through the superior orbital fissure and the common tendinous ring.
   2. **The optic nerve, ophthalmic artery, and central artery of the retina** enter the orbit through the optic canal and the tendinous ring.
   3. **The superior ophthalmic vein and the trochlear, frontal, and lacrimal nerves** enter the orbit through the superior orbital fissure but outside the tendinous ring.

V. Lacrimal Apparatus (Figure 8-37)

A. **Lacrimal gland**
   – lies in the upper lateral region of the orbit on the lateral rectus and the levator palpebrae superioris muscles.
   – is drained by 12 **lacrimal ducts,** which open into the superior conjunctival fornix.

B. **Lacrimal canaliculi**
   – are two curved canals, which begin as a lacrimal punctum (or pore) in the margin of the eyelid and open into the lacrimal sac.

C. **Lacrimal sac**
   – is the upper dilated end of the **nasolacrimal duct,** which opens into the inferior meatus of the nasal cavity.

D. **Tears**
   – are produced by the **lacrimal gland.**
   – pass through excretory ductules into the superior conjunctival fornix.
   – are spread evenly over the eyeball by blinking movements and accumulate in the area of the **lacrimal lake.**

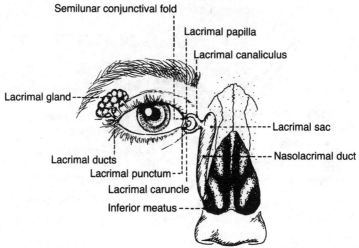

Semilunar conjunctival fold
Lacrimal papilla
Lacrimal canaliculus
Lacrimal gland
Lacrimal sac
Nasolacrimal duct
Lacrimal ducts
Lacrimal punctum
Lacrimal caruncle
Inferior meatus

Figure 8-37. Lacrimal apparatus.

– enter the lacrimal canaliculi through their lacrimal puncta (which is on the summit of the lacrimal papilla) before draining into the lacrimal sac, nasolacrimal duct, and finally the inferior nasal meatus.

VI. Eyeball (Figure 8-38)

A. External white fibrous coat

– consists of the sclera and the cornea.

1. Sclera

– is a tough white fibrous tunic enveloping the posterior five-sixths of the eye.

2. Cornea

– is a transparent structure forming the anterior one-sixth of the external coat.

– is responsible for the **refraction of light** entering the eye.

B. Middle vascular pigmented coat

– consists of the choroid, ciliary body, and iris.

1. Choroid

– consists of an outer pigmented (dark-brown) layer and an inner highly vascular layer, which invest the posterior five-sixths of the eyeball.

– **nourishes the retina** and darkens the eye.

2. Ciliary body

– is a **thickened portion of the vascular coat** between the choroid and the iris, and consists of the ciliary ring, ciliary processes, and ciliary muscle.

a. The **ciliary processes** are radiating pigmented ridges that encircle the margin of the lens.

b. The **ciliary muscle** consists of meridional and circular fibers of smooth muscle innervated by parasympathetic fibers. It contracts to pull the **ciliary ring** and ciliary processes, relaxing the suspensory ligament of the lens and allowing it to increase its convexity.

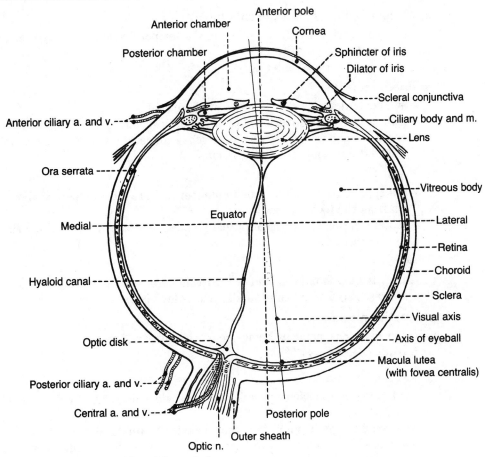

Figure 8-38. Horizontal section of the eyeball.

3. Iris

- is a thin, contractile, circular, pigmented diaphragm with a central aperture, the **pupil.**
- contains circular muscle fibers (**sphincter pupillae**), which are innervated by parasympathetic fibers, and radial fibers (**dilator pupillae**), which are innervated by sympathetic fibers.

C. Internal nervous coat

- consists of the **retina,** which has an outer pigmented layer and an inner nervous layer.
- Its posterior part is photosensitive; its anterior part, which is not photosensitive, constitutes the inner lining of the ciliary body and the posterior part of the iris.

1. Optic disk (blind spot)

- consists of **optic nerve fibers** formed by axons of the ganglion cells. These cells are connected to the rods and cones by bipolar neurons.
- is located nasal (or medial) to the fovea centralis and the posterior pole of the eye, has no receptors, and is insensitive to light.
- has a depression in its center termed the **physiological cup.**

2. **Macula (yellow spot or macula lutea)**
   - is a yellowish area of the retina on the temporal side of the optic disk for the most distinct vision.
   - contains the fovea centralis.

3. **Fovea centralis**
   - is a **central depression** (foveola) in the macula.
   - is avascular and is nourished by the choriocapillary lamina of the choroid.
   - has cones only (no rods), each of which is connected with only one ganglion cell, and functions in detailed vision.

4. **Rods**
   - are approximately 120 million in number and are most numerous about 0.5 cm from the fovea centralis.
   - contain **rhodopsin,** a visual purple pigment, and are specialized for **vision in dim light.**

5. **Cones**
   - are 7 million in number and are most numerous in the foveal region.
   - are associated with **visual acuity** and **color vision.**

D. **Refractive media**
   - consist of cornea, aqueous humor, lens, and vitreous body.

   1. **Cornea** (see VI A 2)

   2. **Aqueous humor**
      - is formed by the ciliary processes and provides nutrients for the avascular cornea and lens.
      - passes through the pupil from the **posterior chamber** (between the iris and the lens) into the **anterior chamber** (between the cornea and the iris) and is drained into the scleral venous plexus through the canal of Schlemm at the iridocorneal angle.
      - Its impaired drainage causes an increased intraocular pressure, leading to atrophy of the retina and blindness.

   3. **Lens**
      - is a transparent **avascular biconvex structure** enclosed in an elastic capsule.
      - is held in position by radially arranged **zonular fibers** (suspensory ligament of the lens), which are attached medially to the lens capsule and laterally to the ciliary processes.
      - flattens to focus on distant objects by pulling the zonular fibers, and it becomes a globular shape to accommodate the eye for near objects by contracting the ciliary muscle and thus relaxing zonular fibers.

   4. **Vitreous body**
      - is a transparent gel called **vitreous humor,** which fills the eyeball posterior to the lens (vitreous chamber between the lens and the retina).
      - holds the retina in place and provides support for the lens.

VII. Clinical Considerations

A. **Horner's syndrome**
   - is caused by **injury to cervical sympathetic fibers.**
   - is characterized by:

1. **Miosis,** pupillary constriction resulting from paralysis of the associated dilator muscle of the pupil

2. **Ptosis,** drooping of an upper eyelid from paralysis of the smooth muscle component of the levator palpebrae superioris

3. **Enophthalmos,** retraction of an eyeball from paralysis of the tarsal muscle

4. **Anhidrosis,** absence of sweating

5. **Vasodilation,** leading to increased blood flow in the facial and cervical regions

B. Crocodile tears syndrome (Bogorad's syndrome)

- is **spontaneous lacrimation during eating,** caused by a lesion of the facial nerve proximal to the geniculate ganglion.
- follows facial paralysis and is due to misdirection of regenerating parasympathetic fibers, which formerly innervated the salivary (submandibular and sublingual) glands, to the lacrimal glands.

C. Glaucoma

- is a condition of **opacity of the crystalline lens.**
- is characterized by **increased intraocular pressure** due to impaired drainage of aqueous humor (which is produced by the ciliary processes) into the venous system through Schlemm's canal and impaired retinal blood flow, producing retinal ischemia or atrophy of the retina, degeneration of the nerve fibers in the retina, particularly at the optic disk, and blindness.

D. Cataract

- is an **opacity (milk-white) of the crystalline eye lens or of its capsule,** necessitating its removal.
- results in little light to be transmitted to the retina, causing blurred images and poor vision.

E. Retinal detachment

- is a **separation of the sensory layer from the pigment layer of the retina.**
- may occur in trauma such as a blow to the head, and can be reattached surgically by photocoagulation by laser beam.

F. Diplopia (double vision)

- is caused by **paralysis of one or more extraocular muscles** due to injury of the nerves supplying them.

G. Myopia (nearsightedness)

- is a condition in which the focus of objects lies in front of the retina, due to elongation of the eyeball.

H. Hyperopia (farsightedness)

- is a condition in which the focus of objects lies behind the retina.

I. Presbyopia

- is a condition in which the **power of accommodation is reduced.**
- is caused by the **loss of elasticity of the crystalline lens.**
- occurs in advanced age and is corrected with bifocal lenses.

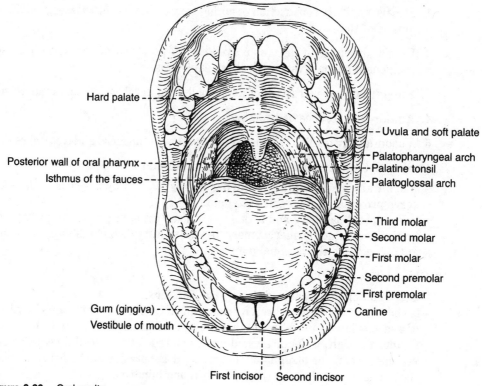

Figure 8-39. Oral cavity.

Oral Cavity and Salivary Glands

I. Oral Cavity (Figure 8-39)
- Its roof is formed by the **palate,** and its floor is formed by the tongue and the mucosa, supported by the geniohyoid and mylohyoid muscles.
- Its lateral and anterior walls are formed by an outer fleshy wall (cheeks and lips) and an inner bony wall (teeth and gums). (The **vestibule** is between the walls, and the **oral cavity proper** is the area inside the teeth and gums.)

II. Palate
- forms the roof of the mouth and the floor of the nasal cavity.

A. Hard palate
- is the anterior four-fifths of the palate and forms a **bony framework covered with a mucous membrane** between the nasal and oral cavities.
- consists of the **palatine processes** of the maxillae and horizontal plates of the palatine bones.
- contains the incisive foramen in its median plane anteriorly and the greater and lesser palatine foramina posteriorly.
- receives sensory innervation through the greater palatine and nasopalatine nerves, and blood from the greater palatine artery.

B. Soft palate
- is a **fibromuscular fold** extending from the posterior border of the hard palate and makes up one-fifth of the palate.

– moves posteriorly against the pharyngeal wall to close the oropharyngeal (faucial) isthmus when swallowing or speaking.
– is continuous with the **palatoglossal and palatopharyngeal folds.**
– receives blood from the greater and lesser palatine arteries of the descending palatine artery of the maxillary artery, the ascending palatine artery of the facial artery, and the palatine branch of the ascending pharyngeal artery.
– receives sensory innervation through the lesser palatine nerves.

C. Muscles of the palate

Muscle	Origin	Insertion	Nerve	Action
Tensor veli palatini	Scaphoid fossa; spine of sphenoid; cartilage of auditory tube	Tendon hooks around hamulus of medial pterygoid plate to insert into aponeurosis of soft palate	Mandibular branch of trigeminal n.	Tenses soft palate
Levator veli palatini	Petrous part of temporal bone; cartilage of auditory tube	Aponeurosis of soft palate	Vagus n. via pharyngeal plexus	Elevates soft palate
Palatoglossus	Aponeurosis of soft palate	Dorsolateral side of tongue	Vagus n. via pharyngeal plexus	Elevates tongue
Palatopharyngeus	Aponeurosis of soft palate	Thyroid cartilage and side of pharynx	Vagus n. via pharyngeal plexus	Elevates pharynx; closes nasopharynx
Musculus uvulae	Posterior nasal spine of palatine bone; palatine aponeurosis	Mucous membrane of uvula	Vagus n. via pharyngeal plexus	Elevates uvula

III. Tongue (Figure 8-40)

– is attached by muscles to the hyoid bone, mandible, styloid process, palate, and pharynx.
– is divided by a V-shaped **sulcus terminalis** into two parts, an anterior two-thirds and a posterior one-third, which differ developmentally, structurally, and in innervation.
– The **foramen cecum** is located at the apex of the V and indicates the site of origin of the embryonic **thyroglossal duct.**

A. Lingual papillae

– are small, nipple-shaped projections on the anterior two-thirds of the dorsum of the tongue.
– are divided into the vallate, fungiform, filiform, and foliate papillae.

1. Vallate papillae

– are arranged in the form of a V in front of the sulcus terminalis.
– are studded with numerous taste buds and are innervated by the glossopharyngeal nerve.

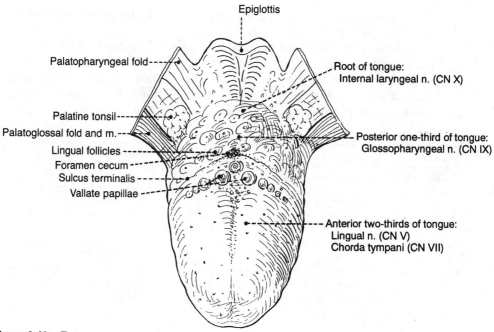

Epiglottis

Palatopharyngeal fold

Palatine tonsil
Palatoglossal fold and m.
Lingual follicles
Foramen cecum
Sulcus terminalis
Vallate papillae

Root of tongue:
Internal laryngeal n. (CN X)

Posterior one-third of tongue:
Glossopharyngeal n. (CN IX)

Anterior two-thirds of tongue:
Lingual n. (CN V)
Chorda tympani (CN VII)

Figure 8-40. Tongue.

2. Fungiform papillae

– are mushroom-shaped projections with red heads and are scattered on the sides and the apex of the tongue.

3. Filiform papillae

– are numerous, slender, conical projections that are arranged in rows parallel to the sulcus terminalis.

4. Foliate papillae

– are found in certain animals but are rudimentary in humans.

B. Lingual tonsil

– is the collection of **nodular masses of lymphoid follicles** on the posterior one-third of the dorsum of the tongue.

C. Lingual innervation

– The extrinsic and intrinsic muscles of the tongue are innervated by the **hypoglossal nerve** except for the palatoglossus, which is innervated by the vagus nerve.
– The anterior two-thirds of the tongue receives general sensory innervation from the **lingual nerve** and taste sensation from the **chorda tympani.**
– The posterior one-third of the tongue and the vallate papillae receive both general and taste innervation from the **glossopharyngeal nerve.**
– The epiglottic region of the tongue and the epiglottis receive both general and taste innervation from the **internal laryngeal branch** of the vagus nerve.

D. Lingual artery

– arises from the external carotid artery at the level of the tip of the greater horn of the hyoid bone in the carotid triangle.

- passes deep to the hyoglossus and lies on the middle pharyngeal constrictor muscle.
- gives rise to the suprahyoid, dorsal lingual, and sublingual arteries and terminates as the **deep lingual artery,** which ascends between the genioglossus and inferior longitudinal muscles.

E. Muscles of the tongue

Muscle	Origin	Insertion	Nerve	Action
Styloglossus	Styloid process	Side and inferior aspect of tongue	Hypoglossal n.	Retracts and elevates tongue
Hyoglossus	Body and greater horn of hyoid bone	Side and inferior aspect of tongue	Hypoglossal n.	Depresses and retracts tongue
Genioglossus	Genial tubercle of mandible	Inferior aspect of tongue; body of hyoid bone	Hypoglossal n.	Protrudes and depresses tongue
Palatoglossus	Aponeurosis of soft palate	Dorsolateral side of tongue	Vagus n. via pharyngeal plexus	Elevates tongue

IV. Teeth and Gums (Gingivae)

A. Structure of the teeth

1. **Enamel** is the hardest substance that covers the crown.

2. **Dentine** is a hard substance that is nurtured through the fine dental tubules of odontoblasts lining the central pulp space.

3. **Pulp** fills the central cavity, which is continuous with the root canal. It contains numerous blood vessels, nerves, and lymphatics, which enter the pulp through an apical foramen at the apex of the root.

B. Parts of the teeth

1. **Crown** projects above the gingival surface and is covered by enamel.

2. **Neck** is the constricted area at the junction of the crown and root.

3. **Root,** embedded in the alveolar part of the maxilla or mandible, is covered with cement, which is connected to the bone of the alveolus by a layer of modified periosteum, the periodontal ligament.

C. Basic types of teeth

1. **Incisors** are chisel-shaped and are used for cutting or biting.

2. **Canines** have a single prominent cone and are used for tearing.

3. **Premolars** usually have two cusps and are used for grinding.

4. **Molars** usually have three cusps and are used for grinding.

D. Two sets of teeth

1. **Deciduous teeth:** two incisors, one canine, and two molars in each quadrant, for a total of 20

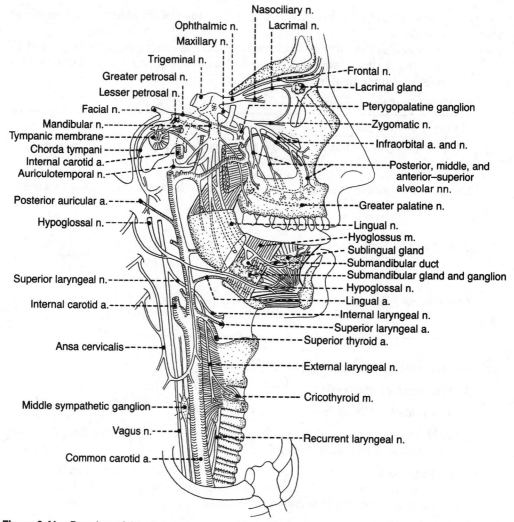

Figure 8-41. Branches of the trigeminal nerve and their relationship with other structures.

2. **Permanent teeth:** two incisors, one canine, two premolars, and three molars in each quadrant, for a total of 32

E. **Innervation of the teeth and gums** (Figure 8-41)

1. **Maxillary teeth** are innervated by the anterior, middle, and posterior–superior alveolar branches of the maxillary nerve.

2. **Mandibular teeth** are innervated by the inferior alveolar branch of the mandibular nerve.

3. **Maxillary gingiva**

   a. **Outer (buccal) surface** is innervated by posterior, middle and anterior–superior alveolar and infraorbital nerves.

   b. **Inner (lingual) surface** is innervated by greater palatine and nasopalatine nerves.

4. Mandibular gingiva

   a. Outer (buccal) surface is innervated by buccal and mental nerves.

   b. Inner (lingual) surface is innervated by lingual nerve.

V. Salivary Glands (see Figure 8-41)

A. Submandibular gland

   – is ensheathed by the investing layer of the deep cervical fascia and lies in the **submandibular triangle.**

   – Its superficial portion is situated superficial to the mylohyoid muscle.

   – Its deep portion is located between the hyoglossus and styloglossus muscles medially and the mylohyoid muscle laterally and between the lingual nerve above and the hypoglossal nerve below.

   – **Wharton's duct** arises from the deep portion and runs forward between the mylohyoid laterally and the hyoglossus medially, where it is crossed laterally by the lingual nerve. It then runs between the sublingual gland and the genioglossus and empties at the summit of the sublingual papilla (caruncle) at the side of the frenulum of the tongue.

   – is innervated by parasympathetic secretomotor fibers from the facial nerve, which run in the chorda tympani and in the lingual nerve and synapse in the submandibular ganglion.

B. Sublingual gland

   – is located in the floor of the mouth between the mucous membrane above and the mylohyoid muscle below.

   – surrounds the terminal portion of the submandibular duct.

   – empties mostly into the floor of the mouth along the sublingual fold by 12 short ducts, some of which enter the submandibular duct.

   – is supplied by postganglionic parasympathetic (secretomotor) fibers from the submandibular ganglion either directly or through the lingual nerve.

VI. Clinical Considerations

A. Abscess or infection of the mandibular teeth

   – might spread through the lower jaw to emerge on the face or in the floor of the mouth.

   – irritates the **mandibular nerve,** causing pain that may be referred to the ear because this nerve also innervates a part of the ear.

B. Abscess or infection of the maxillary teeth

   – irritates the **maxillary nerve,** causing upper **toothache.**

   – may result in symptoms of **sinusitis** with pain referred to the distribution of the maxillary nerve.

C. Lesion of the hypoglossal nerve (hypoglossal paralysis)

   – causes **deviation of the protruded tongue** toward the side of the lesion, due to paralysis of the tongue muscles, especially the genioglossus muscle.

D. Tongue-tie (ankyloglossia)

   – is an **abnormal shortness of frenulum linguae,** resulting in limitation of its movement and thus a severe speech impediment.

   – can be corrected surgically by cutting the frenulum.

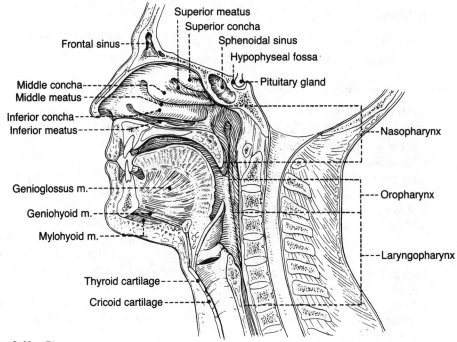

Figure 8-42. Pharynx.

Pharynx and Tonsils

I. Pharynx (Figure 8-42)
– is a **funnel-shaped fibromuscular tube** that extends from the base of the skull to the inferior border of the cricoid cartilage.
– conducts food to the esophagus and air to the larynx and lungs.

II. Subdivisions of the Pharynx

A. Nasopharynx
– is situated behind the nasal cavity above the soft palate and communicates with the nasal cavities through the **nasal choanae.**
– contains the **pharyngeal tonsils** in its posterior wall.
– is connected with the tympanic cavity through the **auditory (eustachian) tube,** which equalizes air pressure on both sides of the tympanic membrane.

B. Oropharynx
– extends between the soft palate above and the superior border of the epiglottis below and communicates with the mouth through the oropharyngeal isthmus.
– contains the **palatine tonsils,** which are lodged in the **tonsillar fossae** and are bounded by the palatoglossal and palatopharyngeal folds.

C. Laryngopharynx (hypopharynx)
– extends from the upper border of the epiglottis to the lower border of the cricoid cartilage.
– contains the **piriform recesses,** one on each side of the opening of the larynx, in which swallowed foreign bodies may be lodged.

III. Muscles of the Pharynx (Figures 8-43 and 8-44)

Muscle	Origin	Insertion	Nerve	Action
Circular muscles:				
Superior constrictor	Medial pterygoid plate; pterygoid hamulus; pterygomandibular raphe; mylohyoid line of mandible; side of tongue	Median raphe and pharyngeal tubercle of skull	Vagus n. via pharyngeal plexus	Constricts upper pharynx
Middle constrictor	Greater and lesser horns of hyoid; stylohyoid ligament	Median raphe	Vagus n. via pharyngeal plexus	Constricts lower pharynx
Inferior constrictor	Arch of cricoid and oblique line of thyroid cartilages	Median raphe of pharynx	Vagus n. via pharyngeal plexus, recurrent and external laryngeal n.	Constricts lower pharynx
Longitudinal muscles:				
Stylopharyngeus	Styloid process	Thyroid cartilage and muscles of pharynx	Glossopharyngeal n.	Elevates pharynx and larynx
Palatopharyngeus	Hard palate; aponeurosis of soft palate	Thyroid cartilage and muscles of pharynx	Vagus n. via pharyngeal plexus	Elevates pharynx and closes nasopharynx
Salpingopharyngeus	Cartilage of auditory tube	Muscles of pharynx	Vagus n. via pharyngeal plexus	Elevates nasopharynx; opens auditory tube

IV. Innervation and Blood Supply of the Pharynx (Figure 8-45)

A. Pharyngeal plexus

- lies on the **middle pharyngeal constrictor.**
- is formed by the **pharyngeal branches** of the glossopharyngeal and vagus nerves and the sympathetic branches from the superior cervical ganglion.
- Its **vagal branch** innervates all of the muscles of the pharynx with the exception of the stylopharyngeus, which is supplied by the glossopharyngeal nerve.
- Its **glossopharyngeal component** supplies sensory fibers to the pharyngeal mucosa.

B. Arteries of the pharynx

- are the ascending pharyngeal artery, ascending palatine branch of the facial artery, descending palatine arteries, pharyngeal branches of the maxillary artery, and branches of the superior and inferior thyroid arteries.

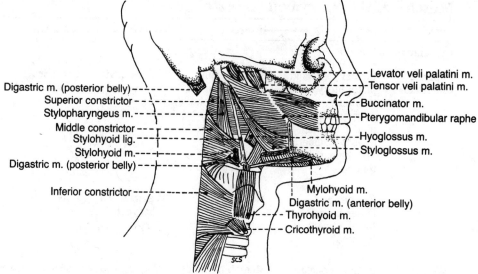

Figure 8-43. Muscles of the pharynx.

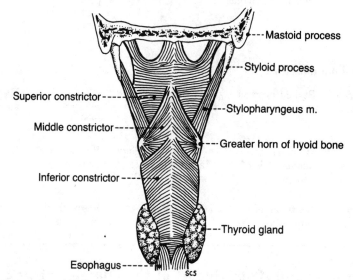

Figure 8-44. Pharyngeal constrictors.

V. Swallowing (Deglutition)

– is described in several stages:

A. The tongue pushes the bolus of food back into the fauces, which is the passage from the mouth to the oropharynx.

B. The palatoglossus and palatopharyngeus muscles contract to squeeze the bolus backward into the oropharynx. The tensor veli palatini and levator veli palatini muscles elevate the soft palate to close the entrance into the nasopharynx.

C. The walls of the pharynx are raised by the palatopharyngeus, stylopharyngeus, and salpingopharyngeus muscles to receive the food. The suprahyoid muscles elevate the hyoid bone and the larynx to close the opening into the larynx, thus preventing the food from entering the respiratory passageways.

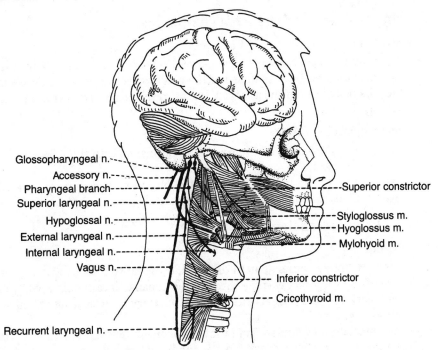

Figure 8-45. Nerve supply to the pharynx.

D. The serial contraction of the superior, middle, and inferior pharyngeal constrictor muscles moves the food through the oropharynx and the laryngopharynx into the esophagus where it is propelled by peristalsis.

VI. Tonsils

A. Pharyngeal tonsil

– is found in the posterior wall and roof of the nasopharynx.

B. Palatine tonsil

– lies on each side of the oropharynx in an interval between the palatoglossal and palatopharyngeal folds.
– receives blood from the ascending palatine and tonsillar branches of the facial artery, the descending palatine branch of the maxillary artery, a palatine branch of the ascending pharyngeal artery, and the dorsal lingual branches of the lingual artery.
– is innervated by branches of the glossopharyngeal nerve and the lesser palatine branch of the maxillary nerve.

C. Tubal tonsil

– lies near the pharyngeal opening of the auditory tube.

D. Waldeyer's ring

– is a **tonsillar ring** at the oropharyngeal isthmus, formed of lingual, palatine, tubal, and pharyngeal tonsils.

VII. Fascia and Space of the Pharynx (see Figure 8-8)

A. Retropharyngeal space

– is a **potential space** between the buccopharyngeal fascia and the prevertebral fascia, extending from the base of the skull to the superior mediastinum.
– permits movement of the pharynx, larynx, trachea, and esophagus during swallowing.

B. Pharyngobasilar fascia

– forms the **submucosa of the pharynx** and blends with the periosteum of the base of the skull.
– lies internal to the muscular coat of the pharynx; these muscles are covered externally by the buccopharyngeal fascia.

VIII. Clinical Considerations

A. Adenoid

– is hypertrophy or **enlargement of the pharyngeal tonsils,** obstructing passage of air from the nasal cavities through the choanae into the nasopharynx.
– may block the pharyngeal orifices of the auditory tube, causing an impairment of hearing.
– The infection may spread from the nasopharynx through the auditory tube to the middle ear cavity, causing **otitis media,** which may result in deafness.

B. Tonsillectomy

– is **surgical removal of a tonsil,** carried out by dissecting the tonsil from its bed.
– may cause much bleeding because the palatine tonsils are highly vascular; severe hemorrhage may also occur after a careless operation because the palatine tonsils are closely related to the internal carotid artery.
– could cause a loss of taste sensation in the posterior part of the tongue from injury to the lingual branches of the glossopharyngeal nerve, and also a loss of general sensation of the anterior two-thirds of the tongue from injury to the lingual nerve.

Nasal Cavity and Paranasal Sinuses

I. Nasal Cavity (Figure 8-46)

– opens on the face through the anterior nasal apertures **(nares, or nostrils)** and communicates with the nasopharynx through a posterior opening, the **choanae.**
– A slight dilatation inside the aperture of each nostril, the **vestibule,** is lined largely with skin containing hair, sebaceous glands, and sweat glands.

A. Roof

– is formed by the nasal, frontal, ethmoid (cribriform plate), and sphenoid (body) bones. The **cribriform plate** transmits the olfactory nerves.

B. Floor

– is formed by the palatine process of the maxilla and the horizontal plate of the palatine bone.
– The **incisive foramen** transmits the nasopalatine nerve and terminal branches of the sphenopalatine artery.

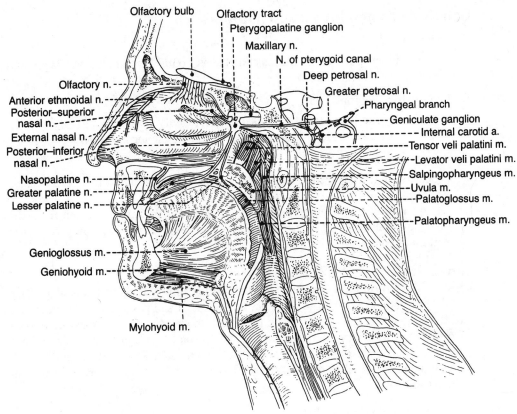

Figure 8-46. Nasal cavity.

C. Medial wall (nasal septum)

– is formed primarily by the perpendicular plate of the ethmoid bone, vomer, and septal cartilage.

– is also formed by processes of the palatine, maxillary, frontal, sphenoid, and nasal bones.

D. Lateral wall

– is formed by the superior and middle conchae of the ethmoid bone and the inferior concha.

– is also formed by the nasal bone, frontal process and nasal surface of the maxilla, lacrimal bone, perpendicular plate of the palatine bone, and medial pterygoid plate of the sphenoid bone.

– contains the following structures, and their openings:

1. **Sphenoethmoidal recess:** opening of the sphenoid sinus

2. **Superior meatus:** opening of the posterior ethmoidal air cells

3. **Middle meatus:** opening of the frontal sinus into the infundibulum; openings of the middle ethmoidal air cells on the **ethmoidal bulla;** openings of the anterior ethmoidal air cells and maxillary sinus in the **hiatus semilunaris**

4. **Inferior meatus:** opening of the **nasolacrimal duct**

5. **Sphenopalatine foramen:** opening into the pterygopalatine fossa; transmits the sphenopalatine artery

II. Subdivisions and Mucous Membranes

A. Vestibule

– is the dilated part inside the nostril that is bound by the alar cartilages and lined by skin with hairs.

B. Respiratory region

– consists of the lower two-thirds of the nasal cavity.
– Its mucous membrane warms, moistens, and cleans incoming air.

C. Olfactory region

– consists of the superior nasal concha and the upper one-third of the nasal septum.
– is innervated by olfactory nerves, which convey the sense of smell from the olfactory cells and enter the cranial cavity through the cribriform plate of the ethmoid bone to end in the olfactory bulb.

III. Blood Supply to the Nasal Cavity

– is from the lateral nasal branches of the anterior and posterior ethmoidal arteries of the ophthalmic artery; the posterior lateral nasal and posterior septal branches of the sphenopalatine artery of the maxillary artery; the greater palatine branch (its terminal branch reaches the lower part of the nasal septum through the incisive canal) of the descending palatine artery of the maxillary artery; the septal branch of the superior labial artery of the facial artery; and the lateral nasal branch of the facial artery.

IV. Nerve Supply to the Nasal Cavity

A. SVA sensation is supplied by the olfactory nerves for the olfactory area.

B. GSA sensation is supplied by the anterior ethmoidal branch of the ophthalmic nerve; the nasopalatine, posterior–superior, and posterior–inferior lateral nasal branches of the maxillary nerve via the pterygopalatine ganglion; and the anterior–superior alveolar branch of the infraorbital nerve.

V. Paranasal Sinuses (Figure 8-47)

A. Ethmoidal sinus

– consists of **ethmoidal air cells,** which are numerous small cavities within the **ethmoidal labyrinth** between the orbit and the nasal cavity.
– can be subdivided into the following groups:

1. **Posterior ethmoidal air cells** drain into the superior nasal meatus.

2. **Middle ethmoidal air cells** drain into the summit of the ethmoidal bulla of the middle nasal meatus.

3. **Anterior ethmoidal air cells** drain into the anterior aspect of the hiatus semilunaris in the middle nasal meatus.

B. Frontal sinus

– lies in the **frontal bone** and opens into the anterior part of the middle nasal meatus by way of the frontonasal duct.
– is innervated by the supraorbital branch of the ophthalmic nerve.

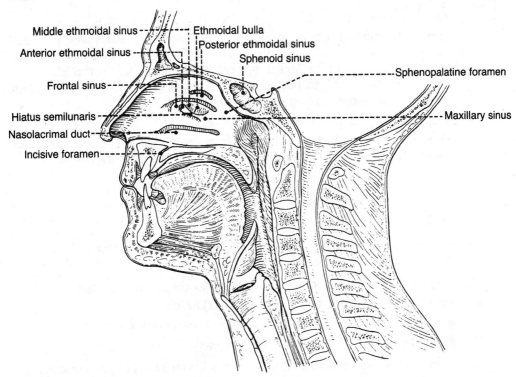

Middle ethmoidal sinus
Ethmoidal bulla
Anterior ethmoidal sinus
Posterior ethmoidal sinus
Sphenoid sinus
Frontal sinus
Sphenopalatine foramen
Hiatus semilunaris
Maxillary sinus
Nasolacrimal duct
Incisive foramen

Figure 8-47. Openings of the paranasal sinuses.

C. Maxillary sinus

- is the largest of the paranasal air sinuses and is the only paranasal sinus that may be present at birth.
- lies in the **maxilla** on each side, lateral to the lateral wall of the nasal cavity and inferior to the floor of the orbit, and drains into the posterior aspect of the hiatus semilunaris in the middle nasal meatus.

D. Sphenoidal sinus

- is contained within the body of the **sphenoid bone.**
- opens into the **sphenoethmoidal recess** of the nasal cavity.
- is innervated by branches from the maxillary nerve and by the posterior ethmoidal branch of the nasociliary nerve.

VI. Clinical Considerations

A. Sneeze

- is an **involuntary, sudden, violent, and audible expulsion of air** through the mouth and nose.
- The afferent limb of the reflex is carried by branches of the maxillary nerve, which convey general sensation from the nasal cavity and palate.

B. Epistaxis

- is a **nosebleed** resulting from rupture of the sphenopalatine artery; also occurs from nose picking, which tears the veins in the vestibule of the nose.

C. Maxillary sinusitis

- mimics the clinical signs of maxillary tooth abscess; in most cases, is related to infected tooth, and infection may spread from the maxillary sinus to the upper teeth and irritate the nerves to these teeth, causing toothache.
- may be confused with toothache since only a thin layer of bone separates the roots of the maxillary teeth from the sinus cavity.

D. Nasal polyp

- is an **inflammatory polyp** developing from the mucosa of the paranasal sinus, which projects into the nasal cavity and may fill the nasopharynx.

Pterygopalatine Fossa

I. Boundaries and Openings

A. Anterior wall: posterior surface of the maxilla or the posterior wall of the maxillary sinus (no openings)

B. Posterior wall: pterygoid process and greater wing of the sphenoid
- Openings and their contents include the following:
1. **Foramen rotundum to middle cranial cavity:** maxillary nerve
2. **Pterygoid canal to foramen lacerum:** nerve of the pterygoid canal
3. **Palatovaginal (pharyngeal or pterygopalatine) canal to choana:** pharyngeal branch of the maxillary artery and pharyngeal nerve from the pterygopalatine ganglion

C. Medial wall: perpendicular plate of the palatine
- Opening is **sphenopalatine foramen to nasal cavity,** which transmits the sphenopalatine artery and nasopalatine nerve.

D. Lateral wall: open (pterygomaxillary fissure to the infratemporal fossa)

E. Roof: greater wing and body of the sphenoid
- Opening is **inferior orbital fissure** to the orbit, which transmits the maxillary nerve.

F. Floor: fusion of the maxilla and the pterygoid process of the sphenoid
- Opening is **greater palatine foramen** to the palate, which transmits the greater palatine nerve and vessels.

II. Contents of the Pterygopalatine Fossa

A. Maxillary nerve (see Figure 8-41)
- passes through the lateral wall of the cavernous sinus and enters the pterygopalatine fossa through the **foramen rotundum.**
- is sensory to the skin of the face below the eye but above the upper lip.
- gives rise to the following branches:

1. **Meningeal branch**
   - innervates the dura mater of the middle cranial fossa.

2. **Pterygopalatine nerves (communicating branches)**
   - are connected to the pterygopalatine ganglion.
   - contain sensory fibers from the trigeminal ganglion.

3. **Posterior–superior alveolar nerves**

    – descend through the pterygopalatine fissure and enter the posterior–superior alveolar canals.

    – innervate the cheeks, gums, molar teeth, and maxillary sinus.

4. **Zygomatic nerve**

    – enters the orbit through the **inferior orbital fissure,** divides into the zygomaticotemporal and zygomaticofacial branches, which supply the skin over the temporal region and over the zygomatic bone, respectively. It joins the lacrimal nerve in the orbit and transmits postganglionic parasympathetic fibers to the lacrimal gland.

5. **Infraorbital nerve**

    – enters the orbit through the inferior orbital fissure and runs through the infraorbital groove and canal.

    – emerges through the infraorbital foramen and divides on the face into the inferior palpebral, nasal, and superior labial branches.

    – gives rise to the middle and anterior–superior alveolar nerves, which supply the maxillary sinus, teeth, and gums.

6. **Branches (sensory) via the pterygopalatine ganglion**

    a. **Orbital branches**

      – supply the periosteum of the orbit and the mucous membrane of the posterior ethmoidal and sphenoidal sinuses.

    b. **Pharyngeal branch**

      – runs in the pharyngeal (palatovaginal) canal and supplies the roof of the pharynx and the sphenoidal sinuses.

    c. **Posterior–superior lateral nasal branches**

      – enter the nasal cavity through the sphenopalatine foramen and innervate the posterior part of the septum, the posterior ethmoidal air cells, and the superior and middle conchae.

    d. **Greater palatine nerve**

      – descends through the palatine canal and emerges through the greater palatine foramen to innervate the hard palate and the inner surface of the maxillary gingiva.

      – gives rise to the posterior–inferior lateral nasal branches.

    e. **Lesser palatine nerve**

      – descends through the palatine canal and emerges through the lesser palatine foramen to innervate the soft palate and the palatine tonsil.

    f. **Nasopalatine nerve**

      – runs obliquely downward and forward on the septum, supplying the septum, and passes through the incisive canal.

B. **Pterygopalatine ganglion** (see Figure 8-28)

    – lies in the pterygopalatine fossa just below the maxillary nerve, lateral to the sphenopalatine foramen and anterior to the pterygoid canal.

    – receives preganglionic parasympathetic fibers from the facial nerve by way of the greater petrosal nerve and the nerve of the pterygoid canal.

– sends postganglionic parasympathetic fibers to the nasal and palatine glands and to the lacrimal gland by way of the maxillary, zygomatic, and lacrimal nerves.

– also receives postganglionic sympathetic fibers (by way of the deep petrosal nerve and the nerve of the pterygoid canal), which are distributed with the postganglionic parasympathetic fibers.

C. Pterygopalatine part of the maxillary artery

– supplies blood to the maxilla and maxillary teeth, nasal cavities, and palate.

– gives rise to the posterior–superior alveolar artery, infraorbital artery (which gives rise to anterior–superior alveolar branches), descending palatine artery (which gives rise to the lesser palatine and greater palatine branches), artery of the pterygoid canal, pharyngeal artery, and sphenopalatine artery.

III. Clinical Considerations: Lesion of the Nerve of Pterygoid Canal

– results in vasodilation and a lack of secretion of the lacrimal, nasal, and palatine glands, and a loss of general and taste sensation of the palate.

Larynx

I. Introduction

– extends from the lower part of the pharynx to the trachea.

– acts as a **compound sphincter** to prevent the passage of food or drink into the airway in swallowing and to close the **rima glottidis** during Valsalva's maneuver (buildup of air pressure during coughing, sneezing, micturition, defecation, or parturition).

– regulates the flow of air to and from the lungs for vocalization (phonation).

– forms a framework of cartilage for the attachment of ligaments and muscles.

II. Cartilages of the Larynx (Figure 8-48)

A. Thyroid cartilage

– is a **single hyaline cartilage** that forms a median elevation called the **laryngeal prominence (Adam's apple).**

– Its **superior horn** is attached to the tip of the greater horn of the hyoid bone, and its **inferior horn** articulates with the cricoid cartilage.

– has an **oblique line** on the lateral surface of its lamina that gives attachment for the inferior pharyngeal constrictor, sternothyroid, and thyrohyoid muscles.

B. Cricoid cartilage

– is a **single hyaline cartilage,** which is shaped like a signet ring.

– Its lower border marks the end of the pharynx and larynx.

C. Epiglottis

– is a **single elastic cartilage** and is a spoon-shaped plate that lies behind the root of the tongue.

– Its lower end is attached to the back of the thyroid cartilage.

D. Arytenoid cartilages

– are **paired elastic and hyaline cartilages.**

– are shaped liked pyramids, and their bases articulate with the cricoid cartilage.

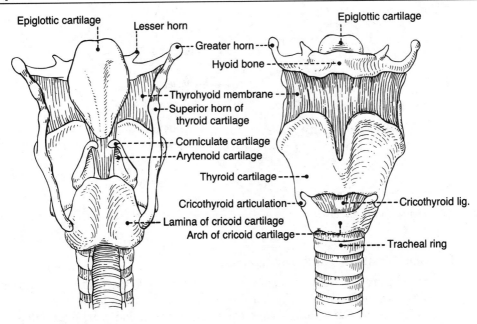

Figure 8-48. Cartilages of the larynx.

    – have **vocal processes,** which give attachment to the vocal ligament and vocalis muscle, and **muscular processes,** which give attachment to the thyroarytenoid, lateral, and posterior cricoarytenoid muscles.

E. Corniculate cartilages

    – are **paired elastic cartilages** that lie on the apices of the arytenoid cartilages.
    – are enclosed within the **aryepiglottic folds** of mucous membrane.

F. Cuneiform cartilages

    – are **paired elastic cartilages** that lie in the aryepiglottic folds anterior to the corniculate cartilages.

III. Ligaments of the Larynx

A. Thyrohyoid membrane

    – extends from the thyroid cartilage to the medial surface of the hyoid bone.
    – Its middle (thicker) part is called the **middle thyrohyoid ligament;** its lateral portion is pierced by the internal laryngeal nerve and the superior laryngeal vessels.

B. Cricothyroid ligament

    – extends from the arch of the cricoid cartilage to the thyroid cartilage and the vocal processes of the arytenoid cartilages.

C. Vocal ligament

    – extends from the posterior surface of the thyroid cartilage to the vocal process of the arytenoid cartilage.
    – is considered the upper border of the **conus elasticus.**

D. Vestibular (ventricular) ligament

    – extends from the thyroid cartilage to the anterior lateral surface of the arytenoid cartilage.

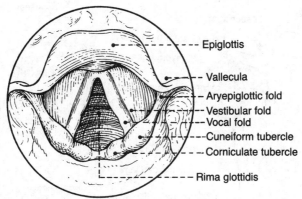

Figure 8-49. Interior view of the larynx.

E. Conus elasticus (cricovocal ligament)

– is the paired lateral portion of the fibroelastic membrane that extends upward from the entire arch of the cricoid cartilage to the vocal ligaments.
– is formed by the cricothyroid, median cricothyroid, and vocal ligaments.

IV. Cavities and Folds of the Larynx (Figure 8-49)

– The larynx is divided into three portions by the vestibular and vocal folds.

A. Vestibule

– extends from the laryngeal inlet to the **vestibular (ventricular) folds.**

B. Ventricles

– extend between the ventricular fold and the vocal fold.

C. Infraglottic cavity

– extends from the rima glottidis to the lower border of the cricoid cartilage.

D. Rima glottidis

– is the space between the vocal folds and arytenoid cartilages.
– is the narrowest part of the laryngeal cavity.

E. Vestibular folds (false vocal cords)

– extend from the thyroid cartilage above the vocal ligament to the arytenoid cartilage.

F. Vocal folds (true vocal cords)

– extend from the angle of the thyroid cartilage to the vocal processes of the arytenoid cartilages.
– contain the **vocal ligament** near their free margin and the **vocalis muscle,** which forms the bulk of the vocal fold.
– are important in **voice production** because they control the stream of air passing through the rima glottidis.
– alter the shape and size of the **rima glottidis** by movement of the arytenoids to facilitate respiration and phonation. (The rima glottidis is wide during inspiration and narrow and wedge-shaped during expiration and sound production.)

V. **Muscles of the Larynx** (Figure 8-50)

Muscle	Origin	Insertion	Nerve	Action on Vocal Cords
Cricothyroid	Arch of cricoid cartilage	Inferior horn and lower lamina of thyroid cartilage	External laryngeal n.	Tenses
Posterior cricoarytenoid	Posterior surface of lamina of cricoid cartilage	Muscular process of arytenoid cartilage	Recurrent laryngeal n.	Abducts
Lateral cricoarytenoid	Arch of cricoid cartilage	Muscular process of arytenoid cartilage	Recurrent laryngeal n.	Adducts
Transverse arytenoid	Posterior surface of arytenoid cartilage	Opposite arytenoid cartilage	Recurrent laryngeal n.	Adducts; closes laryngeal inlet
Oblique arytenoid	Muscular process of arytenoid cartilage	Apex of opposite arytenoid	Recurrent laryngeal n.	Adducts; closes laryngeal inlet
Aryepiglottic	Apex of arytenoid cartilage	Side of epiglottic cartilage	Recurrent laryngeal n.	Adducts
Thyroarytenoid	Inner surface of thyroid lamina	Anterolateral surface of arytenoid cartilage	Recurrent laryngeal n.	Adducts; relaxes
Thyroepiglottic	Anteromedial surface of lamina of thyroid cartilage	Lateral margin of epiglottic cartilage	Recurrent laryngeal n.	Adducts
Vocalis	Angle between two laminae of thyroid cartilage	Vocal process	Recurrent laryngeal n.	Adducts and tenses

VI. **Innervation of the Larynx** (Figure 8-51)

A. **Recurrent laryngeal nerve**

– innervates all of the intrinsic muscles of the larynx except the cricothyroid, which is innervated by the external laryngeal branch of the superior laryngeal branch of the vagus nerve.

– supplies sensory innervation below the vocal cord.

– Its terminal portion above the lower border of the cricoid cartilage is called the **inferior laryngeal nerve.**

B. **Internal laryngeal nerve**

– innervates the mucous membrane above the vocal cord.

– is accompanied by the superior laryngeal artery and pierces the thyrohyoid membrane.

C. **External laryngeal nerve**

– innervates the cricothyroid and inferior pharyngeal constrictor muscles.

– is accompanied by the superior thyroid artery.

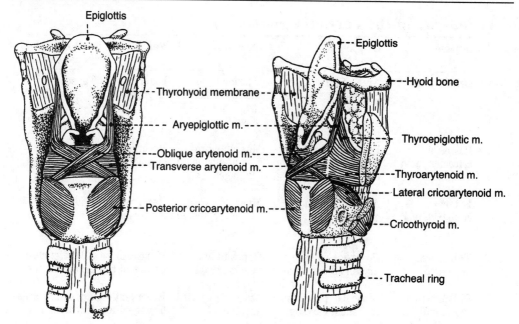

Figure 8-50. Muscles of the larynx.

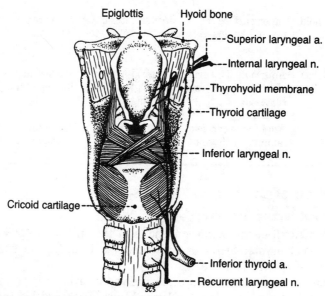

Figure 8-51. Nerve supply to the larynx.

VII. Clinical Considerations

A. Laryngeal obstruction (choking)

- is caused by aspirated foods, which are usually lodged at the rima glottidis.
- could be released by compression of the abdomen to expel air from the lungs and thus dislodge the foods.

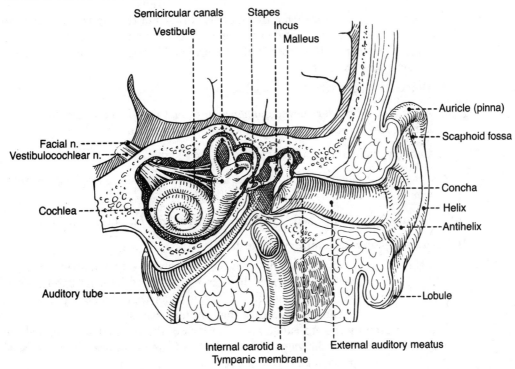

Figure 8-52. External, middle, and inner ear.

B. Laryngitis

– is an **inflammation of the mucous membrane of the larynx.**

– is characterized by dryness and soreness of the throat, hoarseness, cough, and dysphagia.

C. Laryngotomy

– is an **operative opening into the larynx** through the cricothyroid membrane (cricothyrotomy), through the thyroid cartilage (thyrotomy), or through the thyrohyoid membrane (superior laryngotomy).

– is performed when severe edema or an impacted foreign body calls for rapid admission of air into the larynx and trachea.

D. Lesion of the recurrent laryngeal nerve

– could be produced during thyroidectomy or cricothyrotomy or by aortic aneurysm.

– may cause respiratory obstruction, hoarseness, and an inability to speak.

Ear

I. External Ear (Figure 8-52)

– consists of the auricle and the external acoustic meatus, and receives sound waves.

A. Auricle

– consists of cartilage connected to the skull by ligaments and muscles and is covered by skin.

– funnels sound waves into the external auditory meatus.

– receives sensory nerves from the auricular branch of the vagus nerve and the greater auricular, auriculotemporal, and lesser occipital nerves.

– receives blood from the superficial temporal and posterior auricular arteries.

– has the following features:

1. **Helix,** the slightly curved rim of the auricle

2. **Antihelix,** a broader curved eminence internal to the helix, which divides the auricle into an outer scaphoid fossa and the deeper concha

3. **Concha,** the deep cavity in front of the antihelix

4. **Tragus,** a small projection from the anterior portion of the external ear anterior to the concha

5. **Lobule,** a structure made up of areolar tissue and fat but no cartilage

B. External acoustic (auditory) meatus

– is about 2.5 cm long, extending from the concha to the tympanic membrane.

– Its external one-third is formed by cartilage, and its internal two-thirds is formed by bone. The cartilaginous portion is wider than the bony portion and has numerous **ceruminous glands** that produce earwax.

– is innervated by the auriculotemporal branch of the trigeminal nerve and the auricular branch of the vagus nerve, which is joined by a branch of the facial nerve and the glossopharyngeal nerve.

– receives blood from the superficial temporal, posterior auricular, and maxillary arteries (a deep auricular branch).

C. Tympanic membrane (eardrum)

– lies obliquely across the end of the meatus; thus, the anterior and inferior walls are longer than the posterior and superior walls.

– consists of **three layers:** an outer (cutaneous), an intermediate (fibrous), and an inner (mucous) layer.

– has a thickened fibrocartilaginous ring at the greater part of its circumference, which is fixed in the tympanic sulcus at the inner end of the meatus.

– The small triangular portion between the anterior and posterior malleolar folds is called the **pars flaccida,** and the remainder of the membrane is called the **pars tensa.**

– contains the **cone of light,** which is a triangular reflection of light seen in the anterior–inferior quadrant.

– The most depressed center point of the concavity is called the **umbo.**

– conducts sound waves to the middle ear.

– Its lateral (outer) concave surface is covered by skin and is innervated by the auriculotemporal branch of the trigeminal nerve and the auricular branch of the vagus nerve. The auricular branch may contain fibers from the glossopharyngeal and facial nerves.

– Its medial (inner) surface is covered by mucous membrane, is innervated by the tympanic branch of the glossopharyngeal nerve, and serves as an attachment for the handle of the **malleus.**

– Its inner surface is supplied by the auricular branch of the occipital artery and the anterior tympanic artery; its outer surface is supplied by the deep auricular artery.

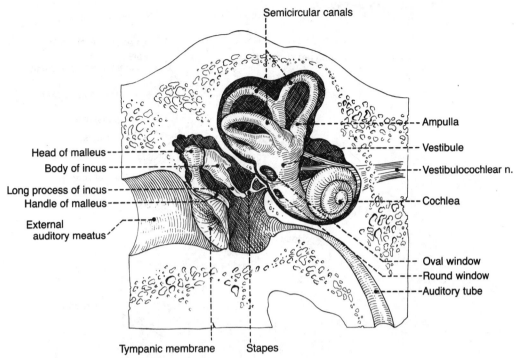

Figure 8-53.   Middle and inner ear.

II. Middle Ear (Figures 8-53 and 8-54)

– consists of the tympanic cavity with its **ossicles,** and transmits the sound waves from air to auditory ossicles and then to the inner ear.

A. Tympanic (middle ear) cavity

– includes the **tympanic cavity proper,** the space internal to the tympanic membrane; and the **epitympanic recess,** the space superior to the tympanic membrane that contains the head of the malleus and the body of the incus.

– communicates anteriorly with the nasopharynx via the **auditory (eustachian) tube** and posteriorly with the mastoid air cells and the mastoid antrum through the **aditus ad antrum.**

– is traversed by the chorda tympani and lesser petrosal nerve.

1. Boundaries of the tympanic cavity

a. **Roof:** tegmen tympani

b. **Floor:** jugular fossa

c. **Anterior:** carotid canal

d. **Posterior:** mastoid air cells and mastoid antrum through the aditus ad antrum

e. **Lateral:** tympanic membrane

f. **Medial:** lateral wall of the inner ear, presenting the promontory formed by the basal turn of the cochlea; the fenestra vestibuli (oval window); the fenestra cochlea (round window); and the prominence of the facial canal

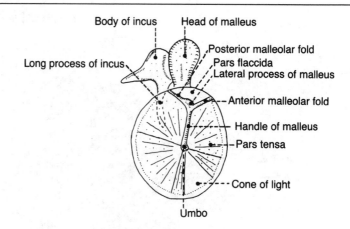

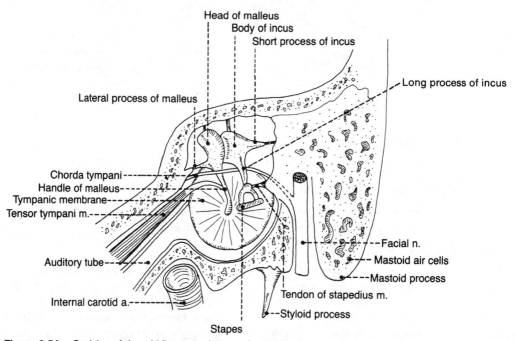

Figure 8-54. Ossicles of the middle ear and tympanic membrane.

2. Oval window (fenestra vestibuli)

– is pushed back and forth by the footplate of the stapes and transmits the sonic vibrations of the ossicles to the perilymph of the scala vestibuli in the inner ear.

3. Round window (fenestra cochlea or tympani)

– is closed by the mucous membrane of the middle ear and accommodates the pressure waves transmitted to the perilymph of the scala tympani.

B. Muscles of the middle ear

1. Stapedius muscle

– is the **smallest of the skeletal muscles** in the human body.
– arises within the pyramidal eminence, and its tendon emerges from the eminence.

– inserts on the neck of the stapes.
– is innervated by a branch of the facial nerve.
– pulls the head of the stapes posteriorly, thereby tilting the base of the stapes in the vestibular window.
– prevents excessive oscillation of the stapes and protects the inner ear from injury during a loud noise.

2. Tensor tympani muscle

– arises from the cartilaginous portion of the auditory tube.
– inserts on the handle (manubrium) of the malleus.
– is innervated by the mandibular branch of the trigeminal nerve.
– draws the manubrium medially, thereby making the tympanic membrane taut.

C. Auditory ossicles

– consist of the malleus, incus, and stapes.
– form a bridge by synovial joints in the middle ear cavity, transmit sonic vibrations from the tympanic membrane to the inner ear, and amplify the force.

1. Malleus (hammer)

– consists of a head, neck, handle (manubrium), and anterior and lateral processes.
– Its rounded head articulates with the incus in the epitympanic recess.
– Its handle is fused to the medial surface of the tympanic membrane and serves as an attachment for the **tensor tympani muscle.**

2. Incus (anvil)

– consists of a body and two processes (crura).
– Its long process descends vertically, parallel to the handle of the malleus, and articulates with the stapes.
– Its short process extends horizontally backward to the fossa of the incus and provides the attachment for the posterior ligament of the incus.

3. Stapes (stirrup)

– consists of a head and neck, two processes (crura), and a base (footplate).
– Its neck provides insertion of the **stapedius muscle.**
– has a hole through which the stapedial artery is transmitted in the embryo; this hole is obturated by a thin membrane in the adult.
– Its base (footplate) is attached by the annular ligament to the margin of the oval window (fenestra vestibuli). Abnormal ossification between the footplate and the oval window (**otosclerosis**) limits the movement of the stapes, causing deafness.

D. Auditory (eustachian) tube

– connects the middle ear to the nasopharynx.
– allows air to enter or leave the middle ear cavity and thus balances the pressure in the middle ear with atmospheric pressure, allowing free movement of the tympanic membrane.
– Its cartilaginous portion remains closed except during swallowing or yawning.
– is opened by the simultaneous contraction of the tensor veli palatini and salpingopharyngeus muscles.

E. Sensory nerve and blood supply to the middle ear

  – is innervated by the auriculotemporal branch of the trigeminal nerve, tympanic branch of the glossopharyngeal nerve, and auricular branch of the vagus nerve.
  – is supplied from the stylomastoid branch of the posterior auricular artery and the anterior tympanic branch of the maxillary artery.

III. Inner Ear (see Figure 8-53)

  – consists of the **acoustic apparatus,** the cochlea housing the cochlear duct for auditory sense, and the **vestibular apparatus,** the vestibule housing the utricle and saccule, and the semicircular canals housing the semicircular ducts for the sense of equilibrium.
  – is the place where vibrations are transduced to specific nerve impulses that are transmitted through the acoustic nerve to the central nervous system.
  – is composed of the bony labyrinth and the membranous labyrinth.

A. Bony labyrinth

  – consists of three parts: the vestibule, the three semicircular canals, and the cochlea, all of which contain the **perilymph,** in which the membranous labyrinth is suspended.
  – The **vestibule** is a cavity of the bony labyrinth communicating with the cochlea anteriorly and the semicircular canals posteriorly.
  – The **bony cochlea** consists of two adjacent ducts, the upper **scala vestibuli,** which begins in the vestibule and receives the vibrations transmitted to the perilymph at the oval window, and the lower **scala tympani,** which communicates with the scala vestibuli through the helicotrema at the apex of the cochlea and ends at the round window, where the sound pressure waves are dissipated.

B. Membranous labyrinth

  – is suspended in perilymph within the bony labyrinth, is filled with **endolymph,** and contains the sensory organs.
  – has comparable parts and arrangement as the bony labyrinth.
  – Its **utricle** and **saccule** are dilated membranous sacs in the vestibule and contain sense organs called maculae.
  – Its **semicircular ducts** consist of anterior (superior), lateral, and posterior, and their dilated ends are called ampullae.
  – Its **cochlear duct (scala media)** is wedged between the scala vestibuli and scala tympani and contains the spiral organ of Corti.

IV. Clinical Considerations

A. Hyperacusis (hyperacusia)

  – is **excessive acuteness of hearing,** due to paralysis of the stapedius muscle (causing uninhibited movements of the stapes), resulting from a lesion of the facial nerve.

B. Otosclerosis

  – is a condition of **abnormal bone formation** around the stapes and the oval window, limiting the movement of the stapes and thus resulting in **progressive conduction deafness.**

C. Conductive deafness

  – is hearing impairment caused by **defect of the sound-conducting apparatus,** such as the auditory meatus, eardrum, or ossicles.

D. Otitis media

– is a condition of **middle ear infection** that may be spread from the nasopharynx through the auditory tube, causing temporary or permanent deafness.

E. Meniere's disease (endolymphatic or labyrinthine hydrops)

– is characterized by **a loss of balance** (vertigo), ringing or buzzing in the ears, and **progressive deafness,** due to edema of the labyrinth or inflammation of the vestibular division of the vestibulocochlear nerve.

Review Test

Directions: Each of the numbered items or incomplete statements in this section is followed by answers or by completions of the statement. Select the ONE lettered answer or completion that is BEST in each case.

1. Damage to the external laryngeal nerve during thyroid surgery could result in the patient's inability to

(A) relax the vocal cords
(B) rotate the arytenoid cartilages
(C) tense the vocal cords
(D) widen the rima glottidis
(E) abduct the vocal cords

2. Upon ligation of the superior laryngeal artery, care must be taken to avoid injury to which of the following nerves?

(A) External laryngeal nerve
(B) Internal laryngeal nerve
(C) Superior laryngeal nerve
(D) Hypoglossal nerve
(E) Vagus nerve

3. In a patient demonstrating a lack of general sensation in the mucosa of the pharyngeal walls, one would expect to find a lesion of the

(A) transverse cervical nerve
(B) phrenic nerve
(C) vagus nerve
(D) glossopharyngeal nerve
(E) ansa cervicalis (descendens hypoglossi)

4. During surgery, a surgeon notices serious bleeding from the deep cervical artery. Which of the following arteries must be ligated immediately?

(A) Inferior thyroid artery
(B) Transverse cervical artery
(C) Thyrocervical trunk
(D) Costocervical trunk
(E) Ascending cervical artery

5. On each side of the body, the phrenic nerve in the neck passes

(A) across the anterior surface of the subclavian vein
(B) across the posterior surface of the subclavian artery
(C) across the deep surface of the anterior scalene muscle
(D) medial to the common carotid artery
(E) across the superficial surface of the anterior scalene muscle

6. Which of the following muscles is a landmark for locating the glossopharyngeal nerve in the neck?

(A) Inferior pharyngeal constrictor muscle
(B) Stylopharyngeus muscle
(C) Posterior belly of the digastric muscle
(D) Longus colli muscle
(E) Rectus capitis anterior muscle

7. The external laryngeal branch of the superior laryngeal nerve innervates the cricothyroid muscle and sends fibers to the

(A) inferior pharyngeal constrictor muscle
(B) middle pharyngeal constrictor muscle
(C) sternothyroid muscle
(D) thyroarytenoid muscle
(E) thyrohyoid muscle

8. A patient with crocodile tears syndrome has spontaneous lacrimation during eating, due to misdirection of regenerating autonomic nerve fibers. Which of the following nerves has been injured?

(A) Facial nerve proximal to the geniculate ganglion
(B) Auriculotemporal nerve
(C) Chorda tympani in the infratemporal fossa
(D) Facial nerve at the stylomastoid foramen
(E) Lacrimal nerve

9. A 13-year-old girl complains of dryness of the nose and the palate, indicating a lesion of which of the following ganglia?

(A) Nodose ganglion
(B) Otic ganglion
(C) Pterygopalatine ganglion
(D) Submandibular ganglion
(E) Ciliary ganglion

10. Bell's palsy can involve corneal inflammation and subsequent corneal ulceration, which results from

(A) sensory loss of the cornea and conjunctiva
(B) lack of secretion of the salivary glands
(C) absence of the corneal blink reflex due to paralysis of the muscles that close the eyelid
(D) absence of the corneal blink reflex due to paralysis of the muscles that open the eyelid
(E) constriction of the pupil due to paralysis of the dilator pupillae

11. Which of the following structures unite to form the straight sinus in the cranial cavity?

(A) Transverse and sigmoid sinuses
(B) Inferior sagittal sinus and great cerebral vein of Galen
(C) Superior sagittal sinus and great cerebral vein of Galen
(D) Superior and inferior petrosal sinuses
(E) Cavernous sinus and basilar plexus

12. Which of the following conditions results from severance of the abducens nerve proximal to its entrance into the orbit?

(A) Ptosis of the upper eyelid
(B) Loss of the pupil's ability to dilate
(C) External strabismus (lateral deviation)
(D) Loss of visual accommodation
(E) Internal strabismus (medial deviation)

13. Death may result from bilateral severance of which of the following nerves?

(A) Trigeminal nerve
(B) Facial nerve
(C) Vagus nerve
(D) Spinal accessory nerve
(E) Hypoglossal nerve

14. When the middle meningeal artery is ruptured but the meninges remain intact, blood enters the

(A) subarachnoid space
(B) subdural space
(C) epidural space
(D) subpial space
(E) dural sinuses

15. Which of the following locations contains preganglionic neurons of the parasympathetic nervous system?

(A) Cervical and sacral spinal cord
(B) Cervical and thoracic spinal cord
(C) Brainstem and cervical spinal cord
(D) Thoracic and lumbar spinal cord
(E) Brainstem and sacral spinal cord

16. Following radical resection of a primary tongue tumor, a patient has lost general sensation on the anterior two-thirds of the tongue. This is probably due to injuries to branches of which of the following nerves?

(A) Trigeminal nerve
(B) Facial nerve
(C) Glossopharyngeal nerve
(D) Vagus nerve
(E) Hypoglossal nerve

17. The pituitary gland lies in the sella turcica, immediately posterior and superior to the

(A) frontal sinus
(B) maxillary sinus
(C) ethmoid cells
(D) mastoid cells
(E) sphenoid sinus

18. After tonsillectomy, a 7-year-old boy is unable to distinguish the sensation of taste on the posterior part of his tongue. Which of the following nerves most likely has been injured?

(A) Internal laryngeal branch of the vagus nerve
(B) Lingual nerve
(C) Lingual branch of the glossopharyngeal nerve
(D) Greater palatine nerve
(E) Hypoglossal nerve

19. Damage to the sella turcica is probably due to fracture of the

(A) frontal bone
(B) ethmoid bone
(C) temporal bone
(D) basioccipital bone
(E) sphenoid bone

20. Which of the following statements concerning accommodation of the eye is true?

(A) The ciliary muscle contracts to make the lens thinner
(B) The ciliary muscle contracts when its parasympathetic fibers are stimulated
(C) During accommodation for objects close to the eye, the lens becomes thinner
(D) Accommodation for objects close to the eye is mediated by sympathetic nerve action
(E) During accommodation, the lens does not change shape; it moves forward or backward

21. A patient has an infectious inflammation of the dural venous sinus nearest the pituitary gland, with secondary thrombus formation. Which of the following is the most likely site of infection?

(A) Straight sinus
(B) Cavernous sinus
(C) Superior petrosal sinus
(D) Sigmoid sinus
(E) Confluence of sinuses

22. Which of the following statements concerning the pterygopalatine fossa is true?

(A) It is located in the petrous portion of the temporal bone
(B) It contains the otic ganglion
(C) It communicates with the middle cranial cavity through the foramen rotundum
(D) It is located directly inferior to the maxillary sinus
(E) It communicates with the infratemporal fossa through the sphenopalatine foramen

23. If a patient is unable to abduct the vocal cords during quiet breathing, which of the following muscles is paralyzed?

(A) Vocalis muscle
(B) Cricothyroid muscle
(C) Oblique arytenoid muscle
(D) Posterior cricoarytenoid muscle
(E) Thyroarytenoid muscle

24. Which of the following pairs of muscles is most instrumental in preventing food from entering the larynx and trachea during swallowing?

(A) Sternohyoid and sternothyroid muscles
(B) Oblique arytenoid and aryepiglottic muscles
(C) Inferior pharyngeal constrictor and thyrohyoid muscles
(D) Levator veli palatini and tensor veli palatini muscles
(E) Uvulae and geniohyoid muscles

25. The veins of the brain are direct tributaries of the

(A) emissary veins
(B) pterygoid venous plexus
(C) diploic veins
(D) dural sinuses
(E) internal jugular vein

26. Which of the following conditions or actions results from stimulation of the parasympathetic fibers to the eyeball?

(A) Enhanced vision for distant objects
(B) Dilation of the pupil
(C) Contraction of capillaries in the iris
(D) Contraction of the ciliary muscle
(E) Flattening of the lens

27. Which of the following statements concerning the tympanic nerve is true?

(A) It is a branch of the facial nerve
(B) It is also known as the deep petrosal nerve
(C) It synapses with fibers of the lesser petrosal nerve
(D) It is a branch of the glossopharyngeal nerve
(E) It is a branch of the vagus nerve

28. Which of the following cavities are separated from the middle cranial fossa by a thin layer of bone?

(A) Auditory tube and bony orbit
(B) Middle ear cavity and sphenoid sinus
(C) Sigmoid sinus and frontal sinus
(D) Sphenoid sinus and ethmoid sinus
(E) Maxillary sinus and middle ear cavity

29. Loss of general sensation in the dura of the middle cranial fossa indicates a lesion of the

(A) vagus nerve
(B) facial nerve
(C) hypoglossal nerve
(D) trigeminal nerve
(E) glossopharyngeal nerve

30. Which of the following statements concerning the carotid sinus is true?

(A) It is located at the origin of the external carotid artery
(B) It is innervated by the facial nerve
(C) It functions as a chemoreceptor
(D) It is stimulated by changes in blood pressure
(E) It communicates freely with the cavernous sinus

31. During a game, a 26-year-old baseball player received a severe blow to the head that fractured his optic canal. Which of the following pairs of structures are most likely to be damaged?

(A) Optic nerve and ophthalmic vein
(B) Ophthalmic vein and ophthalmic nerve
(C) Ophthalmic artery and optic nerve
(D) Ophthalmic nerve and optic nerve
(E) Ophthalmic artery and ophthalmic vein

32. Paralysis of the posterior belly of the digastric muscle would result from a lesion of the

(A) spinal accessory nerve
(B) trigeminal nerve
(C) ansa cervicalis
(D) facial nerve
(E) glossopharyngeal nerve

33. Contraction of the tensor tympani and the stapedius prevents damage to the eardrum and middle ear ossicles. These muscles are most likely controlled by the

(A) chorda tympani and tympanic nerve
(B) trigeminal and facial nerves
(C) auditory and vagus nerves
(D) facial and auditory nerves
(E) trigeminal and accessory nerves

34. If a patient's pupil remains small when room lighting is subdued, this may indicate damage to the

(A) trochlear nerve
(B) superior cervical ganglion
(C) oculomotor nerve
(D) ophthalmic nerve
(E) abducens nerve

35. Which of the following statements concerning the frontal sinus is true?

(A) It extends into the parietal bone
(B) It communicates with the superior nasal meatus
(C) It receives sensory innervation from the maxillary nerve
(D) It is supplied with blood by branches of the ophthalmic artery
(E) It is a cranial dural venous sinus that receives the ophthalmic vein

36. A patient can move his eyeball normally and see distant objects clearly but cannot focus on near objects. This condition may indicate damage to the

(A) ciliary ganglion and oculomotor nerve
(B) oculomotor nerve and long ciliary nerve
(C) short ciliary nerves and ciliary ganglion
(D) superior cervical ganglion and long ciliary nerve
(E) oculomotor, trochlear, and abducens nerves

37. Which of the following structures enters the orbit through the superior orbital fissure and the common tendinous ring?

(A) Frontal nerve
(B) Lacrimal nerve
(C) Trochlear nerve
(D) Abducens nerve
(E) Ophthalmic vein

38. Which of the following groups of cranial nerves innervate the muscles attached to the styloid process?

(A) Facial, glossopharyngeal, and hypoglossal nerves
(B) Hypoglossal, vagus, and facial nerves
(C) Glossopharyngeal, trigeminal, and vagus nerves
(D) Vagus, spinal accessory, and hypoglossal nerves
(E) Facial, glossopharyngeal, and vagus nerves

Directions: Each of the numbered items or incomplete statements in this section is negatively phrased, as indicated by a capitalized word such as NOT, LEAST, or EXCEPT. Select the ONE lettered answer or completion that is BEST in each case.

39. Before excising a large benign tumor over a patient's neck, a surgical resident examined the posterior cervical triangle. All of the following statements concerning the triangle are true EXCEPT

(A) its posterior boundary is the anterior border of the trapezius
(B) it is subdivided by the omohyoid anterior belly
(C) it contains the accessory nerve lying on the levator scapulae
(D) it is traversed by the suprascapular artery, a branch of the thyrocervical trunk
(E) its roof is formed by the platysma

40. A low tracheotomy performed below the isthmus of the thyroid may encounter all of the following vessels EXCEPT

(A) inferior thyroid vein or its tributaries
(B) jugular arch
(C) costocervical trunk
(D) thyroidea ima artery
(E) left brachiocephalic vein

41. If a patient has no cutaneous sensation on the areas supplied by the cervical plexus, it is likely that damage has occurred to all of the following branches of the plexus EXCEPT

(A) phrenic nerve
(B) greater auricular nerve
(C) transverse cervical nerve
(D) supraclavicular nerve
(E) lesser occipital nerve

42. In a patient with swelling of the mucous membranes of the middle nasal meatus, all of the openings of the paranasal sinuses are plugged EXCEPT

(A) middle ethmoidal air cells
(B) maxillary sinus
(C) sphenoid sinus
(D) anterior ethmoidal air cells
(E) frontal sinus

43. All of the following statements concerning the cervical plexus and its branches are true EXCEPT

(A) cervical nerves C1–C4 provide cutaneous nerve fibers to the cervical plexus
(B) the transverse cervical nerve provides sensory innervation to the anterior and lateral parts of the neck
(C) the supraclavicular nerves innervate the skin over the shoulder
(D) the motor nerves for most infrahyoid muscles are branches of the ansa cervicalis
(E) cervical nerves C1–C4 contribute motor fibers to the cervical plexus

44. All of the following statements concerning the larynx are true EXCEPT

(A) the inlet to the larynx is formed by the aryepiglottic folds
(B) the true vocal folds are superior to the ventricle of the larynx
(C) afferent nerve fibers from the larynx are carried by the vagus nerve
(D) the larynx extends inferiorly to the level of the sixth cervical vertebra
(E) the larynx regulates the flow of air to and from the lungs for sound production

45. All of the following nerves pass through the superior orbital fissure EXCEPT

(A) abducens nerve
(B) ophthalmic nerve
(C) oculomotor nerve
(D) trochlear nerve
(E) optic nerve

46. All of the following statements concerning the salivary glands are true EXCEPT

(A) the duct of the sublingual gland empties mostly into the floor of the mouth along the sublingual fold
(B) the duct of the submandibular gland arises from the superficial portion of the gland
(C) the duct of the parotid gland pierces the buccinator muscle
(D) the deep portion of the submandibular gland lies between the mylohyoid and hyoglossus muscles
(E) the sublingual gland is innervated by secretomotor fibers that travel within the chorda tympani

47. All of the following statements concerning the nasal cavity are true EXCEPT

(A) the conchae are attached to its lateral wall
(B) the ethmoid bone contributes to its roof and medial and lateral borders
(C) its septum is partially cartilaginous
(D) most of its floor is formed by the palatine bone
(E) part of its roof is formed by the vomer

48. A horizontal cut through the neck that severs the inferior thyroid arteries will also sever all of the following structures EXCEPT

(A) recurrent laryngeal nerves
(B) external carotid arteries
(C) inferior thyroid veins
(D) vagus nerves
(E) trachea

49. All of the following statements concerning the tensor tympani are true EXCEPT

(A) it inserts on the handle of the malleus
(B) it is innervated by a branch of the facial nerve
(C) it runs parallel to the auditory tube
(D) it arises chiefly from the cartilaginous portion of the auditory tube
(E) it functions to tighten the tympanic membrane

50. Severance of the oculomotor nerve can cause all of the following conditions EXCEPT

(A) partial ptosis
(B) adduction of the eyeball
(C) a dilated pupil
(D) impaired lacrimal secretion
(E) paralysis of the ciliary muscle

51. All of the following nerves supply striated muscles and are of branchiomeric origin EXCEPT

(A) vagus nerve
(B) glossopharyngeal nerve
(C) oculomotor nerve
(D) facial nerve
(E) trigeminal nerve

52. Infection within the carotid sheath may damage all of the following structures EXCEPT

(A) vagus nerve
(B) common carotid artery
(C) internal jugular vein
(D) sympathetic trunk
(E) internal carotid artery

53. During surgery for malignant parotid tumors, the main trunk of the facial nerve is lacerated, resulting in paralysis of all of the following muscles EXCEPT

(A) buccinator muscle
(B) stylohyoid muscle
(C) posterior belly of the digastric
(D) tensor tympani
(E) zygomaticus major

54. All of the following statements concerning the scalenus anterior are true EXCEPT

(A) it divides arbitrarily the subclavian artery into three parts
(B) it inserts on the first rib
(C) the phrenic nerve passes anterior to it
(D) the subclavian artery runs anterior to it
(E) the roots of the brachial plexus lie posterior to it

55. All of the following statements concerning the internal laryngeal nerve are true EXCEPT

(A) it is a branch of the superior laryngeal nerve
(B) it may contain taste fibers from the epiglottis
(C) it provides sensory innervation to the mucosa of the larynx
(D) it provides motor innervation to the cricothyroid muscle
(E) it is accompanied by the superior laryngeal artery

56. All of the following structures are embedded in the wall of the cavernous sinus along part of its course EXCEPT

(A) oculomotor nerves
(B) abducens nerves
(C) trochlear nerves
(D) the mandibular division of the trigeminal nerve
(E) ophthalmic nerves

57. All of the following statements concerning the palatine tonsil are true EXCEPT

(A) it is bounded by the palatoglossal and palatopharyngeal folds
(B) it receives sensory innervation from the glossopharyngeal nerve
(C) it receives blood from branches of the facial artery
(D) it is located on each side of the oropharynx
(E) it is called the adenoid when enlarged

58. An eroded lesion occurring in the jugular foramen may damage all of the following structures EXCEPT

(A) vagus nerve
(B) spinal accessory nerve
(C) internal jugular vein
(D) glossopharyngeal nerve
(E) hypoglossal nerve

59. All of the following statements concerning the orbicularis oculi muscle are true EXCEPT

(A) it is innervated by temporal and zygomatic branches of the facial nerve
(B) its function is to open the eyelids
(C) its paralysis results in spilling of tears
(D) it is branchiomeric in origin
(E) it is one of the muscles of facial expression

60. Severance of the greater petrosal nerve would produce all of the following conditions EXCEPT

(A) decreased lacrimal gland secretion
(B) loss of taste sensation in the palate
(C) dryness in the nose and palate
(D) decreased parotid gland secretion
(E) loss of general sensation in the roof of the mouth

61. A benign tumor in the pterygoid canal could injure all of the following nerve fibers EXCEPT

(A) preganglionic parasympathetic fibers
(B) taste fibers
(C) postganglionic parasympathetic fibers
(D) postganglionic sympathetic fibers
(E) general visceral afferent (GVA) fibers

62. The pupillary light reflex can be eliminated by cutting all of the following nerves EXCEPT

(A) short ciliary nerves
(B) long ciliary nerve
(C) oculomotor nerve
(D) optic nerve
(E) ciliary ganglion

63. A dry corneal surface due to a lack of moistening fluid could indicate damage to all of the following nerves EXCEPT

(A) terminal portion of the lacrimal nerve
(B) zygomatic branch of the maxillary nerve
(C) lesser petrosal nerve
(D) greater petrosal nerve
(E) nerve of the pterygoid canal

64. All of the following statements concerning the tongue are true EXCEPT

(A) its muscles are innervated by the hypoglossal nerve
(B) taste buds in the vallate papillae are innervated by the chorda tympani of the facial nerve
(C) its anterior two-thirds receives general sensory innervation from the trigeminal nerve
(D) it receives taste fibers from the facial and glossopharyngeal nerves
(E) its root near the epiglottis receives taste sensation from the vagus nerve through the internal laryngeal nerve

65. A functional impediment to the act of swallowing can result from damage to any of the following nerves EXCEPT

(A) hypoglossal nerve
(B) spinal accessory nerve
(C) vagus nerve
(D) facial nerve
(E) trigeminal nerve

Directions: Each set of matching questions in this section consists of a list of four to twenty-six lettered options (some of which may be in figures) followed by several numbered items. For each numbered item, select the ONE lettered option that is most closely associated with it. To avoid spending too much time on matching sets with large numbers of options, it is generally advisable to begin each set by reading the list of options. Then, for each item in the set, try to generate the correct answer and locate it in the option list, rather than evaluating each option individually. Each lettered option may be selected once, more than once, or not at all.

Questions 66 and 67

Match each structure below with the appropriate lettered structure in this radiograph of a lateral view of the head.

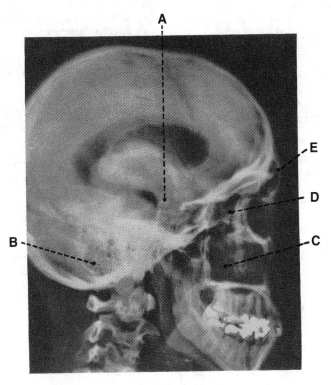

66. Maxillary sinus

67. Mastoid air cells

Questions 68–70

Match each description below with the most appropriate muscle.

(A) Lateral pterygoid muscle
(B) Medial pterygoid muscle
(C) Buccinator muscle
(D) Orbicularis oris
(E) Zygomaticus major

68. These are muscles of smiling

69. The jaws are closed by these muscles

70. The lips are closed by these muscles

Questions 71–75

Match each description below with the most appropriate sinus.

(A) Cavernous sinus
(B) Sigmoid sinus
(C) Superior sagittal sinus
(D) Transverse sinus
(E) Straight sinus

71. Lies at the junction of the falx cerebri and the tentorium cerebelli

72. Curves laterally and forward in the convex outer border of the tentorium cerebelli

73. Lies in the superior convex border of the falx cerebri

74. Communicates directly with the ophthalmic veins

75. Becomes continuous with the internal jugular vein at the jugular foramen

Questions 76–80

Match each description below with the most appropriate nerve.

(A) Hypoglossal nerve
(B) Vagus nerve
(C) Chorda tympani
(D) Lingual nerve
(E) Glossopharyngeal nerve

76. Provides motor innervation to the intrinsic muscles of the larynx

77. Carries general sensation from the anterior two-thirds of the tongue

78. Provides parasympathetic innervation for the parotid gland

79. Provides motor innervation to the intrinsic muscles of the tongue

80. Carries general somatic sensory information to the tympanic cavity (or the middle ear)

Questions 81–85

Match each statement below with the most appropriate opening of the skull.

(A) Foramen ovale
(B) Foramen magnum
(C) Petrotympanic fissure
(D) Cribriform plate
(E) Foramen rotundum
(F) Pterygoid canal
(G) Superior orbital fissure
(H) Sphenopalatine foramen
(I) Internal auditory meatus
(J) Jugular foramen

81. Injury to the nerve passing through this structure causes a loss of general sensation of the maxillary teeth

82. Injury to the nerve passing through this structure causes a loss of sensation of the temporomandibular joints

83. Traversed by sensory fibers from the posterior one-third of the tongue

84. Traversed by sensory fibers from the mucosa of the nasal septum, posterior lateral nasal wall, and anterior portion of the hard palate

85. Traversed by preganglionic parasympathetic fibers to the lacrimal gland

Questions 86–90

For each clinical condition described, select the lesion or other insult most closely related to it.

(A) Lesion of the glossopharyngeal nerve
(B) Destruction of the supraorbital notch
(C) Lesion of the facial nerve
(D) Rupture of cerebral arteries
(E) Lesion of the vagus nerve
(F) Lesion of the trigeminal nerve
(G) Rupture of cerebral veins
(H) Lesion of the oculomotor nerve
(I) Lesion of the trochlear nerve
(J) Laceration of the middle meningeal artery
(K) Lesion of the deep petrosal nerve
(L) Lesion of the greater petrosal nerve

86. A radiograph of a patient's skull reveals a fracture of the squamous part of the temporal bone and a fracture line through the foramen spinosum.

87. After a car accident, a 21-year-old woman is unable to close her lips.

88. A patient presents with a complaint of ptosis, drooping of an upper eyelid.

89. A baseball pitcher who receives a blow on the head is found to have subdural hematoma.

90. A patient develops a vasodilation in the lacrimal gland.

Answers and Explanations

1–C. The external laryngeal nerve innervates the cricothyroid muscle, and damage to this nerve results in inability to tense the vocal cords. The rima glottidis is widened by the posterior cricoarytenoid muscle.

2–B. The internal laryngeal nerve accompanies the superior laryngeal artery.

3–D. The glossopharyngeal nerve supplies sensory innervation to the mucosa of the pharynx and to the stylopharyngeus, whereas the vagus nerve innervates all of the remaining pharyngeal muscles.

4–D. The costocervical trunk divides into the deep cervical and superior intercostal arteries.

5–E. The phrenic nerve descends on the superficial surface of the anterior scalene muscle and passes into the thorax deep to the subclavian vein.

6–B. The glossopharyngeal nerve innervates the stylopharyngeus. This muscle is a landmark for locating the glossopharyngeal nerve because, as the nerve enters the pharyngeal wall, it curves posteriorly around the lateral margin of the stylopharyngeus.

7–A. The external laryngeal branch of the superior laryngeal nerve supplies the cricothyroid and inferior pharyngeal constrictor muscles.

8–A. Crocodile tears syndrome (Bogorad's syndrome) is caused by a lesion of the facial nerve proximal to the geniculate ganglion, due to misdirection of degenerating parasympathetic fibers, which formerly innervated the salivary glands, to the lacrimal glands.

9–C. Postganglionic parasympathetic fibers originating in the pterygopalatine ganglion innervate glands in the palate and nasal mucosa.

10–C. Bell's palsy (facial paralysis) can involve inflammation of the cornea leading to corneal ulceration, which probably is attributable to an absence of the corneal blink reflex due to paralysis of the orbicularis oculi, which closes the eyelid.

11–B. The inferior sagittal sinus and the great cerebral vein of Galen unite to form the straight sinus in the cranial cavity.

12–E. The abducens nerve (CN VI) innervates the lateral rectus muscle, which abducts the eyeball. A lesion of the abducens nerve results in medial strabismus and diplopia (double vision).

13–C. Bilateral severance of the vagus nerve (CN X) causes a loss of reflex control of circulation due to an increase in heart rate and blood pressure; poor digestion due to decreased gastrointestinal motility and secretion; and a difficulty in swallowing, speaking, and breathing due to paralysis of laryngeal and pharyngeal muscles. All of these effects may result in death.

14–C. Rupture of the middle meningeal artery in the cranial cavity causes an epidural hemorrhage.

15–E. Preganglionic neurons of the parasympathetic nervous system are located in the brainstem (cranial outflow) and sacral spinal cord segments S2–S4 (sacral outflow).

16–A. The anterior two-thirds of the tongue is innervated by the lingual nerve, a branch of the mandibular division of the trigeminal nerve (CN V).

17–E. The pituitary gland lies in the hypophyseal fossa of the sella turcica of the sphenoid bone, which lies immediately posterior and superior to the sphenoid sinus and medial to the cavernous sinus.

18–C. The posterior one-third of the tongue receives both general and taste innervation from the lingual branch of the glossopharyngeal nerve.

19–E. The sella turcica is part of the sphenoid bone.

20–B. When parasympathetic fibers are stimulated, the ciliary muscle contracts, making the lens of the eye thicker. Accommodation for objects close to the eye is mediated by parasympathetic nerve action.

21–B. The dural venous sinus nearest the pituitary gland is the cavernous sinus. Cavernous sinus thrombophlebitis is an infectious inflammation of the sinus that may produce meningitis, papilledema, exophthalmos, and ophthalmoplegia.

22–C. The pterygopalatine fossa lies between the pterygoid plates of the sphenoid and palatine bone below the apex of the orbit. It contains the pterygopalatine ganglion and communicates with the middle cranial cavity through the foramen rotundum and with the infratemporal fossa through the pterygomaxillary fissure.

23–D. The posterior cricoarytenoid muscle is the only muscle that abducts the vocal cords during quiet breathing.

24–B. The oblique arytenoid and aryepiglottic muscles act to tilt the arytenoid cartilages and approximate them, assisting in closing off the larynx and preventing food from entering the larynx and trachea during the process of swallowing. The cricopharyngeus fibers of the inferior pharyngeal constrictors act as a sphincter that prevents air from entering the esophagus.

25–D. The veins of the brain are direct tributaries of the dural venous sinuses. The emissary veins connect the dural venous sinuses with the veins of the scalp; the diploic veins lie in channels in the diploë of the skull and communicate with the dural sinuses, the veins of the scalp, and the meningeal veins. The pterygoid venous plexus communicates with the cavernous sinus through an emissary vein.

26–D. When the parasympathetic fibers to the eyeball are stimulated, the pupil constricts and the ciliary muscle contracts, resulting in a thicker lens and enhanced vision for near objects (accommodation). Contraction of capillaries in the iris and enhanced ability to see distant objects (flattening of the lens) result from stimulation of sympathetic nerves.

27–D. Preganglionic parasympathetic fibers run in the tympanic nerve, or Jacobson's nerve (a branch of the glossopharyngeal nerve), and then in the lesser petrosal nerve to reach the otic ganglion, where they synapse with postganglionic neurons.

28–B. The middle ear cavity and the sphenoid sinus are separated from the middle cranial fossa by a thin layer of bone.

29–D. The cranial dura is innervated by the ophthalmic division of the trigeminal nerve in the anterior cranial fossa, the maxillary and mandibular divisions of the trigeminal nerve in the middle cranial fossa, and the vagus and hypoglossal nerves in the posterior cranial fossa.

30–D. The carotid sinus is a spindle-shaped dilatation of the origin of the internal carotid artery. It is a pressoreceptor that is stimulated by changes in blood pressure. The carotid sinus is innervated by the carotid sinus branch of the glossopharyngeal nerve and by a branch of the vagus nerve.

31–C. The optic canal transmits the optic nerve, ophthalmic artery, and the central vein of the retina. The ophthalmic nerve and ophthalmic vein pass through the superior orbital fissure.

32–D. The digastric anterior belly is innervated by the trigeminal nerve; the digastric posterior belly is innervated by the facial nerve.

33–B. The tensor tympani is innervated by the trigeminal nerve; the stapedius is innervated by the facial nerve.

34–B. When the pupil remains small in a dimly lit room, it is an indication that postganglionic sympathetic fibers that originate from the superior cervical ganglion and innervate the dilator pupillae (radial muscles of the iris) are damaged.

35–D. The frontal sinus, a paranasal air sinus, receives blood from branches of the ophthalmic artery. It lies in the frontal bone, communicates with the middle nasal sinus, and is innervated by the supraorbital branch of the ophthalmic nerve.

36–C. Damage to the parasympathetic fibers in the ciliary ganglion or in the short ciliary nerves will impair a patient's ability to focus on close objects (accommodation). The patient will be able to see distant objects clearly because the long ciliary nerve also carries sympathetic fibers to the dilator pupillae. The ability to move the eyeball normally indicates that the oculomotor, trochlear, and abducens nerves are intact.

37–D. The abducens, oculomotor, and nasociliary nerves enter the orbit through the superior orbital fissure and the common tendinous ring. The trochlear, lacrimal, and frontal nerves and the ophthalmic vein enter the orbit through the superior orbital fissure outside the common tendinous ring.

38–A. The stylohyoid muscle is innervated by the facial nerve; the styloglossus muscle by the hypoglossal nerve; and the stylopharyngeus muscle by the glossopharyngeal nerve.

39–B. The posterior cervical triangle is further divided by the omohyoid posterior belly into the occipital and subclavian triangles.

40–C. The costocervical trunk is a short trunk from the subclavian artery that arches to the neck of the first rib, where it divides into the deep cervical and superior intercostal arteries. Consequently, it is not closely associated with the isthmus of the thyroid gland. A low tracheotomy is a surgical incision of the trachea through the neck, below the isthmus of the thyroid gland.

41–A. The phrenic nerve, a branch of the cervical plexus, contains motor and sensory fibers but no cutaneous nerve fibers.

42–C. The sphenoid sinus opens into the sphenoethmoidal recess in the nasal cavity. The maxillary sinus, frontal sinus, and several ethmoid sinuses drain into the hiatus semilunaris, and infection spreads readily among these sinuses.

43–A. Cervical nerve C1 contains no cutaneous nerve fibers.

44–B. The true vocal folds are inferior to the ventricle of the larynx. The afferent nerve fibers from the larynx are carried by the internal laryngeal branch of the superior laryngeal nerve and the inferior or recurrent laryngeal nerve, which are all branches of the vagus nerve.

45–E. The optic nerve and the ophthalmic artery pass through the optic canal.

46–B. The duct of the submandibular gland arises from the deep portion of the gland.

47–E. The vomer forms part of the nasal septum.

48–B. A horizontal cut through the neck will sever the inferior thyroid arteries and the common carotid arteries, but not the external carotid arteries.

49–B. The tensor tympani is innervated by the mandibular branch of the trigeminal nerve.

50–D. The secretomotor fibers for lacrimal secretion come through the pterygopalatine ganglion, and so severance of the oculomotor nerve will have no effect on lacrimal secretion. The oculomotor nerve carries parasympathetic fibers to the constrictor pupillae and ciliary muscles.

51–C. Nerves that supply the muscles of the eyeball and tongue, such as the oculomotor nerve, are not of branchiomeric origin. Special visceral efferent (SVE) nerve fibers in the trigeminal, facial, glossopharyngeal, vagus, and accessory nerves innervate a variety of skeletal muscles in the head and neck, and are branchiomeric (nonsomitic) in origin because they originate from the branchial arches.

52–D. The carotid sheath contains the vagus nerve, the common and internal carotid arteries, and the internal jugular vein.

53–D. The tensor tympani is innervated by the mandibular division of the trigeminal nerve.

54–D. The subclavian artery passes posterior to the scalenus anterior and the subclavian vein anterior to it.

55–D. The internal laryngeal nerve is a branch of the superior laryngeal nerve, provides sensory innervation to the mucosa of the larynx above the vocal cord, and contains taste fibers from the epiglottis and the root of the tongue adjacent to it. The external laryngeal nerve innervates the cricothyroid and inferior pharyngeal constrictor muscles.

56–D. The oculomotor, abducens, trochlear, and ophthalmic nerves all lie in the wall of the cavernous sinus. The mandibular division of the trigeminal nerve does not lie in the wall of the cavernous sinus.

57–E. The pharyngeal tonsil, located in the posterior wall and roof of the nasopharynx, is called the adenoid when enlarged.

58–E. The jugular foramen transmits the glossopharyngeal, spinal accessory, and vagus nerves and the internal jugular vein.

59–B. The orbicularis oculi muscle, derived from the second pharyngeal (or hyoid) arch, is innervated by the facial nerve and is active in eye closure. It is a muscle of facial expression, and its paralysis results in drooping of the lower eyelid and spilling of tears.

60–D. The greater petrosal nerve transmits parasympathetic (preganglionic) fibers, which are secreto-motor fibers, to the lacrimal glands and mucous glands in the nasal cavity and palate; taste fibers from the palate; and general visceral afferent (GVA) fibers from the nasal cavity, palate, and roof of the oral cavity. A lesion of the lesser petrosal nerve causes decreased parotid gland secretion.

61–C. The nerve of the pterygoid canal (vidian nerve) contains taste fibers from the palate, GVA fibers, postganglionic sympathetic fibers, and preganglionic parasympathetic fibers.

62–B. The efferent limbs of the reflex arc concerned in the pupillary light reflex (i.e., constriction of the pupil in response to illumination of the retina) are composed of parasympathetic preganglionic fibers in the oculomotor nerve and parasympathetic postganglionic fibers in the short ciliary nerves. The afferent limbs of this reflex are optic nerve fibers. The long ciliary nerves contain postganglionic sympathetic fibers.

63–C. The secretomotor fibers to the lacrimal gland are parasympathetic fibers that run in the facial, greater petrosal, vidian, maxillary, zygomatic, and lacrimal nerves. The lesser petrosal nerve contains secretomotor (preganglionic parasympathetic) fibers to the parotid gland.

64–B. The vallate papillae are located in the anterior two-thirds of the tongue, but their taste buds are innervated by the glossopharyngeal nerve, which supplies both general and taste sensations to the posterior one-third of the tongue.

65–B. Swallowing involves movements of the tongue to push the food into the oropharynx, elevation of the soft palate to close the entrance of the nasopharynx, elevation of the hyoid bone and the larynx to close the opening into the larynx, and contraction of the pharyngeal constrictors to move the food through the pharynx. The mandibular division of the trigeminal nerve supplies the suprahyoid muscles (e.g., the anterior belly of the digastric and the mylohyoid muscles). The vagus nerve innervates the muscles of the palate, the larynx, and the pharynx. The hypoglossal nerve supplies all of the tongue muscles except the palatoglossus, which is innervated by the vagus nerve.

66–C.

67–B.

68–E. The zygomaticus major muscles act to produce a smile by drawing the angles of the mouth backward and upward.

69–B. The medial pterygoid, temporalis, and masseter muscles are involved in opening the jaws. The lateral pterygoid muscles close the jaws by depressing the mandible.

70–D. The orbicularis oris muscles can close the lips.

71–E. The straight sinus runs along the line where the falx cerebri attaches to the tentorium cerebelli.

72–D. The transverse sinus runs laterally and forward in the convex outer border of the tentorium cerebelli.

73–C. The superior sagittal sinus lies in the superior convex border of the falx cerebri.

74–A. The cavernous sinus communicates directly with the ophthalmic veins.

75–B. The sigmoid sinus becomes continuous with the internal jugular vein at the jugular foramen.

76–B. The vagus nerve provides motor innervation to the intrinsic muscles of the larynx through the recurrent and external laryngeal nerves.

77–D. The lingual nerve carries general sensation from the anterior two-thirds of the tongue.

78–E. The glossopharyngeal nerve carries the preganglionic parasympathetic fibers that run in the tympanic and lesser petrosal nerves and synapse in the otic ganglion with cell bodies of postganglionic parasympathetic fibers, which run in the auriculotemporal nerve and innervate the parotid gland.

79–A. The hypoglossal nerve innervates all intrinsic muscles of the tongue and all extrinsic muscles of the tongue except for the palatoglossus, which is innervated by the vagus nerve.

80–E. The glossopharyngeal nerve carries general somatic sensory information from the tympanic cavity, the auditory tube, and the mastoid air cells.

81–E. The foramen rotundum transmits the maxillary division of the trigeminal nerve. Injury to this nerve causes a loss of general sensation of the maxillary teeth.

82–A. The foramen ovale transmits the mandibular division of the trigeminal nerve. Injury to this nerve causes a loss of sensation of the temporomandibular joint.

83–J. The jugular foramen transmits the glossopharyngeal nerve, which carries sensory fibers from the posterior one-third of the tongue.

84–H. The sphenopalatine foramen transmits the sphenopalatine nerve, which carries sensory fibers from the mucosa of the nasal septum, the posterior lateral nasal wall, and the anterior portion of the hard palate.

85-F. The pterygoid canal transmits the nerve of the pterygoid canal, which contains preganglionic parasympathetic fibers to the lacrimal gland.

86–J. A fracture of the foramen spinosum would result in a laceration of the middle meningeal artery, which it transmits.

87–C. The lips are closed by the orbicularis oris muscles, which are innervated by the facial nerve. However, the mouth or jaws are closed by the temporalis, masseter, and medial pterygoid muscles, which are innervated by the trigeminal nerve.

88–H. The eyes are opened by the levator palpebrae superioris muscles, which are innervated by the oculomotor nerve.

89–G. Subdural hematoma is caused by rupture of cerebral veins as they pass from the brain surface into one of the venous sinuses.

90–K. Sympathetic postganglionic fibers running in the deep petrosal nerve supply blood vessels in the lacrimal glands.

Comprehensive Examination

Directions: Each of the numbered items or incomplete statements in this section is followed by answers or by completions of the statement. Select the ONE lettered answer or completion that is BEST in each case.

1. If a stab wound lacerates the posterior humeral circumflex artery passing through the quadrangular space on the shoulder region, which of the following nerves might be injured?

(A) Radial nerve
(B) Axillary nerve
(C) Thoracodorsal nerve
(D) Suprascapular nerve
(E) Accessory nerve

2. A car accident victim has lost the function of all abductors of the arm. This indicates damage to which of the following parts of the brachial plexus?

(A) Middle trunk and posterior cord
(B) Middle trunk and lateral cord
(C) Lower trunk and lateral cord
(D) Upper trunk and posterior cord
(E) Lower trunk and medial cord

3. Which of the following muscles acts to flex the elbow but is not innervated by the musculocutaneous, median, or ulnar nerves?

(A) Flexor digitorum longus
(B) Brachioradialis
(C) Brachialis
(D) Extensor digitorum longus
(E) Biceps brachii

4. A lesion of the radial nerve in the spiral groove of the humerus would cause which of the following conditions?

(A) Numbness over the medial side of the forearm
(B) Inability to oppose the thumb
(C) Weakness in pronating the forearm
(D) Weakness in abducting the arm
(E) Inability to extend the hand

5. A patient with an intercondylar fracture of the humerus is unable to use the muscle that inserts into the pisiform bone. Which of the following nerves is most likely to have been damaged?

(A) Axillary nerve
(B) Radial nerve
(C) Musculocutaneous nerve
(D) Median nerve
(E) Ulnar nerve

6. Abduction of the fingers would be impaired most by paralysis of the

(A) ulnar nerve
(B) median nerve
(C) radial nerve
(D) musculocutaneous nerve
(E) ancillary nerve

7. When the hand is in a resting supine position, the radius is in articulation at the radiocarpal joint with which of the following bones?

(A) Triquetrum and trapezium
(B) Lunate and trapezium
(C) Lunate and scaphoid
(D) Scaphoid and hamate
(E) Capitate and scaphoid

8. An injury to the thoracodorsal nerve would probably affect the strength of which of the following movements?

(A) Abduction of the arm
(B) Lateral rotation of the arm
(C) Adduction of the scapula
(D) Extension of the arm
(E) Elevation of the scapula

9. The cell bodies of nerve fibers in the lateral antebrachial cutaneous nerve are located in the

(A) collateral ganglia and dorsal root ganglia
(B) sympathetic trunk and anterior horn of spinal cord
(C) dorsal root ganglia and sympathetic trunk
(D) lateral horn of spinal cord and sympathetic trunk
(E) dorsal root ganglia and anterior horn of spinal cord

10. A patient cannot extend the proximal interphalangeal joint of the ring finger. Which of the following pairs of nerves are damaged?

(A) Radial and median nerves
(B) Radial and axillary nerves
(C) Radial and ulnar nerves
(D) Ulnar and median nerves
(E) Ulnar and axillary nerves

11. Which of the following statements concerning the anterior cruciate ligament of the knee joint is true?

(A) It becomes taut during flexion of the leg
(B) It resists posterior displacement of the femur on the tibia
(C) It inserts into the medial femoral condyle
(D) It helps prevent hyperflexion of the knee joint
(E) It is lax when the knee is extended

12. Which of the following actions is most seriously affected by paralysis of the deep peroneal nerve?

(A) Plantar flexion of the foot
(B) Dorsiflexion of the foot
(C) Abduction of the toes
(D) Eversion of the foot
(E) Adduction of the toes

13. The first vascular channel likely to be obstructed or occluded by an embolus from the deep veins of a lower limb would be

(A) tributaries of the renal veins
(B) branches of the coronary arteries
(C) sinusoids of the liver
(D) tributaries of the pulmonary veins
(E) branches of the pulmonary arteries

14. A patient presents with flatfeet. The foot is displaced laterally and everted, and the head of the talus is no longer supported. Which of the following ligaments probably is stretched?

(A) Plantar calcaneonavicular ligament
(B) Calcaneofibular ligament
(C) Anterior tibiofibular ligament
(D) Lateral talocalcaneal ligament
(E) Anterior tibiotalar ligament

15. Which of the following ligaments is important in preventing forward displacement of the femur on the tibia when the weight-bearing knee is flexed?

(A) Anterior meniscofemoral ligament
(B) Fibular collateral ligament
(C) Oblique popliteal ligament
(D) Posterior cruciate ligament
(E) Anterior cruciate ligament

16. Cell bodies of nerve fibers in the ventral root of a thoracic spinal nerve are located in the

(A) dorsal root ganglia and sympathetic trunk
(B) lateral horn of spinal cord and dorsal root ganglia
(C) ventral horn and lateral horn of spinal cord
(D) sympathetic trunk and lateral horn of spinal cord
(E) ventral horn of spinal cord and sympathetic trunk

17. Which of the following structures is found in the right ventricle of the heart?

(A) Fossa ovalis
(B) Septomarginal trabecula
(C) Pectinate muscles
(D) Sinoatrial node
(E) Openings of pulmonary veins

18. Inadequate blood flow in the artery that runs aside the great cardiac vein in the anterior interventricular sulcus of the heart would result from occlusion of the

(A) circumflex branch of the left coronary artery
(B) marginal branch of the right coronary artery
(C) left coronary artery
(D) right coronary artery
(E) marginal branch of the left coronary artery

19. Severely diminished blood flow in the coronary arteries would most likely result from embolization of an atherosclerotic plaque in the

(A) pulmonary trunk
(B) ascending aorta
(C) coronary sinus
(D) descending aorta
(E) aortic arch

20. Which set of conditions describes the respective positions of the pulmonary valve, the aortic valve, and both atrioventricular valves during ventricular systole?

(A) Open, open, closed
(B) Closed, closed, open
(C) Closed, open, closed
(D) Open, closed, open
(E) Open, closed, closed

21. Which of the following statements concerning the internal abdominal oblique muscle is correct?

(A) It forms the floor of the inguinal canal
(B) Its aponeurosis contributes to the posterior wall of the inguinal canal
(C) Its aponeurosis aids in the formation of the falx inguinalis (conjoint tendon)
(D) Its aponeurosis contributes to the formation of the posterior layer of the rectus sheath below the arcuate line
(E) Its muscle fibers travel in the same general direction as those of the external abdominal oblique muscle

22. Which of the following structures forms part of the anterior wall of the inguinal canal?

(A) Transversalis fascia
(B) Aponeurosis of the transverse abdominal muscle
(C) Aponeurosis of the external abdominal oblique muscle
(D) Falx inguinalis
(E) Lacunar ligament

23. An indirect inguinal hernia occurs

(A) lateral to the inferior epigastric artery
(B) between the inferior epigastric artery and the obliterated umbilical artery
(C) medial to the obliterated umbilical artery
(D) between the median umbilical fold and the obliterated umbilical artery
(E) between the median umbilical fold and the inferior epigastric artery

24. Lesions of parasympathetic fibers in the vagus nerve interfere with glandular secretory or smooth muscle functions in which of the following organs?

(A) Bladder
(B) Transverse colon
(C) Sigmoid colon
(D) Prostate gland
(E) Rectum

25. A slowly growing tumor in the head of the pancreas could compress the

(A) duodenojejunal junction
(B) gastroduodenal artery
(C) bile duct
(D) inferior mesenteric artery
(E) common hepatic duct

26. Which of the following statements concerning the deep perineal space is correct?

(A) It is formed superiorly by the perineal membrane
(B) It contains the superficial transverse perineal muscles
(C) In males, it contains a segment of the dorsal nerve of the penis
(D) It is formed inferiorly by Colles' fascia
(E) In females, it contains the greater vestibular glands

27. If the urethra tears distal to the urogenital diaphragm, urine might collect in the

(A) retropubic space
(B) medial aspect of the thigh
(C) ischiorectal fossa
(D) superficial perineal space
(E) paravesical fossa

28. The inferior hypogastric (pelvic) plexus contains parasympathetic nerve fibers from the

(A) lumbar splanchnic nerves
(B) pelvic splanchnic nerves
(C) sacral sympathetic ganglia
(D) vagus nerve
(E) sacral splanchnic nerves

29. Parasympathetic fibers in the distal portion of the inferior mesenteric plexus are branches of the

(A) vagus nerve
(B) pelvic splanchnic nerves
(C) sacral splanchnic nerves
(D) lesser splanchnic nerves
(E) greater splanchnic nerves

30. Carcinoma of the uterus can spread directly to the labia majus through lymphatics that follow the

(A) ovarian ligament
(B) suspensory ligament of the ovary
(C) round ligament of the uterus
(D) uterosacral ligaments
(E) pubocervical ligaments

31. Which of the following groups of locations best describes the normal position of the dens of the axis in relation to the anterior arch of the atlas, the cruciform ligament, and the body of the axis?

(A) Anterior, posterior, superior
(B) Anterior, anterior, inferior
(C) Posterior, anterior, superior
(D) Anterior, posterior, inferior
(E) Posterior, posterior, superior

32. A patient shows numbness of the skin over the anterior triangle of the neck. Which of the following nerves might be injured?

(A) Great auricular nerve
(B) Transverse cervical nerve
(C) Superior ramus of the ansa cervicalis
(D) Inferior ramus of the ansa cervicalis
(E) Superior laryngeal nerve

33. If a benign tumor is found where the common carotid artery usually bifurcates, it would be located at the level of the

(A) thyroid isthmus
(B) cricoid cartilage
(C) angle of the mandible
(D) superior border of the thyroid cartilage
(E) jugular notch

34. Damage to the articular disk of the temporomandibular joint would result in paralysis of which of the following muscles?

(A) Masseter
(B) Temporalis
(C) Medial pterygoid muscle
(D) Lateral pterygoid muscle
(E) Buccinator

35. A patient with ptosis of the left eyelid probably has a damaged left

(A) trochlear nerve
(B) abducens nerve
(C) oculomotor nerve
(D) ophthalmic nerve
(E) facial nerve

36. The superior petrosal sinus lies in the margin of the

(A) tentorium cerebelli
(B) falx cerebri
(C) falx cerebelli
(D) diaphragma sellae
(E) straight sinus

37. Which of the following statements best explains the importance of arachnoid granulations?

(A) They allow cerebrospinal fluid (CSF) to pass from the subarachnoid space into the venous sinuses of the dura mater
(B) They are a storage area for CSF
(C) They increase the surface area available for the production of CSF
(D) They provide a shunt that allows CSF to return to the ventricles of the brain
(E) They receive blood from the diploë of the skull

38. The great cerebral vein of Galen drains into the

(A) superior sagittal sinus
(B) inferior sagittal sinus
(C) cavernous sinus
(D) transverse sinus
(E) straight sinus

39. While a surgical intern is making a local excision for a tumor in the palate, he accidentally cuts a tendon that loops around the pterygoid hamulus. Which of the following muscles would most likely be paralyzed as a result?

(A) Tensor tympani
(B) Tensor veli palatini
(C) Levator veli palatini
(D) Superior pharyngeal constrictor
(E) Stylohyoid

40. If a vertical stab wound lacerates the pterygomandibular raphe, which of the following muscles would be paralyzed?

(A) Superior and middle pharyngeal constrictors
(B) Middle and inferior pharyngeal constrictors
(C) Superior pharyngeal constrictor and buccinator muscles
(D) Medial and lateral pterygoid muscles
(E) Tensor veli palatini and levator veli palatini

41. A patient presents with a large benign tumor that blocks a communication between the infratemporal fossa and the pterygopalatine fossa. The tumor would be located in or at which of the following structures?

(A) Pharyngeal canal
(B) Pterygopalatine canal
(C) Sphenopalatine foramen
(D) Pterygomaxillary fissure
(E) Petrotympanic fissure

42. Which of the following muscles indents the submandibular gland and divides it into superficial and deep parts?

(A) Hyoglossus
(B) Genioglossus
(C) Styloglossus
(D) Superior constrictor muscle
(E) Mylohyoid muscle

43. An abscess in the auditory tube may block a communication between the nasopharynx and which of the following structures?

(A) Vestibule of the inner ear
(B) Middle ear
(C) Semicircular canals
(D) External ear
(E) Inner ear

44. A patient is unable to open his jaw, due to paralysis of the

(A) medial pterygoid muscle
(B) masseter
(C) temporalis
(D) lateral pterygoid muscle
(E) buccinator

45. Which of the following conditions could result from damage to parasympathetic fibers in the lesser petrosal nerve?

(A) Lack of lacrimal secretion
(B) Lack of submandibular secretion
(C) Lack of parotid secretion
(D) Constriction of the pupil
(E) Ptosis of the upper eyelid

Directions: Each of the numbered items or incomplete statements in this section is negatively phrased, as indicated by a capitalized word such as NOT, LEAST, or EXCEPT. Select the ONE lettered answer or completion that is BEST in each case.

46. All of the following statements concerning the porta hepatis are true EXCEPT

(A) it is a transverse fissure where the hepatic portal vein enters the liver
(B) it lies between the caudate and quadrate lobes of the liver
(C) it is the area where the right and left hepatic ducts exit the liver
(D) it is the separation point between the fissure for the ligamentum teres hepatis and the fissure for the ligamentum venosum
(E) it is the portal fissure where the common hepatic artery is divided into the proper hepatic and gastroduodenal arteries

47. All of the following statements concerning the medial epicondyle of the humerus are correct EXCEPT

(A) it provides an attachment for many of the wrist flexors
(B) it is more prominent than the lateral epicondyle
(C) it is closer to the basilic vein than to the cephalic vein
(D) it is grooved posteriorly by the ulnar nerve
(E) it is the point at which the brachial artery divides into the ulnar and radial branches

48. In Erb-Duchenne paralysis, the nerve fibers in the roots of C5 and C6 of the brachial plexus usually are damaged. Paralysis is likely to occur in all of the following muscles EXCEPT

(A) biceps brachii
(B) flexor carpi ulnaris
(C) brachioradialis
(D) brachialis
(E) coracobrachialis

49. A patient with carpal tunnel syndrome could have compression of all of the following structures EXCEPT

(A) flexor pollicis longus tendon
(B) ulnar nerve
(C) median nerve
(D) flexor digitorum superficialis tendon
(E) flexor digitorum profundus tendon

50. All of the following bones are associated with the medial longitudinal arch of the foot EXCEPT

(A) talus
(B) medial three metatarsal bones
(C) navicular (scaphoid) bone
(D) calcaneus
(E) cuboid bone

51. Dislocation of the hip results in a vascular necrosis of the hip joint due to injuries to all of the following arteries EXCEPT

(A) lateral femoral circumflex artery
(B) obturator artery
(C) medial femoral circumflex artery
(D) inferior gluteal artery
(E) deep iliac circumflex artery

52. All of the following statements characterize the medial meniscus of the knee joint EXCEPT

(A) it is nearly circular
(B) it is attached to the tibial collateral ligament
(C) it is larger than the lateral meniscus
(D) it lies outside the synovial cavity
(E) it is more frequently torn in injuries than the lateral meniscus

53. The manubrium of the sternum articulates with all of the following structures EXCEPT

(A) body of the sternum
(B) first rib
(C) second rib
(D) third rib
(E) clavicle

54. All of the following statements concerning the ribs are correct EXCEPT

(A) the intercostal nerves, arteries, and veins are associated with the costal grooves
(B) the tubercles articulate with the spinous processes of the vertebrae
(C) the costal cartilages attach the upper 10 pairs of ribs to the sternum
(D) the upper 7 pairs of ribs are called true ribs
(E) the lower 2 pairs of ribs are called floating ribs

55. General visceral afferent (GVA) fibers are contained in all of the following structures EXCEPT

(A) sympathetic chain
(B) dorsal root
(C) greater splanchnic nerve
(D) gray rami communicantes
(E) white rami communicantes

56. Thrombosis in the coronary sinus might cause dilation of all of the following veins EXCEPT

(A) great cardiac vein
(B) middle cardiac vein
(C) anterior cardiac vein
(D) small cardiac vein
(E) oblique cardiac vein

57. The mediastinum contains all of the following structures EXCEPT

(A) thymus gland
(B) esophagus
(C) trachea
(D) lungs
(E) heart

58. All of the following statements concerning the respiratory system are correct EXCEPT

(A) the left lung has a lingula
(B) cartilaginous rings are found in the main bronchi
(C) the left lung has a smaller volume than the right lung
(D) the main bronchus branches into the lobar bronchi at the carina
(E) the right lung usually receives a single bronchial artery

59. An occlusive lesion in the celiac trunk may result in ischemia of all of the following organs EXCEPT

(A) liver
(B) spleen
(C) pancreas
(D) gallbladder
(E) stomach

60. The portal venous system contains all of the following veins EXCEPT

(A) left colic vein
(B) splenic vein
(C) superior rectal vein
(D) appendicular vein
(E) hepatic vein

61. All of the following statements concerning the scrotum are true EXCEPT

(A) it is homologous to the labia majora in females
(B) most of its lymphatic drainage enters superficial inguinal lymph nodes
(C) it is supplied with blood by the testicular artery
(D) it has a dartos layer of fascia and muscle that is continuous with Colles' layer of superficial fascia in the perineum
(E) it is innervated anteriorly by the ilioinguinal nerve

62. The pelvic outlet is formed by all of the following structures EXCEPT

(A) sacrotuberous ligament
(B) inferior pubic ramus
(C) pubic tubercle
(D) ischial tuberosity
(E) coccyx

63. All of the following statements contrasting male and female pelvic structures are true EXCEPT

(A) in females, the pelvic inlet is oval; in males, it is more heart-shaped
(B) the depth of the entire pelvis is generally greater in females than in males
(C) the angle formed by the inferior pubic rami is greater in females than in males
(D) the sacrum is wider in females than in males
(E) the lips of the pubis are more everted in males than in females

64. During palatine tonsillectomy, a surgeon would ligate branches from all of the following arteries EXCEPT

(A) lesser palatine artery
(B) facial artery
(C) lingual artery
(D) superior thyroid artery
(E) ascending pharyngeal artery

65. The nasal cavity is chronically dry due to a lack of glandular secretions, indicating lesions of all of the following structures EXCEPT

(A) pterygopalatine ganglion
(B) nerve of the pterygoid canal
(C) facial nerve
(D) greater petrosal nerve
(E) deep petrosal nerve

Directions: Each set of matching questions in this section consists of a list of four to twenty-six lettered options (some of which may be in figures) followed by several numbered items. For each numbered item, select the ONE lettered option that is most closely associated with it. To avoid spending too much time on matching sets with large numbers of options, it is generally advisable to begin each set by reading the list of options. Then, for each item in the set, try to generate the correct answer and locate it in the option list, rather than evaluating each option individually. Each lettered option may be selected once, more than once, or not at all.

Questions 66–70

Match each statement below with the muscle it describes.

(A) Pectoralis major
(B) Latissimus dorsi
(C) Anterior serratus
(D) Infraspinatus
(E) Long head of triceps brachii

66. Forms the anterior axillary fold; functions to flex and adduct the arm

67. Arises from the scapula; is innervated by branches from the radial nerve

68. Arises from the thoracodorsal fascia; with the teres major, forms the posterior axillary fold

69. Forms the anterior wall of the axilla; is innervated by the lateral and medial cords of the brachial plexus

70. Helps stabilize the glenohumeral joint; is innervated by a branch from the suprascapular nerve

Questions 71–75

Match each statement below with the muscle it describes.

(A) Interossei
(B) Lumbrical muscles
(C) Flexor digitorum profundus
(D) Flexor digitorum superficialis
(E) Extensor digitorum

71. Flexes interphalangeal joints; is innervated by the median and ulnar nerves

72. Originates from the radial side of tendons of another muscle

73. Flexes interphalangeal joints; is innervated solely by the median nerve

74. Extends interphalangeal joints when metacarpophalangeal joints are flexed

75. Inserts into extensor expansion; adducts and abducts the fingers

Questions 76–80

Match each statement below with the nerve it describes.

(A) Femoral nerve
(B) Obturator nerve
(C) Pudendal nerve
(D) Superior gluteal nerve
(E) Sciatic nerve

76. Innervates a muscle that also receives innervation from the sciatic nerve

77. Enters the gluteal region through the greater sciatic foramen and exits this region at the inferior border of the gluteus maximus

78. Enters the gluteal region through the greater sciatic foramen and exits this region through the lesser sciatic foramen in close proximity to the ischial spine

79. Innervates the tensor fascia lata

80. Innervates the gracilis

Questions 81–85

Match each statement below with the artery it describes.

(A) Right coronary artery
(B) Superior intercostal artery
(C) Left coronary artery
(D) Internal thoracic artery
(E) Pulmonary artery

81. A branch of the costocervical trunk

82. Gives rise to the superior epigastric artery

83. Carries the major blood supply for the anterior portion of the interventricular septum

84. Gives rise to anterior intercostal arteries

85. Usually carries the major blood supply for the posterior portion of the interventricular septum

Questions 86–90

Match each statement below with the ligament it describes.

(A) Broad ligament
(B) Round ligament of the uterus
(C) Ovarian ligament
(D) Suspensory ligament of the ovary
(E) Cardinal ligament

86. Enters the deep inguinal ring

87. A double layer of mesentery that attaches to the lateral surface of the uterus

88. Homologous to the most superior portion of the gubernaculum in males

89. Extends from the ovary to the dorsolateral body wall

90. A uterine support that extends from the cervix and the lateral fornices of the vagina to the pelvic wall

Questions 91–95

Match each statement below with the nerve it describes.

(A) Hypoglossal nerve
(B) Recurrent laryngeal nerve
(C) Chorda tympani
(D) Lingual nerve
(E) Glossopharyngeal nerve

91. Provides motor innervation to the intrinsic muscles of the larynx

92. Carries general sensation from the anterior two-thirds of the tongue

93. Carries special visceral (taste) sensation from the anterior two-thirds of the tongue

94. Provides motor innervation to the intrinsic muscles of the tongue

95. Carries sensation from pressure receptors in the carotid sinus

Questions 96–99

Match each description below with the appropriate lettered site or structure in this radiograph of the bones of the hand.

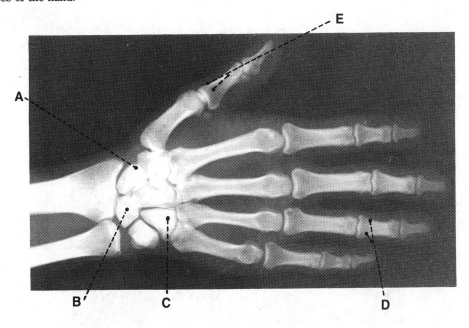

96. Lunate bone

97. Hamate bone

98. Site of attachment of the muscles that form the thenar eminence

99. Site of tendinous attachment of the flexor digitorum superficialis

Questions 100–103

Match each description below with the appropriate lettered structure in this computed tomogram showing a sectional view of the abdomen.

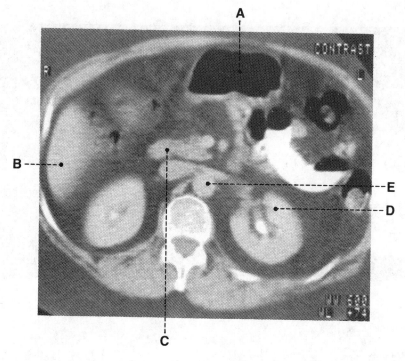

100. Pyloric portion of the stomach

101. Left kidney

102. Portal vein

103. Right lobe of the liver

Questions 104 and 105

Match each description below with the appropriate lettered structure in this radiograph showing a frontal view of the head.

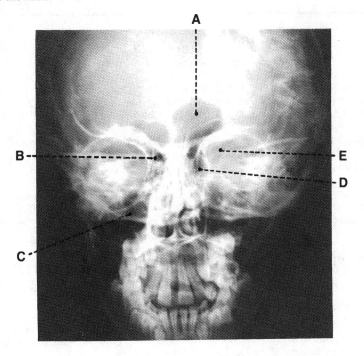

104. Frontal sinus

105. Superior orbital fissure

Answers and Explanations

1–B. The axillary nerve runs posteriorly to the humerus, accompanying the posterior humeral circumflex artery through the quadrangular space, and innervates the teres minor and deltoid muscles.

2–D. The abductors of the arm are the deltoid and supraspinatus muscles. The deltoid is innervated by the axillary nerve, which arises from the posterior cord of the brachial plexus; the supraspinatus is innervated by the suprascapular nerve, which arises from the upper trunk of the brachial plexus.

3–B. The brachioradialis is innervated by the radial nerve and functions to flex the elbow. The biceps brachii and brachialis muscles flex the elbow and are innervated by the musculocutaneous nerve. The flexor digitorum longus and extensor digitorum longus do not act in the elbow.

4–E. The radial nerve innervates the extensor muscles of the hand. The skin on the medial side of the forearm is innervated by the medial antebrachial nerve. The opponens pollicis, pronator teres, and quadratus muscles are innervated by the median nerve. The abductor of the arm (the deltoid muscle) and the teres minor muscle are innervated by the axillary nerve.

5–E. The flexor carpi ulnaris inserts on the pisiform bone and is innervated by the ulnar nerve.

6–A. The ulnar nerve innervates the dorsal interossei, which are the abductors of the second, third, and fourth fingers.

7–C. The radius and the articular disk articulate with the scaphoid, lunate, and triquetrum at the radiocarpal (wrist) joint; however, the triquetrum does not articulate with the radius, but instead articulates with the articular disk on the head of the ulna.

8–D. The thoracodorsal nerve innervates the latissimus dorsi, which adducts, extends, and medially rotates the arm.

9–C. The lateral antebrachial cutaneous nerve contains general somatic afferent (GSA) fibers, which have cell bodies located in the dorsal root ganglia, and sympathetic postganglionic (general visceral efferent [GVE]) fibers, which have cell bodies located in the sympathetic chain ganglia.

10–C. The proximal and distal interphalangeal joints of the ring finger are extended by the extensor digitorum, which is innervated by the radial nerve. When the metacarpophalangeal joint of the ring finger is extended by the extensor digitorum, the interphalangeal joints can be extended by the dorsal and palmar interossei and lumbrical muscles, which are innervated by the ulnar nerve.

11–B. The anterior cruciate ligament of the knee joint prevents posterior displacement of the femur on the tibia and limits hyperextension of the knee joint. This ligament is lax when the knee is flexed and becomes taut when the knee is extended. It inserts into the lateral femoral condyle posteriorly within the intercondylar notch.

12–B. The deep peroneal nerve innervates the dorsiflexors of the foot, which include the tibialis anterior, extensor hallucis longus, extensor digitorum longus, and peroneus tertius muscles.

13–E. An embolus from the deep veins of the lower limb would travel through the femoral vein, the external and common iliac veins, the inferior vena cava, the right atrium, the right ventricle, the pulmonary trunk, and into the pulmonary arteries, where it could obstruct and occlude these vessels.

14–A. Flatfoot is characterized by disappearance of the medial portion of the longitudinal arch, which appears completely flattened. The plantar calcaneonavicular (spring) ligament supports the head of the talus and the medial side of the longitudinal arch.

15–D. The posterior cruciate ligament prevents forward displacement of the femur on the tibia when the knee is flexed; the anterior cruciate ligament prevents backward dislocation of the femur on the tibia when the knee is extended.

16–C. The ventral root of a thoracic spinal nerve contains sympathetic preganglionic fibers, which have cell bodies located in the lateral horn of the gray matter of the spinal cord, and general somatic efferent (GSE) fibers, which have cell bodies located in the ventral horn of the gray matter of the spinal cord.

17–B. The right ventricle contains the septomarginal trabecula (moderate band). The right atrium contains the fossa ovalis, sinoatrial node, and pectinate muscles. The pulmonary veins open into the left atrium.

18–C. The great cardiac vein is accompanied by the anterior interventricular artery, which is a branch of the left coronary artery.

19–B. The right and left coronary arteries branch from the ascending aorta.

20–A. During ventricular systole (contraction of both ventricles), the pulmonary valve is open, the aortic valve is open, and both atrioventricular valves are closed.

21–C. The falx inguinalis (conjoint tendon) is formed by the aponeuroses of the internal abdominal oblique and transverse abdominal muscles. The internal abdominal oblique muscle contributes to the formation of the anterior wall and the roof of the inguinal canal and to the cremaster fascia. The posterior wall of the inguinal canal is formed by the aponeurosis of the transverse abdominal muscle and the transversalis fascia, and the floor is formed by the inguinal and lacunar ligaments. The external abdominal oblique muscle travels inferiorly and medially. The internal abdominal oblique muscle runs superiorly and medially.

22–C. The anterior wall of the inguinal canal is formed by the aponeuroses of the external and internal abdominal oblique muscles. The falx inguinalis is formed by the aponeuroses of the internal abdominal oblique and transverse abdominal muscles.

23–A. An indirect inguinal hernia occurs lateral to the inferior epigastric vessels. A direct inguinal hernia arises medial to these vessels.

24–B. The vagus nerve supplies parasympathetic fibers to the thoracic and abdominal viscera, including the transverse colon. The descending and sigmoid colons and other pelvic viscera are innervated by the pelvic splanchnic nerves.

25–C. The bile duct traverses the head of the pancreas and thus a tumor located there could compress this structure.

26–C. The deep perineal space, a space between the superior and inferior fasciae of the urogenital diaphragm, contains a segment of the dorsal nerve of the penis in males. The superficial transverse perineal muscles and the greater vestibular glands are found in the superficial perineal space.

27–D. Extravasated urine can pass into the superficial perineal space. The urine could spread inferiorly into the scrotum, anteriorly around the penis, and superiorly into the abdominal wall, but it could not spread into the thigh because the superficial fascia of the perineum is firmly attached laterally to the ischiopubic rami and connected with the deep fascia of the thigh (the fascia lata).

28–B. The inferior hypogastric (pelvic) plexus contains preganglionic parasympathetic fibers from the pelvic splanchnic nerves. The sacral splanchnic nerves contain preganglionic sympathetic fibers.

29–B. The pelvic splanchnic nerves contain parasympathetic preganglionic fibers, which join the inferior mesenteric plexus to supply the descending and sigmoid colons. The vagus nerve provides parasympathetic fibers up to the transverse colon. The greater, lesser, lowest, lumbar, and sacral splanchnic nerves contain sympathetic preganglionic fibers.

30–C. Carcinoma of the uterus can spread directly to the labium majus through the lymphatics that follow the round ligament of the uterus. The round ligament of the uterus extends from the uterus, enters the inguinal canal at the deep inguinal ring, emerges from the superficial inguinal ring, and merges with the subcutaneous tissue of the labium majus.

31–C. The dens (odontoid process) of the axis is located posterior to the anterior arch of the atlas, anterior to the cruciform ligament, and superior to the body of the axis.

32–B. The transverse cervical nerve innervates the skin over the anterior cervical triangle; the great auricular nerve innervates the skin behind the auricle and over the parotid gland. The ansa cervicalis innervates the infrahyoid muscles, including the sternohyoid, sternothyroid, and omohyoid muscles.

33–D. The common carotid artery usually bifurcates into the external and internal carotid arteries at the level of the superior border of the thyroid cartilage.

34–D. The lateral pterygoid muscle is inserted on the articular disk and capsule of the temporomandibular joint.

35–C. Damage to the oculomotor nerve results in ptosis (drooping) of the eyelid because the levator palpebrae superioris is innervated by the oculomotor nerve. The facial nerve innervates the orbicularis oculi, which functions to close the eyelids.

36–A. The superior petrosal sinus lies in the margin of the tentorium cerebelli.

37–A. Arachnoid granulations are tuft-like collections of highly folded arachnoid that project into the superior sagittal sinus and other dural sinuses. They absorb the cerebrospinal fluid into dural sinuses and often produce erosion or pitting of the inner surface of the calvaria.

38–E. The great cerebral vein of Galen and the inferior sagittal sinus unite to form the straight sinus.

39–B. The tendon of the tensor veli palatini curves around the pterygoid hamulus.

40–C. The pterygomandibular raphe serves as a common origin for the superior pharyngeal constrictor and buccinator muscles.

41–D. The pterygopalatine fossa communicates laterally with the infratemporal fossa by way of the pterygomaxillary fissure, medially with the nasal cavity through the sphenopalatine foramen, posteriorly with the foramen lacerum through the pterygoid canal, superiorly with the skull through the foramen rotundum, and anteriorly with the orbit through the inferior orbital fissure. The petrotympanic fissure transmits the chorda tympani.

42–E. The submandibular gland is indented by and divided into superficial and deep parts by the mylohyoid muscle.

43–B. The auditory (eustachian) tube connects the nasopharynx with the middle ear cavity.

44–D. The action of the lateral pterygoid muscles opens the jaws. The medial pterygoid, masseter, and temporalis muscles are involved in closing the jaws.

45–C. Parasympathetic preganglionic fibers in the lesser petrosal nerve enter the otic ganglion, where they synapse, and postganglionic parasympathetic fibers join the auriculotemporal nerve to supply the parotid gland.

46–E. The porta hepatis is a transverse fissure where the hepatic portal vein enters the liver and the hepatic ducts leave the liver. It lies between the caudate and quadrate lobes and marks the separation point between the fissure for the ligamentum teres hepatis (round ligament of the liver), which lies to the left of the quadrate lobe, and the fissure for the ligamentum venosum, which lies to the left of the caudate lobe.

47–E. The brachial artery divides into the radial and ulnar arteries at the level of the radial neck, in the cubital fossa about 1 cm before the bend of the elbow.

48–B. In Erb-Duchenne paralysis (or upper trunk injury), the nerve fibers in the roots of C5 and C6 of the brachial plexus (roots of anterior primary rami of spinal nerves C5 and C6) are damaged. The biceps brachii, which is innervated by the musculocutaneous nerve (C5–C7), and the brachioradialis, which is innervated by the radial nerve (C5–T1), usually are paralyzed. The flexor carpi ulnaris and adductor pollicis muscles are not paralyzed because they are innervated by the ulnar nerve, which is formed by the roots of C8 and T1.

49–B. Structures entering the palm deep to the flexor retinaculum are compressed in carpal tunnel syndrome; these include the median nerve and the tendons of the flexor pollicis longus, flexor digitorum profundus, and flexor digitorum superficialis muscles.

50–E. The medial longitudinal arch of the foot is formed by the talus, calcaneus, navicular bone, cuneiform bones, and medial three metatarsal bones, whereas the lateral longitudinal arch is formed by the calcaneus, cuboid bone, and lateral two metatarsal bones.

51–E. The hip joint receives blood from branches of the medial and lateral femoral circumflex, superior and inferior gluteal, and obturator arteries.

52–A. The medial meniscus is C-shaped or forms a semicircle.

53–D. The third rib articulates with the body of the sternum rather than the manubrium.

54–B. The tubercles of the ribs articulate with the transverse processes of the vertebrae.

55–D. The gray rami communicantes contain sympathetic postganglionic (GVE) fibers, which have their cell bodies located in the chain ganglia, rather than general visceral afferent (GVA) fibers.

56–C. The anterior cardiac vein drains directly into the right atrium, whereas all other cardiac veins drain into the coronary sinus.

57–D. The mediastinum contains the heart, trachea, esophagus, and the thymus gland; it does not contain the lungs.

58–D. The carina is the point where the trachea divides into the right and left main bronchi. The main bronchi contain cartilaginous rings. The right lung usually receives one bronchial artery, and the left lung receives two bronchial arteries.

59–C. The arterial supply to the pancreas is from both the celiac and superior mesenteric distributions. The pancreas receives blood from the superior pancreaticoduodenal branch of the gastroduodenal artery and from the dorsal pancreatic and pancreatic branches of the splenic artery, which arise from the celiac artery. The pancreas also receives blood from the inferior pancreaticoduodenal artery, which branches from the superior mesenteric artery.

60–E. The hepatic veins are systemic veins that drain hepatic blood into the inferior vena cava.

61–C. The scrotum receives blood from the posterior scrotal branch of the internal pudendal artery and the external pudendal artery.

62–C. The pelvic outlet (lower pelvic aperture) is bounded posteriorly by the sacrum and coccyx; laterally by the ischial tuberosities and sacrotuberous ligaments; and anteriorly by the pubic symphysis, the arcuate ligament, and the rami of the pubis and ischium.

63–E. Due to its everted ischial tuberosities, the female pelvic outlet is larger than the male pelvic outlet.

64–D. The palatine tonsil receives blood from the lesser palatine branch of the maxillary artery, the ascending palatine branch of the facial artery, the ascending pharyngeal artery, and the dorsal lingual branches of the lingual artery.

65–E. The parasympathetic secretomotor fibers for mucous glands in the nasal cavity run in the facial nerve, the greater petrosal nerve, the nerve of the pterygoid canal, and the pterygopalatine ganglion. The deep petrosal nerve contains sympathetic postganglionic fibers which supply blood vessels in the lacrimal gland and the nasal and oral mucosa.

66–A. The pectoralis major adducts and medially rotates the arm. The clavicular part rotates the arm medially and flexes it; the sternocostal part depresses the arm and shoulder. The lateral border of the pectoralis major forms the anterior axillary fold.

67–E. The long head of the triceps brachii originates from the infraglenoid tubercle of the scapula and is innervated by branches from the radial nerve.

68–B. The latissimus dorsi arises from the thoracodorsal fascia and, with the teres major, forms the posterior axillary fold.

69–A. The pectoralis major is innervated by the lateral and medial pectoral nerves and forms the anterior wall of the axilla.

70–D. The tendon of the infraspinatus forms the rotator (musculotendinous) cuff and thus helps to stabilize the glenohumeral joint. It is innervated by a branch from the suprascapular nerve.

71–C. The flexor digitorum profundus can flex the distal interphalangeal joints. It is innervated by the median and ulnar nerves.

72–B. The lumbricals arise from the radial side of the tendon of the flexor digitorum profundus. They are innervated by the median and ulnar nerves.

73–D. The flexor digitorum superficialis flexes the proximal interphalangeal joints. It is innervated solely by the median nerve.

74–E. The extensor digitorum communis extends the proximal and distal interphalangeal joints when the metacarpophalangeal joints are flexed by the interossei and the lumbricals.

75–A. The dorsal and palmar interossei insert into the extensor expansion and are innervated by the ulnar nerve. The dorsal interossei abduct the finger; the palmar interossei adduct the finger.

76–B. The adductor magnus is innervated by the obturator and sciatic nerves.

77–E. The sciatic nerve enters the gluteal region through the greater sciatic foramen, has no branches in the gluteal region, and exits this region at the inferior border of the gluteus maximus.

78–C. The pudendal nerve enters the gluteal region through the greater sciatic foramen and exits this region through the lesser sciatic foramen in close proximity to the ischial spine.

79–D. The superior gluteal nerve innervates the gluteus medius, gluteus minimus, and tensor fascia lata muscles.

80–B. The obturator nerve innervates the medial muscles of the thigh.

81–B. The superior intercostal and deep cervical arteries branch from the costocervical trunk.

82–D. The internal thoracic artery gives rise to the anterior intercostal arteries and then terminates at the sixth intercostal space by dividing into the superior epigastric and musculophrenic arteries.

83–C. The anterior interventricular artery branches from the left coronary artery and supplies the anterior portion of the interventricular septum.

84–D. The anterior intercostal arteries are branches of the internal thoracic artery.

85–A. The right coronary artery gives rise to the posterior interventricular artery, which provides the major blood supply of the posterior portion of the interventricular septum.

86–B. The round ligament of the uterus enters the deep inguinal ring, runs through the inguinal canal, emerges from the superficial inguinal ring, and becomes lost in the labium majus.

87–A. The broad ligament is a double layer of mesentery that attaches to the lateral surface of the uterus.

88–C. The ovarian (proper) ligament is homologous to the most superior portion of the gubernaculum in males.

89–D. The suspensory ligament of the ovary extends from the ovary to the dorsolateral body wall and is composed of the connective tissue around the ovarian vessels.

90–E. The cardinal (lateral cervical) ligament is an important uterine support. It is composed of fibromuscular condensations of pelvic fascia from the cervix and the lateral fornices of the vagina that extend to the pelvic wall.

91–B. The recurrent laryngeal nerve provides motor innervation to the intrinsic muscles of the larynx.

92–D. The lingual nerve carries general sensation from the anterior two-thirds of the tongue.

93–C. The chorda tympani carries special visceral sensation from the anterior two-thirds of the tongue.

94–A. The hypoglossal nerve provides motor innervation to the intrinsic muscles of the tongue.

95–E. The glossopharyngeal nerve carries sensation from pressure receptors in the carotid sinus.

96–B. The lunate bone.

97–C. The hamate bone.

98–E. The site of attachment for the adductor pollicis brevis and the flexor pollicis brevis, which, along with the opponens pollicis, form the thenar eminence.

99–D. The site of attachment for the flexor digitorum superficialis.

100–A. The pyloric portion of the stomach.

101–D. The left kidney.

102–C. The portal vein.

103–B. The right lobe of the liver.

104–A. The frontal sinus.

105–E. The superior orbital fissure.

Index

Note: Page numbers in italics denote illustrations, those followed by t denote tables, those followed by Q denote questions, and those followed by E denote explanations.